Sparkling Wines

Sparkling Wines

The Equinox Guide for Wine Professionals and Connoisseurs

PIOTR NAGÓRKA
with
MARTA RĘGOWSKA

equinox

SHEFFIELD UK BRISTOL CT

Published by Equinox Publishing Ltd.

UK: Office 415, The Workstation, 15 Paternoster Row, Sheffield, South Yorkshire S1 2BX
USA: ISD, 70 Enterprise Drive, Bristol, CT 06010

www.equinoxpub.com

First published 2026

British Library Cataloguing-in-Publication Data

A catalogue record for this book is available from the British Library.

ISBN-13 978 1 80050 684 8 (hardback)
978 1 80050 686 2 (ePDF)
978 1 80050 713 5 (ePub)

Library of Congress Cataloging-in-Publication Data

Names: Nagórka, Piotr author | Ręgowska, Marta author | Equinox Publishing (Great Britain) issuing body
Title: Sparkling wines : the equinox guide for wine professionals and connoisseurs / Piotr Nagórka and Marta Ręgowska.
Other titles: Equinox guide for wine professionals and connoisseurs
Description: Sheffield, South Yorkshire ; Bristol, CT : Equinox Publishing Ltd, 2026. | Includes bibliographical references and index. | Summary: "Sparkling wines belong to the group of special wines, which owe their origin to human ingenuity. The history of sparkling wines reaches back to 17th-century English taverns and 18th-century wineries of Champagne. While there are a number of books on champagne, this book is one of the few published so far with the intention to cover the entire spectrum of the sparkling wine trade"-- Provided by publisher.
Identifiers: LCCN 2025020365 (print) | LCCN 2025020366 (ebook) | ISBN 9781800506848 hardback | ISBN 9781800506862 ePDF | ISBN 9781800507135 ePub
Subjects: LCSH: Sparkling wines | Champagne (Wine)
Classification: LCC TP548.6.S65 N34 2025 (print) | LCC TP548.6.S65 (ebook) | DDC 663/.224--dc23/eng/20250717
LC record available at https://lccn.loc.gov/2025020365
LC ebook record available at https://lccn.loc.gov/2025020366

Typeset by Scribe Inc.

CONTENTS

FIGURES AND TABLES

FIGURES:

TABLES:

ILLUSTRATIONS

AUTHOR'S PREFACE

Sparkling wines belong to a group of special wines that owe their origin to human ingenuity. The history of sparkling wines reaches back to 17th-century English taverns and to 18th-century wineries in Champagne. While there are a number of books on champagne, this book is one of the few published so far covering the entire spectrum of the sparkling wine trade. Here, wine connoisseurs can learn about sparkling wines — which are currently made in all continents except Antarctica — through the many voices and opinions of wine professionals, based on their wide and various experience. For instance, there is a variety of philosophies on the merits of lees contact — for one thing, preferences differ about the duration of the process — and this can result in a story being told from different, sometimes contrasting, angles. The tools of the science of terminology have enabled an extensive study of concepts, and allowed claims to be fact-checked and data to be integrated. The result is a distillation of nuanced knowledge on sparkling wines into a compendium, which you are now holding, and which provides its readers with guidance in many areas, from early historical background to the attitudes and practices of the third decade of the 21st century.

My terminological study of concepts in the sparkling wine industry dates back to the late 2000s, but it wasn't until the latter half of the 2010s that I began to undertake advanced research in this area. The idea for this book owes much to the success of *Madeira, Port, Sherry: The Equinox Companion to Fortified Wines*, which received the OIV Award in 2022. When the Equinox Publishing managing director Janet Joyce expressed her interest in a follow-up, I was immediately thinking in terms of a co-author. In late 2021 I invited wine educator Marta Ręgowska to contribute her rich expertise to this enterprise, not only because journeys — intellectual ones included — are usually best taken with others, but also because I knew Marta's in-depth knowledge of sparkling wine — which she has always regarded as much more than a wine for celebrations — would prove invaluable for the wine and food pairing chapter.

The scope of this venture has exceeded by far that of my previous studies of concepts in the fortified wine industries

— which underlay *Madeira, Port, Sherry*. This is very much to do with both the reach of and interest in these types of wine. Suffice to say that the global annual consumption of sparkling wines in 2018 was about 19 million hectolitres, whereas the combined sales of madeira, port and sherry in the same period was under 110 million litres. Notably, the market for champagne alone is now over twice the size of that for the classic fortified wines combined. Considering the ever-increasing advancements in the field of sparkling wines — such as sustainable practices or experimental ageing of sparkling wine underwater — I am all the more grateful to my co-author for her unwavering dedication to keeping up with new developments in this fast-changing industry, even during difficult and personally challenging times.

While the major part of the primary data has resulted from desk research, the field studies I have been conducting since 2015 at a number of sparkling winemaking estates, notably those in Champagne, Lower Austria, England's East Sussex, and western Poland, confirmed my conceptual schemes and completed the photographic record. Photographs from the region of Penedès (in Cataluña) and the areas of Conegliano and Valdobbiadene in the northeastern part of Italy were taken by Marta, who also visited the Pfalz region in southwestern Germany. In taking both of us on a tour of Europe's vineyards, this research has resulted in friendships as well as valuable relationships with many in the sparkling wine trade, for which we are particularly grateful.

This encyclopaedia is designed to serve as an educational tool. It enables an appreciation of the rich, multifaceted history of the sparkling wine trade, from the mid-17th century to the present, and offers unique educational journeys through winemaking activities, from grape selection to bottle ageing. It covers classifications, serving temperatures, sales dynamics in the most important export markets, while also exploring flavour compounds responsible for notes identified in wine bouquets. Guidelines on wine and food pairings are included. This book addresses a wide audience, from wine professionals and connoisseurs to gourmet chefs and home cooks.

I am indebted to all those who have given of their time to assist me in my research. I have received a lot of support from both my colleagues at the Faculty of Applied Linguistics and in the wine world, including oenologists, winemakers, sommeliers, historians and wine educators. A debt of deep gratitude is owed to my consultants, especially to the Leclerc Briant *chef de cave* Hervé Jestin; owner, CEO and head winemaker of Adoria Vineyards, Mike Whitney; Ridgeview Wine Estate's head winemaker Simon Roberts; as well as the owner and winemaker of Winnica Gostchorze, Guillaume Dubois. I would also like to thank Jacek Turnau, CEO of Turnau Vineyard; Artur Kojder, co-owner of Winnice Kojder; Mardi Roberts, Ridgeview Wine Estate's director of communications; and Tomasz Kowalski, the curator at Wine Museum in Zielona Góra, for their hospitality, and their endless patience with my never-ending lists of questions.

My sincere thanks go to the head of the Institute of Ukrainian Studies, Katarzyna Jakubowska-Krawczyk, and the Dean of the Faculty of Applied Linguistics, Magdalena Olpińska-Szkiełko. Their decisions helped me fund the field research and carry it out. Moreover, I am grateful to representatives of wine producers Graham Beck Estate,

Adoria Vineyards, Spier Wine Farm, Schloss Gobelsburg, Woodfine (Wine), and Soumah of Yarra Valley, for granting me permission to reproduce some of their photographs in this book. Special thanks are due to Casey Scumaci, Małgorzata Kucikiewicz, Aldona Szafrańska, and also to Chris Bielecki and his wife Sara. Their photographs represent an immense contribution to the aesthetic quality of this volume. The photographic documentation in the area of champagnes would have been incomplete without the invaluable assistance of my brother Paweł, who helped post-process some of the photographs I took during my field studies in Champagne.

Finally, I want to express my heartfelt appreciation to my family and friends, whose enormous contribution — albeit indirect, more personal than professional, and (for this reason) difficult to measure — can hardly be overestimated. When photographing food, I drew my inspiration from images once shared with me by my sister-in-law Wioletta Kieblesz. Still very vivid in my memory is the favour granted me by Joanna Raszczuk, who carried a heavy volume of Frederick J. Ryan Jr's book *Wine and the White House: A History* all the way from London so that I could include it in my bibliography. I would also like to thank my parents, Irena and Ryszard, for providing me with a quiet work space in their home during the final stages of writing. I owe a debt of appreciation that cannot be repaid to my wife Beata. All the exquisite meals that she prepared and served during the course of my work on this book were small works of art that not only managed to bring the family together around the table but also became healthy distractors, each time providing me with fresh energy and inspiration.

Piotr Nagórka
25 October 2024

NOTE TO THE READER

SPARKLING WINES represent a vital part of the field of special wines. This companion has been written for those seeking to understand the sparkling wine trade as a whole: its history, regions, producers, crafting methods, and the practical aspects of enjoying these wines. It is designed to accompany the reader on their personal journeys of discovery in the world of wine, which I encourage them to undertake. My intended audience comprises a large group of beneficiaries: from the production sector, to wine experts and traders, to the market for sparkling wines. This book is intended as both a point of reference and a learning stimulus for: beginners in the sparkling wine industry and winemaking trainees; sommeliers; wine business investors and marketeers; wine educators and wine journalists; connoisseurs; gourmet restaurant chefs; regular wine customers; and home cooks. It could especially be of use to degree students in viticulture, oenology or related fields, or to those who run, or participate in, a wine (and food pairing) course (or workshop).

As with the book *Madeira, Port, Sherry: The Equinox Companion to Fortified Wines*, which initiated the series on special wines, the aspect of wine education is reflected in the design of this volume. Displays of professional content have been organized based on the principle of progressive complexity. This guide has the following sections:

- A factual guide briefly discussing the history of the sparkling wine trade
- A map of terms, serving as an overview of the production of sparkling wines
- An A–Z display giving definitions of all wine terms found on the map
- Wine and food pairing, which lays out the principles of matching a chosen sparkling wine with foods, while offering full explanations with numerous examples
- An order of events, providing a chronological list of developments that have shaped the trade from its origins to the present day

The terminological research underpinning this book gives the reader access to a treasure trove of structured information. Students of oenology can learn how the makers of sparkling wines differ from one another, not only in terms of preferred crafting methods but also approaches to the various stages of the production process. Those interested in the arcane will learn what honey can be used for and at what stage of production it is applied. Those specializing in wine chemistry can learn which chemical compounds are responsible for flavour differences between sparkling wines that have matured on lees (in bottles as well as in pressure tanks) for varying periods.

This guide is also for those who have only recently begun their career in the sparkling wine trade and are eager to learn. They can learn about the extent to which sustainable practices are now found in the sparkling wine business; which of the major sparkling wine industries are leaders in mitigating the effects of climate change? It also serves as a snapshot of the dynamics of the sparkling wine business: it discusses not only the world's largest sparkling wine producers but also the largest consumers of wines from this category. It shows the multiple relationships that sparkling wine has developed with other industries over the years, most notably with various agricultural sectors, the hospitality industry and the cultural and creative economy, with some producers — both large and small — crossing the natural boundaries of their own industry to enhance customer experience.

Finally, this encyclopaedia has also been conceived as a rich and reliable resource for wine educators preparing intellectual games for their workshops. To reveal just a few points of general interest, the reader will learn:

- Why are the English credited with the invention of fully sparkling wines?
- When did champagne first appear in a painting, laying the groundwork for a lasting relationship between the sparkling wine business and the world of visual arts?
- Which champagne houses were honoured with Royal Warrants by Queen Victoria in the 19th century, becoming official suppliers of fine wine to the British (royal) court?
- Which champagne brands were Sir Winston Churchill's favourites?
- Which sparkling wines were most often served at state events hosted by US presidents in the latter half of the 20th century and the first quarter of the 21st century?
- Which Spanish sparkling wine appellation has been recognized since 2017 as the world's first to be applied to wines made using traditional cultivation methods and exclusively organic techniques?
- Which champagne was served on 19 May 2018 at a luncheon reception hosted by Queen Elizabeth II at St George's Hall in Windsor Castle following the wedding of HRH Prince Henry of Wales and Ms Meghan Markle?
- Which is currently the largest sparkling wine-producing country in the Southern Hemisphere?

Readers who wish to see the results of quantitative terminological research on professional communication in the sparkling wine industry are referred to 'Appendix: The Language of Wine'.

SPARKLING WINES

FACTUAL GUIDE[1]

MOST SPARKLING wines draw their roots from European soils but production now takes place on all continents bar Antarctica. The grapes come from vineyards that enjoy diverse climatic conditions. The world's largest sparkling wine-producing regions are located in the Northern Hemisphere and include major viticultural areas of Italy, France, Germany, Spain (Catalonia mostly) and the United States, with the greatest share of American sparkling wines crafted in the Napa Valley. The Southern Hemisphere is also represented, albeit to a lesser extent, with vineyards scattered as far as Brazil, South Africa and Australia.

Grapes destined for sparkling wines are grown in various terroirs. The impact of natural factors on viticulture varies from one wine-growing region to another, with some regions being internally differentiated. For example, the Prosecco zone is located in the northeastern part of Italy, a climate with minor influences from the Italian Alps in the Valdobbiadene area, summer temperatures averaging 23°C, and with moderate influences from the Adriatic Sea in the Conegliano area, where summer temperatures are about 25°C.

1 Merret [1662] 1669: 227; MacCulloch [1816] 1829: 118–20; PPH 1841: 2; Maumené and Jaunay 1859: No. 1578; Heidsieck 1860; Haraszthy 1862; Shaw 1863: 248–9; McCabe 1871: 333–73; Vizetelly 1882: 205–8, 259–65; Sheridan 1888: 411–30; Wheatley 1891: 524–5; *Leader* 1901: 58; Marsden 1905; Simon 1905: 118–36, 184; *The Age* 1905: 5; Crawford 1910: 1,011, 1,164; McMillan 1943: 240–2; Baldwin 1945: 527–9; Cole [1950] 1993: 2; Forbes [1967] 1989: 151–8, 172–3, 190–212, 467–74; Rosberg [1994] 2015: 97–8. De la Serna [1994] 2015: 212–13; Platter and Fridjhon [1994] 2015; Speller [1994] 2015a: 357; Thomases [1994] 1999: 300; Peters 1996: 63–4; Guy 1999; Kladstrup and Kladstrup 2001: 55–62, 129–35; Fielden [2001] 2010: 45–6; Sumners 2002: 64–6; Kladstrup and Kladstrup [2005] 2006; Bargmann 2005; Jeffs 2006; Staniforth 2007; Braatz *et al.* [2007] 2014; Dungworth 2012: 37–8; Charters and Spielmann 2014; Jones *et al.* 2014; Dessauve [2014] 2022; Phillips 2016: 276–7; Tardi 2016: 40–52; Liem 2017: 24–6, 84–7; Chevet *et al.* 2018: 61–5; Shepro 2018: 346–8; Harding 2018: 101–5; Becuwe *et al.* [2018] 2019: 47; Bielka-Vescovi 2019; Krebiehl 2019; Stevenson 2019: 45; Coppola 2020; OIV 2020a: 6–8; Ryan 2020: 289–91, 381–439; Tomasi and Gaiotti 2020: 51–65; Deceurs 2020; Crowden 2021: 72–87; Rose 2021; Skelton 2021: 59–105; Antonsson 2022; Lanson-BCC 2022; LP 2022: 6–8; Mazzeo 2022; McCoy 2022: 30–8; WA 2022; Perrin and Péters 2022; Sheppard 2023; WA 2023; Ruth 2024; Gilinsky *et al.* 2021; Green 2023a.

From April to September, 15–20 mm less rain falls each month in Conegliano, with total precipitation in this period lower than 680 mm.

Similarly, vignerons in Champagne need to adapt to varying climate influences: for the most part, being semi-oceanic in the north and mainly semi-continental in the south. Temperatures in the region rarely fall below −5°C in the winter, or exceed 30°C in the summer, but the Champenois must be alert to occasional hard frosts, especially from mid-December to late February, and temperature spikes in the warm season, which may at times culminate in violent storms. While rainfall in the whole region throughout the year is steady and moderate, Champagne's subregions differ markedly in subsoil types, with Montagne de Reims and Côte des Blancs dominated by chalk, Vallée de la Marne located on clays and sands, and vineyards in Côte des Bar known for their marl soils.

Photo 1. Vineyards of Champagne Augustin at Avenay Val d'Or (Vallée de la Marne), a five-minute walk from the train station of Avenay.
Photo: author, July 2024

Diverse terroirs are also typical of Spain's cava-producing region, a phenomenon explained by the region's division into wine-growing zones spread throughout the country. One needs only compare the Mediterranean climate of Serra de Mar, a subzone of Comtats de Barcelona, with its annual rainfall of more than 470 mm, and the continental climate of Valle del Cierzo, part of the Valle del Ebro zone, which receives less than 300 mm of rain. Moreover, the sandy and hence highly permeable soils of Serra de Mar, capable of retaining solar heat, provide a marked contrast to calcareous soils of the Levante zone, which provide a limited but regular water supply to the region's grapes, grown at an altitude of 600–900 metres.

Germany may be Europe's most varied sparkling wine producing country in terms of terroir, given that the grapes for the lowest-category sparkling wines made in this country's wineries can be sourced from any part of the EU, with the quality sekt wines, a showpiece in each of Germany's wine-producing regions, crafted using grapes grown exclusively on German soils. These soils vary in type and structure, and include clay, chalk, granite, keuper, limestone, loam, loess, quartzite, sand, sandstones, slate and shell limestone. The latter is found primarily in Franconia, Saale-Unstrut, Württemberg, the Mosel Valley — which has some of Europe's steepest vineyards with gradients in excess of 60° — and the country's warmest wine-growing region, Baden, where growers benefit from regular rainfall in the vegetation period, ranging from 50–100 mm per month.

Producing around 1 million hectolitres of sparkling wine in 2018, which accounts for nearly 80% of the US's total production, the Napa Valley boasts a wide range of soils and microclimates. Containing half of the world's soil orders, this winegrowing region numbers among the most geologically diverse. Grapes typically destined for sparkling wines such as Chardonnay and Pinot Noir are grown mainly in the southern wine districts of Los Carneros and Wild Horse Valley, where mid-summer temperatures rarely exceed 27°C. While annual rainfall in Calistoga and Diamond Mountain District reaches between 1,400 mm and 1,500 mm, growers in Los Carneros can expect about 610 mm of rain, which is offset to some extent by cooling breezes from San Pablo Bay.

The largest sparkling wine producing country in the Southern Hemisphere, Australia, has produced more than 700 thousand hectolitres of sparkling wine a year since 2018. The industry is mainly concentrated in Australia's cool-temperature regions, notably the Adelaide Hills (South Australia), Yarra Valley (Victoria) and Tasmania. The low-lying vineyards in the Adelaide Hills are dominated by heavy soils, with well-drained soils found at higher elevations. Although rainfall in the growing season is less than 270 mm, growers can rely on the region's robust underground water supplies. Yarra Valley's northern parts are composed of sand and clay loam, which are relatively acidic and permeable, while the southern side of the valley features fertile red volcanic soils. In the growing season, lasting from autumn to spring, the region vines receive around 560 mm of rain.

Tasmania is Australia's coolest winegrowing region, with mean temperatures in January, the warmest month, not exceeding 15.6°C. The vast majority of Tasmania's vineyards are located at altitudes lower than 100 metres. On low slopes, the vineyard soils consist of sandstones and mudstones, as well as igneous rocks of volcanic origin. Peaty alluvial soils can be found in Coal River Valley; deep, friable, free-draining soils are typical of the Pipers River area; schists are present in Derwent Valley; while the vines growing in the area of Tamar Valley emerge from gravelly basalt, clay and limestone. In the growing season, October to April, rainfall in the areas under vines — essentially Riesling, Chardonnay and Pinot Noir — is in the range of 450–480 mm.

Also in the Southern Hemisphere, South America, particularly Brazil and Argentina, further adds to the world's already large variety of viticultural conditions suitable for the creation of sparkling wines. The area of Altos de Pinto Bandeira, part of Brazil's southern wine-producing region Serra Gaúcha, is known for its unique terroir, with vineyards, planted predominantly with Chardonnay, Pinot Noir and Riesling Italico, located at an average altitude of over 600 metres. Growers benefit from the area's acidic soils with elements of weathered basalt, which contain varying proportions of water-retaining clay. The annual rainfall in the area reaches up to 1,800 mm, of which usually a little less than half falls in the growing season: September to February.

Vineyards in Valle de Uco (Uco Valley), part of Mendoza — Argentina's most prolific sparkling winegrowing province — located in close proximity to the Andes, are among the world's highest. Typical altitudes at which vines, mainly Chardonnay and Pinot Noir, are grown in this region range from 1,000 to 1,650 metres, which keeps temperatures rather low, a circumstance compensated by over 300 days of sunshine a year. With an average annual rainfall of just 200 mm, the mountains provide a vital source of water for the vines. Soils in these high-altitude vineyards are described as particularly calcareous, consisting predominantly of limestone, gravel, clay and permeable sandy loam.

This wealth of natural factors that favour the cultivation of grapes for sparkling wines is further complemented by those found in the Western Cape regions of South Africa. With their rich limestone deposits, the soils of the Robertson district, part of the Breede River Valley, allow the production of high-acid grapes. With annual rainfall in the district of less than 500 mm, and day temperatures exceeding 30°C, the River Breede and the cooling southeasterly winds are considered the lifeblood of the local vineyards. By contrast, districts in the Coastal Region, such as Stellenbosch and Franschhoek, which benefit from a Mediterranean climate with maritime influences, have acidic and potassium-rich soils derived from the weathering of granites, resulting in the presence of kaolin clay.

Photo 2. Graham Beck Estate's vineyard in the Robertson area, in the Breede River Valley (South Africa's Western Cape). The vineyard is located within seven miles of the town of Robertson.
Photo: courtesy of Graham Beck Estate, February 2017

While the monks of the Abbey of Saint-Hilaire in the south of France in the second quarter of the 16th century may indeed have been the first to deliberately produce mildly effervescent wines, fully sparkling wines must have been created much later. One reason for this observation is that the wood-based glassmaking technology of the time did not allow the production of bottles capable of holding pressures of up to six atmospheres without breaking. Methods for producing glass robust enough for the purpose, i.e. using coal-burning furnaces, only began to be developed (in England) between the last quarter of the 16th and the mid-17th centuries.

In the first quarter of the 17th century, the depletion of Britain's timber resources was becoming a problem, particularly because of a need to retrofit and expand the Royal Navy fleet. On 23 May 1615, King James I banned the use of wood and its products in the manufacture of glass, thereby laying the foundation for the creation of an industrial economy based on coal, the combustion of which enabled pressure-resistant bottles to be produced. The nascent glass industry attracted numerous entrepreneurs but was dominated for several decades by Sir Robert Mansell, whose glassworks were established in various parts of England.

By the mid-17th century English glassmakers had developed thick-walled, narrow-necked bottles, the invention of which is ascribed to Sir Kenelm Digby, a courtier and man of numerous talents, who was later to become one of the earliest members of the Royal Society. These bottles had collars of thick glass at the top of the neck, which served as anchors for the piece of string holding the cork. This invention proved indispensable for the birth and development of sparkling wines as we know them today.

On 17 December 1662 Christopher Merret presented a paper to the Royal Society — *Some Observations Concerning the Ordering of Wines* — evidence that in the mid-17th century English coopers were initiating a second fermentation in the still wines imported to England in barrels, with the intention of producing sparkling wines. This was enabled by English glass-makers and their robust bottles crafted in coal-fired furnaces, which could withstand the pressure deriving from the refermentation of base wines, notably those from Champagne.

English wine bottles gained a reputation for strength, indispensable for effervescent beverages, but it was not until the beginning of the 18th century that coal-powered bottle factories built in the English manner began to appear in France. Among the first French-based bottle houses using coal to meet growing demands of the French wine-producing market was one launched in Picardy by Gaspard Thévenot in 1709 and one in Bordeaux by Peter Mitchell in 1723.

The sparkling wines of Champagne were appreciated by Louis XV, and a royal edict of 25 May 1728 granted producers and merchants the right to transport these wines in bottles. Soon after, in 1729, the cloth merchant Nicolas Ruinart established a winery dedicated to sparkling wines. Ruinart's winery evolved over time into a respected champagne house, today part of the business group Moët Hennessy–Louis Vuitton (commonly referred to as LVMH), based in Paris.

By the mid-1730s, sparkling alcoholic beverages made from Champagne grapes and sold in cork-stoppered glass bottles were enjoying the status of a drink that was served with meals. We can see this in Jean-François de Troy's painting *Le Déjeuner d'Huîtres* ('Oyster Lunch'), commissioned by the King of France in 1734 and unveiled in 1735. Currently in the collection of the Musée Condé in Château de Chantilly, the painting depicts a bottle of fizzy wine being opened during a festive meal, a champagne cork captured in flight.

Some of today's most renowned champagne houses were founded in that period, including Chanoine, set up by the young merchant Pierre Chanoine in 1730, and Forest-Fourneaux, now known as Taittinger, founded on the initiative of Jacques Fourneaux in 1734. Wine merchant Claude Moët established his own business dedicated to producing and selling sparkling wine in 1743. Seeds were soon planted (and the foundations laid) for firms currently known by the names Henri Abelé, Lanson, and Louis Roederer, their roots dating back to 1757, 1760 and 1765 respectively (although the official year of Louis Roederer's birth is 1776). They were soon to share the market with firms run by the Clicquot and Heidsieck families.

A banker and vineyard owner, Philippe Clicquot began to commercialize his wines in 1772, initially selling champagne mainly to business acquaintances in France. With a solid grounding in commerce, Florens-Louis Heidsieck, who first visited Reims in 1777, had an ambitious vision for his business and adopted a different strategy. In 1785 he married Agathe Perthois and established a wine company called Heidsieck & Cie. In order to acquire significant name recognition, this German-born merchant would present his wines to prominent figures: Queen Marie-Antoinette was considered one of Heidsieck & Cie's first ambassadors.

Napoleon Bonaparte's rise to power in 1799 marked the beginning of a prosperous era for champagne producers, a time

Photo 3. Very old wooden casks in the underground cellars of the champagne house Veuve Clicquot, once used for the purpose of maturation of base wines before the operation of bottling.
Photo: author, July 2024

also characterized by technical progress in the industry. In 1805, after the death of François, Phillipe Clicquot's son, his widowed wife Barbe Nicole Clicquot, née Ponsardin, took over the business. In 1810, the company, now known as Veuve Clicquot, made what seems to be the first vintage champagne in the trade's history. Meanwhile, under the leadership of Jean-Rémy Moët, who took control of the family business in the 1790s, the company currently known by the name Moët & Chandon continued to grow in international prominence, for which he was awarded the title Chevalier de la Légion d'Honneur in March 1814.

Towards the second decade of the 19th century, the firm Veuve Clicquot was at the forefront of innovation, with Antoine Müller, the company's cellar master at the time, credited with introducing the first riddling racks to the sparkling wine industry, around 1816. The racks solved a problem whereby sediment was formed by dead yeast cells in the bottle following the second fermentation: they eliminated the need to decant the wine into another bottle, which was not only time-consuming and costly but resulted in a loss of effervescence. Müller's invention is still used by quality-conscious winemakers committed to traditional practices.

In 1826, Georg Christian von Kessler, who by that time had worked for almost two decades at the champagne house Veuve Clicquot, for the most part as the company's foreign trade manager and the owner's business partner, moved from Reims to Esslingen am Neckar to found a wine company under his own name. Now known as current-day Germany's oldest sparkling wine producer, G.C. Kessler & Co. quickly gained recognition, not only in the Kingdom of Württemberg and other German states, but also beyond their borders. Towards the end of the 1820s the firm began exporting its sparkling wines to Great Britain, Russia and the United States.

In the years that followed, a new wave of producers emerged in Champagne, especially as entrepreneurs began to realize that the champagne industry was strong enough to survive not only difficult economic times but also political turmoil, owing much to the unrelenting demand for its wine. The industry's good prospects attracted investment from both France and abroad. The company known today by the name G.H. Mumm & Cie was established in 1827 in Reims by a German winemaking family. In February 1829, at Aÿ, the foundations were laid by Athanase-Louis-Emmanuel de Villermont, and his partners (Joseph Bollinger and Paul Renaudin) for a new champagne-trading firm, now called Bollinger. In 1837, Swiss merchant Henri-Marc de Venoge and his son Joseph set up a wine company in Épernay named De Venoge & Cie.

There has been a long-standing dispute about which firm was first to produce sparkling wines — by the method used in Champagne — on German soil. The dispute involved the firms Häusler, Förster & Grempler from Grünberg (now the city of Zielona Góra, western Poland) and G.C. Kessler & Co. from Esslingen am Neckar (the current-day Würtemberg region in southwestern Germany) both founded in 1826. It appears to have been settled in favour of the former. First of all, the Grünberg company was registered at the initiative of the winemaker Carl Samuel Häusler in February 1826, predating the foundation of G.C. Kessler & Co. by over four months; secondly, Carl Samuel Häusler began producing his first sekts around two years before his company was officially established. At the time of the company's registration, its cellars were already full of wine made from grapes from the 1824 harvest — the collection of which had been personally supervised by the founder.

Photo 4. The rock formation known as London Bridge, near Point Nepean (Victoria's Mornington Peninsula), around four miles from the entrance to Port Phillip Bay, where the barque *William Salthouse*, carrying a cargo of champagne baskets, struck a submerged rock and sank on 28 November 1841.
Photo: Chris Bielecki and Sara Bielecka, January 2024

Sparkling champagne's reputation as one of the world's most desirable wines is confirmed as early as the first half of the 19th century by the trade between Montreal (Canada) and Australia. The incoming cargo manifest of the barque *William Salthouse*, which sank on 28 November 1841 at the entrance to Port Phillip Bay, near Melbourne, was published in the *Port Phillip Herald* on 10 December 1841, telling of a cargo that included 20 baskets of champagne, a brand most probably owned by the Gosset family. Production and sales continued to grow, and more companies began to operate in Champagne in the 1840s.

The renowned champagne producer Krug, now part of the LVMH group, was established in Reims by Joseph Krug in 1843. The French sparkling wine sector saw very little economic impact either from France's February Revolution of 1848 or from the political upheavals in other European countries from 1846 to 1849, known collectively as the Springtime of Nations. Trusting that the champagne industry would weather these turbulent times, in 1849 Pol Roger founded a wine company bearing his name, which, like the industry itself, prospers to this day. Suffice it to say that in the period 1847–1850 annual sales of sparkling wine made in the Champagne region amounted to around seven million bottles, climbing to nearly eight million between April 1850 and March 1851.

The following decade saw the establishment of two champagne houses whose founders became famous for their marketing skills. Charles Heidsieck established his wine company in 1851. The growing number of producers sharing European markets for champagne provided the spur of increased competition, which may have been among the reasons that prompted Heidsieck to embark on a series of business trips around the United States the following year in search of reliable markets for his company's wines. After he had set up his champagne house in 1858, Eugène Mercier authored some of the most creative promotional campaigns in the industry's rich history, leaving a legacy of commitment to customer communication that made wines previously considered the preserve of the elite accessible to everyone without compromising on their quality.

In the late 1850s, cooperation between Edme-Jules Maumené, one of the pioneers of sparkling wine science, and Louis Brunet Jaunay, an inventor and wine merchant operating in Reims, resulted in the development of a relatively new method of creating effervescent wines using closed tanks equipped with adjustable valves, to allow regulation of the amount of mousse in the final wine. They laid both the scientific and technical foundations for the bulk production of (tank-fermented) sparkling wines, although the commercial face of their invention was not to emerge until the beginning of the following century.

In the 1860s, the significant presence of counterfeit champagnes in wine markets on both sides of the Atlantic was met with determined reactions from the affected producers. In a letter to the editor of the *New York Dispatch*, published on 21 January 1860, Charles Heidsieck declared that the amount of wine imitating the sparkling wines of Champagne and sold to the US market might be several times the quantity of genuine champagne imported to that country per year by authorized agents.

Over the following years, sparkling winemaking expertise, as applied for generations in Champagne, began to spread to the Italian region of Piedmont. Having studied champagne-making techniques during his time at Reims, Carlo Gancia returned to his native Piedmont in 1850. His numerous experiments with locally grown grapes, notably Moscato, and with winemaking techniques observed in Champagne, which he tried to adapt to the winegrowing conditions of his homeland, led to his creation, in the mid-1860s, of the first quality bottle-fermented sparkling wines in Piedmont, considered the forerunners of today's Asti Spumante wines.

Photo 5. Attached to the walls of Champagne Boizel's underground cellars in Épernay are plaques with annual dates and city names, indicating when the company's champagne was first shipped to a particular foreign market. The first shipment of Boizel's champagne to London was in 1851. Deliveries of the firm's champagnes to addresses in New York were not until 1908.
Photo: author, July 2024

In the mid-1860s, the export of champagne, which had already been climbing steadily since the early Victorian era, began to undergo a rapid expansion. This is largely attributable to significant reductions in English tariffs on French wines, a result of the carefully negotiated Cobden–Chevalier Treaty, signed on 23 January 1860 between France and the United Kingdom, one of the champagne industry's top export markets. The progressive conquest of world markets by the Champenoise is evident in the statistics: the number of bottles of champagne sold abroad rose from around 9.8 million in 1863–1864 to over 13.8 million in 1869–1870.

Producers in Champagne were soon to be severely affected by the Franco-Prussian War, declared by the French emperor Napoleon III against Prussia on 19 July 1870 and concluded with the Treaty of Frankfurt, signed on 10 May the following year. The military occupation of the Champagne region by the Prussian army, from early September 1870, resulted in a sharp decline in the region's wine sales, from over 17.4 million bottles sold in 1869–1870 to fewer than 9.2 million bottles commercialized in 1870–1871. When trade between champagne producers and their trading partners was quickly restored, the industry's sales performance in the following months exceeded expectations.

The post-war years were the heyday of the champagne trade. Figures for 1871–1874 showed record results, with annual sales exceeding 20 million bottles. Champagne producers responded to various demands in the market, tailoring their wines to their customers' tastes. While the French had a preference for light and moderately sweet wines, in the United States medium-light champagnes with intermediate sweetness were appreciated. The sweetest champagnes were requested by customers in Russia and Germany, the strongest bottlings purchased by those in Australia and South Africa, and light dry wines expected in China and India.

Meanwhile, creators of sparkling wines in Champagne found followers among winemakers across France's southern border. In 1872, Josep Raventós i Fatjó, a leading winegrower in Spain and head of the family firm Codorníu, located in the region of Penedès in Catalonia, made his first bottles of sparkling wine in the likeness of wines produced in Champagne, the only difference being that he used grapes native to Catalan soils. These bottlings, released a few years later, laid the foundations for the development of the sparkling wine trade in Penedès, and eventually throughout Spain.

Britain, one of France's most significant export markets in Europe at that time, demonstrated a growing demand for dry champagnes. Champagne manufacturers, such as Moët & Chandon and Pommery & Greno, capitalized on the outstanding harvest of 1874, launching vintage and non-vintage champagnes marked by varying degrees of dryness. Prominent among them was Pommery's Brut Nature Champagne 1874, a wine that Adolphe Hubinet, Pommery & Greno's agent in London, introduced to the British market in the latter half of the 1870s, an event widely noted as the birth of fully dry vintage champagnes.

Sparkling wines of Champagne were increasingly regarded by consumers in the UK as a symbol of wealth and high social standing. Perception of champagne as a status symbol in Britain was consolidated and reinforced in 1877, when champagne house Pol Roger & Co. received its first Royal Warrant from Queen Victoria, becoming the official supplier of sparkling wines to the British (royal) court. The premium sector of the champagne trade in Britain, represented by producers such as Heidsieck & Co., Pommery & Greno and Veuve Cliquot, was keen to capitalize on the existing sentiment by raising prices on their non-vintage champagnes in subsequent years.

Champagne producers continued to protect their businesses from fraud while advertising their wines in both the domestic market and export markets, where they operated through sales agents. On 4 November 1882, a joint initiative of a number of champagne houses, primarily Heidsieck & Cie, Giesler & Cie and G.H. Mumm & Cie, led to the creation of Syndicat du Commerce des Vins de Champagne. The organization was established to support champagne manufacturers in promoting their wines, defend them against unfair competition, and put pressure on legislators to modify laws and regulations in their favour.

Towards the end of the 19th century, with the assistance of Syndicat du Commerce des Vins de Champagne, the most powerful champagne houses managed to lay enduring foundations for the development of one of the world's first widely recognized territorial brands. The judgement of the Angers Court of Appeal of 1887 declared that the name 'champagne' must not be used as belonging to the public domain, being a designation only for wines from the Champagne region, thus indicating where they were produced; this was considered the earliest major success of champagne producers united by the common goal of defending their master brands.

With the arrival of the phylloxera plague in Champagne's vineyards at the beginning of the 1890s, the economic situation among the region's vine-growers began to deteriorate. Worse still, faced with shortages in the supply of grapes, some of the less scrupulous among champagne producers resorted to sourcing grapes, juice and base wine for their sparkling wines from outside the region, to the growing distress of local suppliers and also to the detriment to an image of the champagne trade that the region's leading houses had worked for many years to build. In these circumstances, legal protection of the champagne brand posed a considerable challenge.

Meanwhile, the method of creating sparkling wines used in Champagne found fertile ground in Great Western Vineyard, an Australian wine company based in the state of Victoria, owned by the passionate entrepreneur and politician Hans William Henry Irvine since 1888. In the early 1890s, during one of his business trips around France, Irvine hired Charles Pierlot, a French winemaker who had previously worked for Pommery & Greno. By the early 1900s, Great Western Vineyard had established itself as Australia's leader in the sparkling wine industry, producing both sparkling wines in the likeness of French champagnes and red sparkling wine, a speciality of Australian winemakers, then commonly known as 'sparkling burgundy'.

A growing number of sparkling wine producers, particularly those in Champagne, continued to learn to make use of cultural and technical advances to showcase the fruits of their labour. Considered as the first portrayal of champagne in a fictional film, the use of the name 'Mercier' on a giant-sized champagne bottle in the 1901 film *La Barbe Bleue*, directed by Georges Méliès, was also one of the first product placements in the history of cinematography. The appearance of the champagne house's name in *La Barbe Bleue* predates by over 20 years the placement of a bottle of Louis Roederer 'Grand Vin'-Sec in Charlie Chaplin's film *A Woman of Paris* (1923), in which restaurant guests seated at a table were served truffles cooked in champagne.

The beginning of the first decade of the 20th century saw many poor harvests in Champagne, which champagne producers tried to compensate for by trying to sell the wines that had been ageing in their cellars. The marketing efforts of major champagne houses at the time can be exemplified by those of Pommery, whose 75,000-litre *foudre* with carved decorations made by master woodworker Emile Gallé — symbolizing amicable trade relations between the Champagne region and the United States — enjoyed pride of place at the St Louis World's Fair. Organized to commemorate the United States' purchase of the Louisiana territory from France in 1803, the exposition was held from 30 April to 1 December 1904, the year after the centenary of the purchase.

Effective protection against unauthorized use of the name 'champagne' required from lawmakers in France a precise delimitation of the area of origin to which the name referred, a measure that proved a bone of contention among Champagne's growers. In October 1910, in the wake of a failed grape harvest, and divided over the issue of Champagne's borders, the region's vignerons were soon to have the limits of their endurance tested. Their misery was compounded by rumours of further fraud by some merchants. This sparked major protests, which culminated in the Marne riots of April 1911, bringing the Champagne region to the brink of civil war.

Further work on the delimitation of the region's grape-growing and production areas was interrupted by the outbreak of

Photo 6. A reproduction of the head of Pommery's *grand foudre* on exhibition in the Musée du vin de Champagne et d'Archéologie régionale in Chateau Perrier in Épernay. The decorations of the original barrel, exhibited at the St Louis World's Fair in 1904, were made by Emile Gallé and represent amicable commercial ties between Champagne and the United States.
Photo: author, June 2023

the Great War (1914–1918). Providing shelter from shellfire and air raids, which had been continuing intermittently from mid-September 1914, wine cellars and underground caves (*crayères*) served as hospitals, schools and even concert venues. Owing much to the effort of women working in vineyards and cellars, and the support offered at harvests by refugees from Belgium and northern France, champagne production was maintained at half pre-war levels. In the grape-growing villages of Champagne, the memory of those who lost their lives during the wartime harvests in the bombing and shelling has been kept alive, despite the passage of time.

The scale of the damage inflicted on the Champagne region during those four years of war was unprecedented. By the end of the war, German artillery shelling and bombardments had razed the city of Reims to the ground, not even sparing Reims Cathedral, traditional site of the coronation of French monarchs. Many maisons, notably those based in Reims, such as Henri Abelé, Lanson, Pommery & Greno and Ruinart Père et Fils, were either turned to rubble or obliterated completely. Air raids over Épernay inflicted extensive damage to the buildings of Moët & Chandon. Determined to fight for their livelihoods amid the devastation, the Champenois were soon looking to rebuild their industry.

But the 1920s were fraught with challenges for vine-growers and producers. The Russian market for luxury goods had already been lost as a consequence of the Bolshevik Revolution, and the US alcohol market was closed once the requisite number of American states had ratified the Eighteenth Amendment. To make matters worse, in the early 1920s champagne sales fell considerably as a result of both rising prices in the domestic market and trade barriers in France's other traditional export markets. Champagne consumption plummeted from over 21 million bottles sold in 1920–1921 to fewer than 13 million in 1921–1922.

Sales of champagnes began to improve from 1923 onwards, owing much to consumption in France, which exceeded demand in foreign markets for four consecutive years, from April 1923 to March 1927. Meanwhile, reconstitution of Champagne's ruined vineyards was nearing completion, achieved by grafting French vines to American (phylloxera-resistant) rootstocks, supported by the AVC (Association Viticole Champenoise). Of note was the French law of 22 July 1927 which ended a long-standing conflict between vignerons from the departments of Marne and Aube by recognizing the Aube as part of the wine-producing area of Champagne.

No sooner had the Champenois regained relative economic stability when they faced yet another serious challenge. The global economic downturn, known as the Great Depression, which began in the United States in 1929, led to a drastic decline in sales of luxury products, champagne included. The situation in export markets was particularly difficult: annual exports dropped from over 15 million bottles sold to foreign markets during a mid-1920s peak to fewer than 4.5 million in the early 1930s. With cellar stocks filled to the brim, producers were forced to increase sales in France by means of low prices, thereby enabling the industry to survive. Still, the crisis continued for some time: at least until the period April 1936–March 1937, in which around 40 million bottles of champagne were sold, including nearly 12 million to external markets.

On 10 May 1940 France was invaded by Germany and soon surrendered, signing the Armistice Agreement on 22 June that year. Three-fifths of mainland France, including the region of Champagne, fell under German military occupation. The occupiers recognized champagne as a valuable commodity and producers were made to supply copious quantities to the German army and German-controlled restaurants in France. Oppressed and exploited economically, in April 1941 the Champenois established Comité Interprofessionnel du Vin de Champagne (CIVC), a body whose aim was to defend the interests of winegrowers and producers against unrealistic demands from the German authorities of the Champagne region, represented by Otto Klaebisch.

Over the course of 1942 champagne houses became vital centres of the French Resistance. Producers would inform the Allies about the locations to which large consignments of their wine were being sent at the behest of the German military, thus indicating the likely direction of

Photo 7. A vista of Épernay from vineyards above the west side of the town. Soldiers of the US Third Army might have had a similar view when approaching the town for the purpose of liberating Épernay and Reims from German occupation during the last days of August 1944.
Photo: author, June 2023

German war campaigns. After the Second Battle of El Alamein in November 1942, thousands of empty champagne bottles were found by the Allies in German camp sites. Any sign of assistance offered to the Allies was severely punished, many Champenois being arrested. Among them was the then incumbent CIVC president representing champagne houses, Robert-Jean de Vogüé, who, following his arrest in November 1943, spent the remainder of the war as a prisoner.

The last months of 1943 in Champagne were marked by severe persecution. Claude Fourmon, director of the CIVC, was deported to the concentration camp Mittelbau. Gaston Poittevin, former president of *Syndicat Général des Vignerons de la Champagne* (SGV), in office from 1938 to 1941, was taken to the concentration camp Buchenwald, from which he never returned. The terror in the Champagne region, including German air raids, continued until the end of August 1944, when Épernay and Reims were liberated by the US Third Army, commanded by Lieutenant General George Smith Patton.

In February 1945, Supreme Headquarters of the Allied Expeditionary Force (SHAEF), headed by General Dwight David Eisenhower, moved to Reims. On 7 May that year, the authorized representative of the German High Command, Alfred Jodl, signed the Act of Military Surrender at the commanding general's headquarters, thus putting an end to World War II in Europe. General Eisenhower's stay at Reims must have left a lasting impression on his predilection for fine beverages, as was to become apparent in the fairly frequent choice of champagne for White House dinners held during his presidency in the period 1953–1961.

On 5 May 1945, part of the Fifth Tactical Group of General Philippe Leclerc's Second Armoured Division climbed the Kehlstein summit to reach Eagle's Nest, a building that had been used frequently by Adolf Hitler and the Nazi Party for official and social meetings. Overlooking the Bavarian town of Berchtesgaden, Eagle's Nest proved to have a secret wine cellar, containing an estimated half-million bottles, many of them confiscated from champagne houses during the war. Among the group of French soldiers making the discovery was Bernard de Nonancourt, the youngest son of Marie-Louise Lanson de Nonancourt — head of Maison Laurent-Perrier.

In the latter half of the 1950s, the then president of Moët & Chandon, Robert-Jean de Vogüé, had the groundbreaking vision of taking the concept of producing quality sparkling wines considered as unique to Champagne to territories outside France, thereby taking advantage of their local terroirs. With the assistance of experienced winemaker Renaud Poirier, he found a suitable site for growing Chardonnay and Pinot Noir for sparkling wines in Argentina, in the Andes foothills, specifically in the Luján de Cuyo region in the northwestern part of the winemaking province of Mendoza. The year 1959 marks the birth of Chandon Argentina, Moët & Chandon's first vineyard holding in the New World, originally headed by Baron Bertrand de Ladoucette.

Two years later and we saw the birth of the now world-famous Franciacorta sparkling wines. Crafted using methods modelled on those applied in the production of champagne, Pinot di Franciacorta, Brut: 1961 was as far as we know the first traditional-method sparkling wine created in the Franciacorta region at the northern tip of Italy — the fruit of a concept that had begun a few years earlier as a conversation between Guido Berlucchi and the forward-thinking oenologist Franco Ziliani. Nowadays, the house Guido Berlucchi & C. shares the market with other companies commercializing their sparkling wines under the same territorial brand: franciacorta, with the leading producers including Ca' del Bosco and Bellavista, both based in Erbusco, near the Lombardian city of Brescia.

The mid-1960s were pivotal both for the Davies family and for US quality sparkling wines. In 1965, Jack and Jamie Davies bought a long-abandoned wine estate by the name of Schramsberg, located in the Napa Valley's Diamond Mountain District, which had been founded by German immigrant Jacob Schram in the early 1860s and run by him until his death in 1905. They replanted the historic vineyards with Chardonnay and Pinot Noir with the intention of producing high-end sparkling wines, thus restoring the winemaking house to its former glory. Since the early 1970s, sparkling wines produced at Schramsberg have been consistently served at state occasions hosted by US presidents.

Having been developed for centuries in Champagne, the concept of traditional sparkling winemaking continued to expand worldwide. In 1971, Frans Malan, owner of Simonsig Wine Estate, located in the Stellenbosch district of the Western Cape's Coastal Region, used Chenin Blanc grapes to create his first natural bottle-fermented sparkling wine. Released in 1973 under the name Kaaps Vonkel, the wine was probably also the first traditional-method sparkling wine in South Africa's winemaking history. In the same year, Moët & Chandon, already part of Moët Hennessy, created two winemaking outposts in the Americas: Chandon Brazil, in the Serra Gaúcha region of south Brazil, and Chandon California, in the Napa Valley on the United States Western Seaboard.

In the following decades, the Champagne region continued to serve as a reference point for enterprising vignerons and vintners in remote parts of the world seeking new ventures in sparkling

Photo 8. This Napa Valley welcome sign can be found on State Route 29 between Yountville and Oakville (northern California). The wine region of Napa Valley is home to a great number of producers of quality traditional-method wines. Among the most recognized of those located from Yountville northwards are: Domaine Chandon (also known as Chandon California), Mumm Napa and Schramsberg.
Photo: Casey Scumaci, May 2011

wine production. Andrew Pirie's search for cool-climate areas similar to Champagne resulted in his purchase of land in northern Tasmania, called Pipers Brook Vineyard, in December 1973. The oenologist Tony Jordan was employed by the Moët Hennessy group to find a site for Chandon Australia. On his advice, in 1986 Moët & Chandon's new sparkling wine-producing estate was founded in Victoria's Yarra Valley, the choice dictated by the great diversity of the terroir, guaranteeing winemakers a wide range of options for blending their wines.

For the champagne industry, the 1980s were a time of significant mergers, the champagne house Maison Laurent-Perrier being a prominent example. The company's then head Bernard de Nonancourt created Laurent-Perrier Group having acquired shares in Maison Champagne de Castellane in 1983, and integrated the brands of Delamotte and Salon into the group's portfolio in 1988. This strategic move, aimed at covering the entire spectrum of the market, also represented the fulfilment of a lifelong dream for de Nonancourt. He must have remembered his own discovery of stolen bottles of Champagne Salon 1928 in the Nazi cellar at Eagle's Nest in May 1945, a wine that he had known well from his apprenticeship at Delamotte, a champagne house located at Le Mesnil-sur-Oger.

The 1990s saw major advancements in sparkling winemaking in New World countries, notably South Africa and Australia. In 1992 Cap Classique Producers Association (CCPA) was formed, an organization of South African wine producers making quality bottle-fermented sparkling wines, since then commonly referred to as cap classique wines. This step can be interpreted as the producers' attempt to secure their interests by differentiating themselves from manufacturers of tank-fermented and carbonated wines. Around that time, in Tasmania foundations were being laid for the now award-winning sparkling wine producer House of Arras. Established in 1995, this firm capitalizes on the concept of multi-regional blending; its winemaker Ed Carr sources fruit from Tasmania's six growing areas: Tamar Valley, Pipers River, Coal River Valley, East Coast, Huon Valley and Derwent Valley.

In the early 2000s, Comité Champagne (CIVC) initiated regular environmental assessments of the champagne industry, which resulted in Champagne becoming the world's first winegrowing region to measure its carbon footprint. The initial efforts by champagne producers to reduce their industry's environmental impact focused on sustainable viticultural and winemaking practices, responsible use of electric equipment and vehicles, energy-efficient management of buildings, and packaging. Sensitive to the needs of individual vine-growing sites, the earliest proponents of biodynamic practices in Champagne began to receive attention, including, among others, the houses of Larmandier-Bernier, Françoise Bedel, and Leclerc Briant.

The signs of the business consolidation era in the sparkling wine industry, which had begun in the 1980s, were still evident in the mid-2000s. A notable example is the acquisition in 2006 of the Lanson International firm by the group of champagne-making businesses known as Boizel Chanoine Champagne (BCC). Having tripled in size, the group became one of the largest champagne producers. Known since 2010 by the name Lanson-BCC, it exploits synergies by bringing together eight champagne houses: Champagne Lanson, Champagne Philipponnat, Champagne De Venoge, Champagne

Besserat de Bellefon, Champagne Boizel, Champagne Chanoine Frères, Maison Burtin and Domaine Alexandre Bonnet.

In the second decade of the 21st century, the concept of sparkling wines continued its worldwide expansion to include sites culturally associated more with tea and spices than with wine. In 2013, an heir to Robert-Jean de Vogüé's idea of growing grapes on several continents, Moët Hennessy, part of the LVMH group, created Chandon China, the first winegrowing operation in Ningxia Hui Autonomous Region (northern China) to make premium sparkling wines, using mainly Chardonnay, Pinot Noir, Marselan and Syrah. In 2014, the same producer established Chandon India, a winemaking estate located at Nashik (in the state of Maharashtra), around 200 km northeast of Mumbai; the grapes used by the estate's chief winemaker, Kaushal Khairnar, include Chenin Blanc, Chardonnay, Pinot Noir and Shiraz.

Ever since the champagne industry's concerted efforts towards sustainability — which in the mid-2010s focused on conserving Champagne's biodiversity and reducing its carbon and water footprints — there has been a growing awareness among sparkling wine producers

Photo 9. A used Möet & Chandon champagne cork. The LVMH-owned Möet & Chandon brand continues to be viewed as the epitome of luxury in Asian markets, most notably in Japan. Other champagne brands with high recognition in the Japanese market in the early 2020s are Dom Pérignon, Louis Roederer, Pommery, Laurent-Perrier, Veuve Clicquot, Taittinger and Perrier-Jouët.

Photo: author, October 2023

across the world of the need to conduct their businesses in a way that takes into account the long-term interests of society and the environment. In 2017, Catalonia's DO Penedès became the world's first appellation to provide a category of premium sparkling wines (known as Classic Penedès) made using a combination of traditional cultivation methods and exclusively organic techniques. By 2020, a commitment to crafting wine sustainably was already widely shared by producers of quality sparkling wines intended for medium- to high-end consumers, even though some — such as Louis Roederer (Champagne) and Apogee (Tasmania) — choose not to use their sustainability credentials in their marketing.

At the end of the first quarter of the 21st century, relations between the film industry and the sparkling wine business are closer than they've ever been; many prominent film-industry figures currently represent both trades. Director Francis Ford Coppola's California winemaking operations include an expanding line of sparkling wines named Sofia (after his daughter, Sofia Carmina Coppola). Around 2015, actor and producer Brad Pitt, in cooperation with the Perrin and Péters families, set up the first and only champagne house exclusively devoted to crafting sparkling rosés. In 2022, actor, producer and environmentalist Leonardo DiCaprio became a minority shareholder in the Damery-based champagne house Telmont, encouraged by the environmental credentials of its owner, the alcoholic beverages group Rémy Cointreau.

MAP OF TERMS

My terminological research in the sparkling wine field has led me to produce a concept map of winemaking terms. This map (see Figure 2 overleaf) illustrates the current understanding of the sparkling winemaking process. It constitutes an ordered collection of terms (concept markers), which should be treated as entry points to bodies of advanced professional knowledge provided in the alphabetic display that follows.

This map is designed as a concept organizer to help the user navigate the system of terms relevant to the field of sparkling wines. It is a snapshot of this field's knowledge that covers winemaking methods, tasks, and frameworks of choices and consequences: the options available to the winemaker that influence both the flavour profile and individuality of the final wine. As such, this graphic display can be used as both a lead-in and a companion to the articles in the alphabetic display and in the wine and food pairing sections.

A legend explains the symbols used for the conceptual pathways typical of communication in the sparkling wine trade. A generic parent–child relationship is marked by a green serpentine arrow (with rounded curves), its head pointing at the concept symbol of child (or species). The markers of child concepts that refer to a single parent concept should be understood as indicators of alternative ideas that the winemaker can use. A whole–part relationship is indicated by a brown zigzagged arrow (with sharp angles), the arrow's head pointing at the symbol for part. Because all of the partonomic concepts used enter into sequential relationships, they are numbered according to their functions. Concept links of companionship are marked by means of light lavender lines. A causative relationship uses a grey dashed arrow, the arrow's head pointing at the symbol for effect. A link between performer and activity is symbolized by a solid grey arrow, indicating the concept symbol of activity. The system can ultimately be reduced to three top terms, shown in the centre: 'sparkling wine producers', 'sparkling winemaking' and 'sparkling wine'.

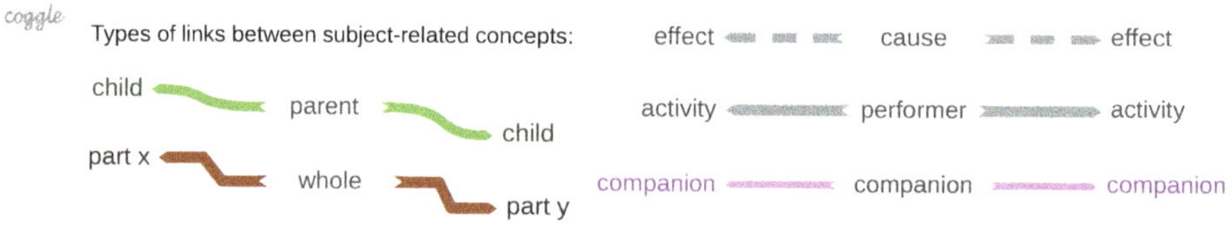

Fig. 1. Symbols of conceptual paths shown on the map of sparkling wine terms (see Figure 2 overleaf).

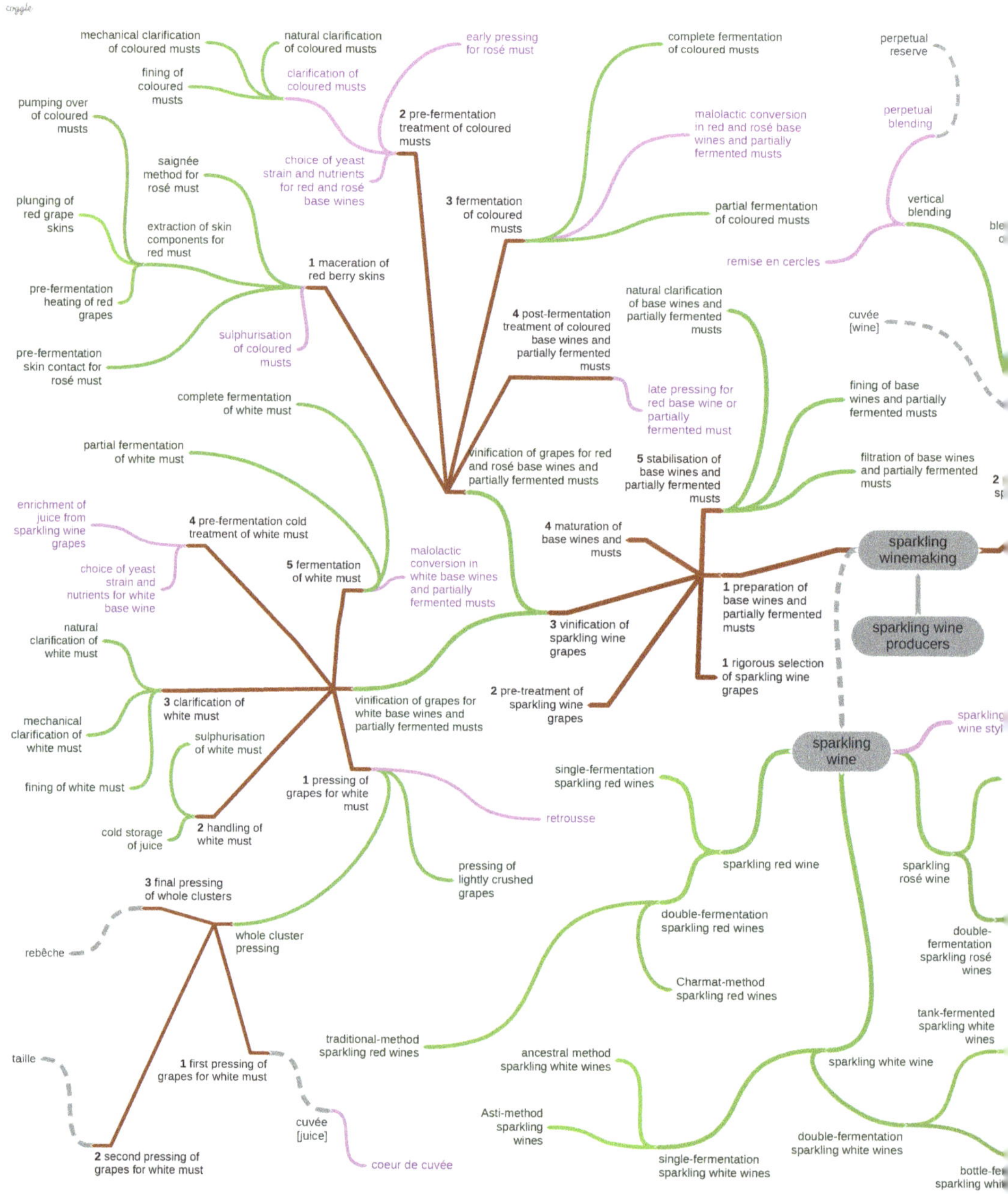

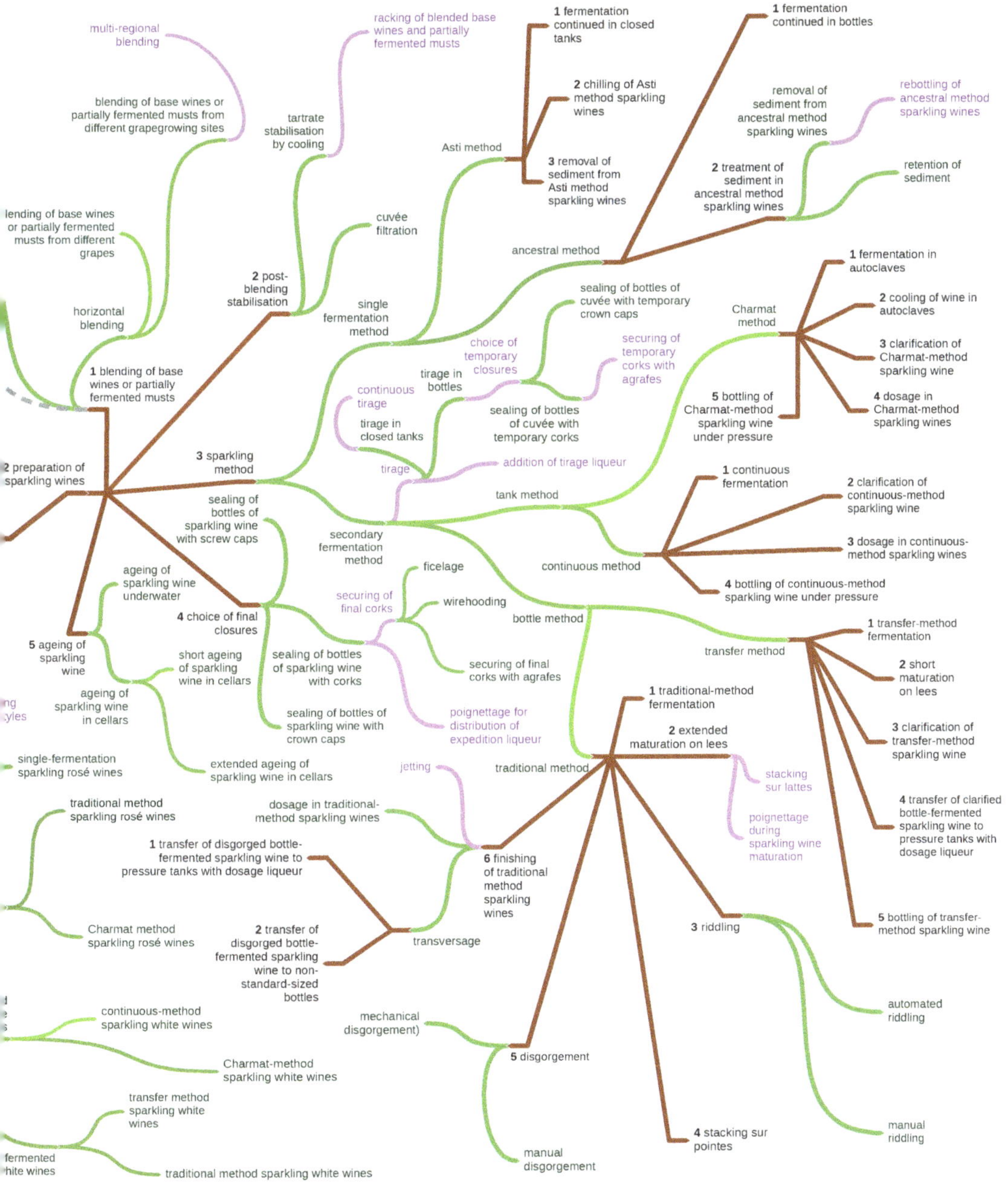

Figure 2. Map of sparkling wine terms.

A–Z DISPLAY

A

addition of tirage liqueur

A winemaking practice that involves supplementing cuvée — a base wine or a blend of base wines destined for secondary fermentation — with a special solution, known as tirage liqueur or *liqueur de tirage*, which typically comprises wine and a sweetening agent in the form of either sucrose extracted from sugar cane or sugar beet, or sugar obtained from grapes, or sweet grape must, or (rectified) concentrated must. The liqueur's sugar content normally ranges between 20 g/l and 25 g/l. This winemaking operation, preceding that of *tirage*, and conducted with the objective of facilitating the second fermentation needed to achieve effervescence, is normally understood as accompanied (or followed) by inoculation of the cuvée with a selected yeast culture, and addition of yeast nutrients, such as nutritive salts (usually ammonium sulphate or diammonium hydrogen phosphate) and growth factors (typically thiamine hydrochloride). Usually, clarifying agents (such as bentonite, tannins or alginates) are also added at this stage to aid the flocculation process and the subsequent removal of yeast cells.

Carrascosa *et al.* 2011: 37; CIVC 2023b; Garofalo *et al.* 2016: 114; Grainger and Tattersall [2005] 2016: 139; Liem 2017: 55–6; OIV 2022: II.4.1–10; Rose 2021: 35–6; SL 2018/2019: 42, 54

maisons-champagne.com/en
www.champagne.fr/en

ageing of sparkling wine

Storage of finished sparkling wines following the sealing of bottles, the aim of which is for the wine to retain and acquire desired characteristics. There are currently two methods of storage, making use of different ageing environments. Traditionally, sparkling wines are stored for various periods in conditions offered by cellars. By contrast, the experimental method, but only currently applied by an insignificant number of producers, involves storing the bottles underwater. Among the main aspects in which these ageing methods differ are storage temperatures, external pressures acting on bottles and bottle closures, and exposure to light and oxygen.

📖 Jeandet *et al.* 2015; Liger-Belair *et al.* 2023: 22,844–52; Mariotti 2023; Rountree 2023; Stevenson [2014] 2022: 244–7, [2015] 2022: 84–7

ageing of sparkling wine in cellars

Storage of finished sparkling wines in cellars, a standard winemaking practice in the creation of sparkling wines with the aim of preserving or assisting the development of the wine's desirable features. The ageing periods tend to be limited in the case of those sparkling wines that do not usually benefit from ageing, such as single-fermentation sparkling wines, a significant proportion of tank-method wines, and also some traditional-method wines. Extended ageing is usually considered beneficial in the case of many traditional-method sparkling wines.

The length of ageing depends on several additional factors. The key ones are the wine's overall quality and the size of the bottle, the latter having an impact on the level of the wine's exposure to oxygen, present in the headspace of the bottle as a result of its ingress at disgorgement (when creating sparkling wines using the traditional method) and in the process of

Photo 10. A box with used champagne corks at the Musée du vin de Champagne et d'Archéologie régionale in Épernay. Following their release, champagnes can continue to age in customers' cellars. Tasters at Moët & Chandon recommend opening standard bottles of the company's non-vintage *brut* champagnes up to around 24 months after purchase. They also suggest that the customer could continue to cellar their vintage champagnes for at least another seven years. Similar recommendations come from professional tasters at Veuve Clicquot; they estimate that most of their vintage-dated cuvées have an ageing potential of over 15 years.

Photo: author, June 2023

(post-disgorgement) ageing. The periods in which sparkling wines are allowed to rest in bottles prior to consumption range from a few weeks to months, or much longer: this depends on the winemaking method used, with particular reference to the time the wine has rested in contact with lees during its sparkling fermentation and maturation, the conditions in which the wine has matured, and the choice of bottle for ageing. Most pétillant naturel wines will be allowed to rest for several decades, which is the case for quality sparkling wines, e.g. champagnes (especially those in large bottles, such as magnums or jéroboams), provided that optimal conditions (including low exposure to light) have been created for their long-term storage.

📖 Avellan 2022b; Douglas 2020; Heiligenthal 2012; Liger-Belair *et al.* 2023: 22,844–52; Rountree 2023; Stevenson [2014] 2022: 244–7, [2015] 2022: 84–7; Teeter 2023

💻 www.moet.com/en-gb
www.veuveclicquot.com/en-int

ageing of sparkling wine underwater

Storage of sparkling wines below water surface, an experimental method of ageing inspired by fortuitous discoveries of well-preserved champagne bottles rescued from shipwrecks between the late 1990s and the end of the first decade of the 21st century. Two of these discoveries received particular media attention. In July 1998, the wreck of the Swedish schooner *Jönköping* was found at the eastern end of the Baltic Sea with around 2,400 bottles of well-preserved Heidsieck & C° Monopole 'Gôut Americain' champagne from the 1907 vintage. The schooner carrying this precious commodity, presumably destined for the Russian Imperial Army, was sunk by a German submarine on 3 November 1916. Similarly, many of the bottles discovered by divers in a shipwreck on the floor of the Baltic Sea near the Finnish Åland archipelago in July 2010 were in particularly good condition, with 47 of the total 168 bottles containing Veuve Clicquot wines from the first half of the 19th century, sealed after 1841. A tasting of some of these wines held in November that year in Mariehamn encouraged the owners of Veuve Clicquot to embark in 2014 on a long-term scientific experiment called 'Cellar in the Sea', designed to last several decades. This involved placing the company's non-vintage Yellow Label champagnes, Vintage Rosé 2004 and *demi-sec* champagnes in an underwater wine cellar called Åland Vault, located 40 metres below the surface of the Baltic Sea, close to where the ~170-year-old wines had been found.

An event organized by champagne producer Veuve Clicquot in late June 2023 can be described as taking wine tours to a new level, unprecedented in the history of wine tourism in Champagne. The company offered oenophiles an immersive four-day experience, lasting from 22–25 June. It combined a visit to Veuve Clicquot's wine cellars at Reims, a gala dinner at the firm's private residence Hôtel du Marc, gourmet meals, and a chartered flight to the Åland Islands for a comparative tasting of Veuve Clicquot's 'Cellar in the Sea' wines.

An emerging trend, this ageing method involves recreating the conditions in which the sparkling wines found in shipwrecks had aged. This involves submerging bottles following disgorgement and storing them in underwater cellars at a depth of 30–60 metres, which guarantees an absence of light, constant temperatures of around 2–4°C and a high pressure corresponding to the pressure inside the bottle. While the managers of Veuve Clicquot are planning to store wines in the depths for several decades, other producers are submerging them for shorter periods. An example is the champagne house Leclerc Briant using the services of Amphoris, a company specializing in running underwater cellars, to store their wines for about a year in the Atlantic Ocean, near the Breton island of Ouessant. Champagne Leclerc Briant 'Cuvée Abyss' 2013, created from Pinot Noir, Meunier and Chardonnay grapes, was bottled in July 2014, aged on lees for three years in the cellars of Épernay, disgorged in March 2017 with no dosage, and submerged in June the same year for 12 months of underwater ageing.

Archeochemical methods are providing new insights about sparkling wines found in shipwrecks. Analyses of the sugar content of the ~170-year-old Veuve Clicquot champagnes found in a shipwreck near the Åland Islands in 2010, combined with evidence from Veuve Clicquot's archives, demonstrate that the popular view that — given the site of the discovery — the wine from the shipwreck was destined for the Russian market may be difficult to defend. Samples of the champagnes were found to contain 144 g/l sugar, half the amount requested by the customers of Madame Clicquot's agent in St Petersburg in the mid-19th century, but close to the demands of the French and German markets at the time. It is therefore likely that the wines were intended for customers in the German Confederation.

A number of other champagne producers have embraced the underwater ageing method, including: Drappier, offering sea-aged champagnes in a range known as 'Les Immersions'; André Heucq, with its H–60 champagnes, created from Chardonnay grapes, aged for 12 months 60 metres below the surface of the sea; and André Chemin, whose portfolio includes sea-aged vintage champagnes such as Champagne Brut 'Millésime' 2015 and Champagne Brut 'Excellence' 2007.

Carr 2021; Edwards 2022; Forshaw 2023; Jeandet *et al.* 2015; Jestin 2023; Jiménez 2014; Juhlin 2023; Kermode 2019; LVMH 2014; Tagliabue 2010; TFB 2019a, 2019b

www.amphoris.fr/en
champagneandrechemin.fr
www.champagne-drappier.com/en
champagne-heucq.com/en
www.heidsieckandco-monopole.com
www.leclercbriant.fr/en
maisons-champagne.com/en

agrafage

≈ securing of final corks with *agrafes*

≈ securing of temporary corks with *agrafes*

Photo 11. A bottle of UJVR Clairette de Die Tradition 'Cuvée César', a lightly sparkling wine made using the ancestral method from grapes grown in the Diois region (central part of the Rhône Valley, southeastern France).
Photo: author, September 2024

ancestral method

A sparkling winemaking method that involves putting partly (or almost fully) fermented must into bottles and sealing the bottles tightly. The remaining sugar is then gradually transformed into alcohol and carbon dioxide, which — when released in contact with fresh air — causes the resultant wine to effervesce.

Interestingly, the concept is sometimes extended — especially by Italian producers — to include methods of making sparkling (mostly lightly sparkling) wines by means of refermentation facilitated by adding a small amount of natural grape must to partly or — in some cases — almost completely fermented base wines prior to bottling.

The method of creating lightly sparkling wines using a single fermentation (also known by the French terms *méthode ancestrale* and *méthode rurale*, or by their Italian equivalent *metodo ancestrale*) has its roots mainly in old winemaking traditions nurtured in several wine regions of France and Italy. These include, among others, the winemaking areas of Bugey, Diois, Gaillac and Limoux (southern France) and the Prosecco zone (northeastern Italy).

Various forms of the ancestral method have developed, modern incarnations resulting in pétillant naturel wines, commonly known as pét-nats, a term currently used mainly for slightly sparkling wines created using some form of the ancestral method in regions with little or no tradition of producing them. Drawing inspiration from the French and Italian traditions, a growing number of winemakers make pétillant naturel wines, with some of them introducing modifications to the ancestral method at different stages of the production process. At Benjamin Bridge, a Canadian (Nova Scotian) wine producer of low-intervention pétillant naturel wines from the Gaspereau Valley, the head winemaker Jean-Benoit Deslauriers uses indigenous yeast to create effervescence in the wine. The company's Benjamin Bridge Pét Nat Rosé 2021, made from Geisenheim, Gamay and Cabernet Franc grapes, has been released with the yeast sediment, as have *col fondo* wines in the Prosecco zone. An interesting example of a variation on the ancestral method is provided by Chandon Australia's chief winemaker Dan Buckle. The production of the company's Chandon Ancestrale Rosé 2019 involves the use of cultured yeast and the addition of a small amount of dosage liqueur — processes not allowed in Montlouis-sur-Loire when making the region's signature pétillant originel wines, such as La Grange Tiphaine's 'Nouveau Nez' 2021 and Domaine de la Croix Mélier's 'La Bulapapa'.

📖 Buglass 2022: 227; Collin [2004] 2009: 207–8; Fielden [1994] 1998: 294; Funke and Parla 2019: 83; Goode 2023; Hooke 2021; Lorch [1994] 2015: 175–6; Maker 2021; Sussman 2020b

💻 benjaminbridge.com
borgoluce.it/en
clairette-de-die.com/en/
winesofnovascotia.ca
www.chandon.com.au
www.prosecco.it/en
www.renardat-fache.com

ancestral method sparkling red wines

≈ single-fermentation sparkling red wines

ancestral method sparkling rosé wines

≈ single-fermentation sparkling rosé wines

ancestral method sparkling white wines

The largest category of effervescent wines made using a single fermentation of grape juice continued and finishing in the bottle. The category comprises (mostly) lightly sparkling wines. Wines of this category include those obtained by bottling partly fermented white juice as well as those created by refermenting almost completely fermented white juice, a process facilitated by adding to the base wine a small amount of (sweet) grape juice before the bottles containing this wine are sealed. Notable examples of lightly sparkling wines obtained by bottling partly fermented grape juice are those made in the Diois region (central part of the Rhône Valley, southeastern France): Union Jeunes Viticulteurs Récoltants' Clairette de Die Tradition 'Cuvée César' and Patrick Lantheaume's Clairette de Die Tradition. By contrast, lightly sparkling wines obtained by refermenting base wines using sweet grape juice can be found in the portfolios of many Italian producers. One example is Costadilà Bianco '280', made from Glera, Bianchetta Trevigiana, Verdiso, Perera and Boschera grapes by refermenting the base wine in bottles aided by the addition of juice obtained from dried grapes. Bottled without filtration, the wine contains yeast sediment, a feature responsible for its cloudy appearance.

Gerling *et al.* 2023; Maker 2021; Parzen 2014, 2018; Quaile 2023: 23

clairette-de-die.com/en/
www.clairette-de-die-lantheaume.fr
www.costadila.it/en

ancestral method sparkling wines

≈ ancestral method sparkling white wines
≈ single-fermentation sparkling red wines
≈ single-fermentation sparkling rosé wines

Asti method

A sparkling winemaking method that involves completing the fermentation of (partly or almost fully fermented) musts in pressure-resistant tanks which are tightly sealed. The remaining sugar is gradually transformed into alcohol and carbon dioxide, which — when released in contact with fresh air — causes the resultant wine to effervesce.

The method of creating semi- or almost fully sparkling wines using a single fermentation taking place in closed tanks has its roots in winemaking traditions nurtured in the Piedmont region (northwestern Italy). Practices associated with crafting Asti wines in Piedmont serve as a benchmark for many winemakers in the rest of the world.

While the original Asti wines, including the fully sparkling Asti wines known as Asti Spumante and the semi-sparkling Asti wines called Moscato d'Asti, are crafted using Moscato Bianco grapes, producers using the Asti method in regions other than Piedmont (including winemaking regions outside Italy) use locally grown grapes.

Prominent examples include the Slovakian producer Hubert J.E. and the producers of sparkling ice wine in Canada. Winemakers applying the Asti method at Hubert J.E. use half-fermented musts made from three different grape varieties: Irsay Oliver, Muscat Ottonel and Moravian Muscat. Marco Zamuner, a winemaker at Magnotta Winery, a family winery located in Ontario (the southernmost province of Canada), crafts the company's sparkling ice wines using Vidal Blanc grapes.

Caliari *et al.* 2015; Clarke [2000] 2003: 20, 28, 41; Cravero 2023; WSET 2020: 65–8; Wu 2015; Zamuner 2021

www.astidocg.it/en
www.hubertsekt.sk/en
www.magnotta.com

Asti-method sparkling white wines

Effervescent white wines crafted using the method of a single fermentation, the final phase of which finishes in a closed fermentation tank. This category encompasses primarily single-fermentation wines made in the region of Piedmont (northwestern Italy). Nonetheless, Piedmontese producers of Asti wines have their followers not only in Europe, most notably Slovakia, but also in winemaking countries as far afield as Brazil and Canada.

Asti-method wines can range from extremely sweet (as in the case of single-fermentation semi-sparkling wines crafted in Piedmont, referred to as Moscato d'Asti, or in the case of sparkling ice wines made by producers in the Canadian province of Ontario) to very dry (as exemplified by the driest Asti Spumante wines). One example of very sweet Asti wine from Piedmont is Moscato d'Asti produced by Cascina Bo (a winemaking estate in Canelli), which boasts a sugar content of 135 g/l sugar.

Sweet Asti-method wines made in Brazil's southern wine-producing region Serra Gaúcha may not be as sweet as Moscato d'Asti made in Piedmont, with their typical sugar contents similar to those expected in sweet styles of Asti Spumante. An illustrative example is provided by Moscatel Espumante, a wine in the portfolio of Brazilian producer Cooperativa Vinícola Garibaldi. Made from Moscatel Branco and Moscato Giallo grapes, the wine boasts a sugar content of 65 g/l. By contrast, the Piedmontese wine Tosti 1820 Asti Secco has a sugar content of 17 g/l.

Very dry styles of Asti-method wines are currently gaining ground. The winemaker Matteo Soria of the family-run winemaking company Azienda Agricola Soria Matteo, located in Castiglione Tinella (in the southern part of Piedmont), has managed to craft Asti Brut 'Bric Prima Bella', a single-fermentation wine from the Moscato Bianco grapes with a sugar content of 9 g/l. Interestingly, after fermenting in steel tanks at 20°C for 10 days, this Asti Spumante wine was allowed to rest for six months on the yeast sediment before bottling.

Araujo *et al.* 2019; Caliari *et al.* 2015: 393–8; Mileham 2022; WSET 2020: 64–9; Wu 2015; Zamuner 2021

cascinabo.it/en
www.astidocg.it/en
www.matteosoria.it
www.vinicolagaribaldi.com.br

automated riddling

A mechanized process in the traditional method of transforming cuvée into effervescent wine that involves rotating bottles of sparkling wine with the aim of dislodging the yeast deposit formed in the bottle during the second fermentation and extended maturation. Bottles are stacked in machines known as *gyropalettes* or *girasols*, large (usually) cube-shaped metal crates, each with the capacity to rotate up to 504 bottles at a time. The machines turn and shake the bottles of sparkling wine (containing lees) at regular intervals progressively towards an inverted vertical position, causing the lees to move to the neck of the increasingly inverted bottle, which facilitates their subsequent removal.

The use of *gyropalettes* reduces the time needed for removing the yeast deposit from bottle-fermented wine from several weeks (required when the operation is performed by hand) to several days, usually around a week. The invention of the device in the late 1960s, fully developed by the mid-1970s, is credited to Claude Cazals and Jacques Ducoin.

Among its earliest adopters (on an industrial scale) have been cava producers and champagne houses, notably Piper Heidsieck and Taittinger. Initially applied by manufacturers of traditional-method wines in France, Italy and Spain, riddling machines, currently ranging from semi-automatic (hand-operated) to fully automated (computer-controlled) devices, are now in widespread use among producers of high-quality sparkling wines around the world.

📖 CIVC 2023b; Edwards 2009: 202; Edwards L. 2023; Epstein 2011: 65; Fielden [1994] 1998: 290; Halliday and Johnson [1992] 2013: 116; Kunc 2012: 142; Melnik 2023; Pessey 2002: 93–4; Quaile 2023: 180; Shevela 2020; WSET 2020: 9, 51

💻 maisons-champagne.com/en
www.champagne-cazals.fr/en

B

blending of base wines or partially fermented musts

A winemaking operation that involves combining base wines from grapes collected in one year's harvest, in which case the activity can be referred to as horizontal blending, or base wines created in different vintages, in which case one can speak about multi-vintage (or vertical) blending. One of the hallmarks of sparkling wine production, the blending operation, sometimes also referred to as *assemblage*, can serve various purposes subsumed under the general purpose of obtaining cuvée with desired characteristic. Typical reasons for blending are to: achieve balance (a harmonious relationship between various components of the wine's profile); ensure consistency (important with non-vintage blends produced to overcome vintage variations); develop a particular style (for instance, creating wines made in the early-drinking style, requiring ripe fruit flavours with moderate acidity, as opposed to wines created in the age-worthy style, which — usually — require concentrated flavours and relatively high acidity), add complexity to the final blend; and minimize faults identified in the wine. The activity of combining base wines comprises: extensive tasting, assessments, bench blending and final blending.

📖 Atkins [1994] 2015b; CIVC 2013: 27, 2023b; Halliday and Johnson [1992] 2013: 114–15; Hubble 2023; Liem 2017: 27–31; Machard de Gramont 2022; Nagórka 2013: 29; Phillips [2012] 2014; OIV 2022: II.4.1–6; WSET 2020: 6–7

💻 maisons-champagne.com/en
www.domperignon.com

Photo 12. A statue of the Benedictine monk Dom Pérignon in the courtyard outside Moët & Chandon's premises in Épernay. An innovative cellar master at the Hautvillers Abbey (Abbaye Saint-Pierre d'Hautvillers), working there from around 1668 until his death in 1715, Dom Pérignon is remembered today mainly for perfecting the art of blending to obtain wines characterized by great harmony and balance.
Photo: author, July 2024

blending of base wines or partially fermented musts from different grape-growing sites

A winemaking procedure of combining two or more base wines created from grapes grown in various sites, which implies differences in these wines' sensory attributes, the aim being to produce a wine with an increased complexity of flavours. Meticulous blending of wines from different vineyards requires that the identities of these wines be preserved, with base wines from each grape-growing site created separately in barrels or vats before being considered for their suitability as components of the final blend.

> A blend of wines from various vineyards, Mumm's Brut 'Cordon Rouge' Champagne has been among the world's best-recognized champagnes ever since its first launches, with the brand's history dating back to the mid-1870s. Its recognition was further cemented by its appearance in the 1942 film *Casablanca*, numbered among Hollywood's most successful productions. Based on Murray Burnett and Joan Alison's play *Everybody Comes to Rick's*, and directed by Michael Curtiz, the film features a flashback scene, where, in June 1940, at a Parisian café called La Belle Aurore, Rick Blaine, portrayed by Humphrey Bogart, and Ilsa Lund, played by Ingrid Bergman, share glasses of G.H. Mumm's 'Cordon Rouge' Champagne as the German army approaches Paris.

Gosset Champagne Celebris Vintage 2012 has been created using 70% Chardonnay and 30% Pinot Noir, with these grapes sourced from a large number of Champagne's villages, including (among others): Ambonnay, Avize, Aÿ, Bouzy, Cramant, Cuis, Grauves, Le Mesnil-sur-Oger, Louvois and Trépail. Another notable multi-vineyard sparkling wine made in Champagne is G.H. Mumm's Brut 'Cordon Rouge' Champagne, a blend of Pinot Noir grapes sourced from Montagne de Reims and Aube with the grapes of Chardonnay collected in Côte des Blancs, Côte de Sézanne, Vitryat and Aube, and Meunier, grown in Vallée de la Marne, Vallée de l'Ardre and Montagne de Reims.

Production of multi-vineyard sparkling wines is by no means exclusive to the Champagne region. Chandon's 50th Anniversary Winemakers' Blend is an interesting multi-vineyard sparkling wine developed at Chandon California by four winemakers: Dawnine Dyer, Wayne Donaldson, Tom Tiburzi and Pauline Lhote, with grapes used for this particular wine (69% Chardonnay, 21% Pinot Noir, and 10% Meunier) sourced from three wine districts: Los Carneros, Mt Veeder and Yountville.

CIVC [2014] 2022: 10; Edwards 2009: 82; Liem 2017: 11–13, 27–31; SL 2018/2019: 6, 14

www.champagne-gosset.com/en
www.chandon.com
www.mumm.com/en-ww

blending of base wines or partially fermented musts from different grapes

A winemaking procedure of combining two or more base wines or partially fermented musts, created from different grape varieties, with the grapes used for the production process — in many cases coming from a single grape-growing site. Notable single-vineyard sparkling wines are produced by Domaine Carneros (founded by the Taittinger family in 1987) and Artesa (Estate Vineyard and Winery) (established by the family business Raventós Codorníu in 1991), winemaking companies representing the wine district Los Carneros, shared between California's two famous wine regions: Napa Valley and Sonoma County. Their wines include: Domaine Carneros' Single Vineyard Brut 2018, created from the grapes of Chardonnay and Pinot Noir, grown in the Tula Vista vineyard, cooled by breezes and fogs from San Pablo Bay; and Artesa's 'Estate Vineyard' Grand Reserve Demi-Sec 2015, composed of 55% Chardonnay and 45% Pinot Noir, sourced from separate blocks of the producer's 150-acre vineyard.

CIVC [2014] 2022: 10; Coates 2000: 547; Frerson 2021; Liem 2017: 27–31, 55–6, 277, 302; SL 2018/2019: 6; WSET 2020: 6–7, 23–4, 49

www.artesawinery.com
www.domainecarneros.com
www.gobelsburg.at/en
raventoscodorniu.com/en
www.taittinger.com/en

Photo 13. Ried Lamm, a south-facing vineyard in the wine region Kamptal, in Lower Austria (Austria's northeastern part), is a source of Grüner Veltliner grapes. Base wine from these grapes has been used as the main component of the cuvée prepared for the traditional-method sparkling wine Schloss Gobelsburg Langenloiser Sekt Extra Brut 2012. The other grapes used for the production of this blend were Riesling and Pinot Noir.
Photo: courtesy of Schloss Gobelsburg (Michael Moosbrugger), October 2013

blending of vins clairs

≈ blending of base wines or partially fermented musts

bottle method

A method of transforming base wine into a finished sparkling wine that involves management of the processes of second fermentation in the prepared cuvée and the resultant wine's maturation using bottles, each of which can withstand more than six atmospheres of pressure building up during the fermentation process. Winemakers distinguish between two general types of bottle method: the traditional method and the transfer method. Among the main differences between these two methods are: the duration of maturation on lees following the second fermentation process (prolonged contact with lees is normally required for traditional-method wines while brief contact with lees is usually the norm in the transfer method); and the method for the removal of sediment from sparkling wine prior to bottling (decanting of sparkling wine in the transfer method from bottles into pressure tanks following maturation eliminates the need for the time-consuming riddling process, viewed as essential in the traditional method of sparkling wine production).

CIVC 2023b; Jackson [1994] 2008: 539, [2002] 2009: 374; Kunkee and Ough 1966: 643; Nagórka 2013: 32; SL 2018/2019: 6–8; Skelton 2021: 151–2; WSET 2020: 7–10

bottle-fermented sparkling white wines

A category of double-fermentation sparkling white wines encompassing wines that have been made effervescent as a result of the second fermentation in sealed bottles. These wines can be divided into traditional-method sparkling white wines, including (among others) cap classiques, cavas, champagnes, franciacortas, etc., and sparkling white wines made using the transfer method. Bottles with sparkling wine crafted by the transfer method can be labelled 'fermented in the bottle' or 'bottle-fermented', which — by implication — distinguishes them from bottles of sparkling wine crafted by the traditional method, the latter undergoing secondary fermentation in the same bottles in which they are sold.

CIVC [2014] 2022: 14–15; Howe 2003: 149; Liem 2017: 297; O'Connor 2024; OIV 2022: II.4.2–1; WSET 2020: 10; Zoecklein 2002: 3

maisons-champagne.com/en
beykush.com/en
www.capclassique.co.za
www.cava.wine/en
www.champagne.fr/en

bottling of transfer-method sparkling wine

Rebottling of the cuvée after it has been transformed into a finished effervescent wine through a sequence of winemaking processes, including: second fermentation of the wine in the cellar bottle, short maturation on lees, and clarification of the wine followed by its transfer to pressure tanks containing *liqueur de dosage*. This operation, often preceded directly by sterile filtration, consists of pumping the transfer-method sparkling wine under pressure into clean bottles in which the wine is to be sold.

Grainger and Tattersall 2005: 97–8; Jackson [2002] 2009: 373; LaVilla 2010: 264; Nagórka 2013: 112–13; Skelton 2021: 152

bouchage

≈ choice of final closures

Photo 14. A cork removed from the bottle of Beykush Winery Beresagne Blanc de Blancs Méthode Classique Brut 2015, a bottle-fermented sparkling white wine made from 65% Chardonnay, 26% Rkatsiteli and 9% Aligoté, grown in the Beykush Winery estate vineyards located on the Black Sea coast of the Mykolaiv region.

Photo: author, September 2024

bulk method

≈ tank method

C

Charmat method

A method of sparkling wine production that involves provoking a second fermentation in base wine placed in a tightly sealed pressure-resistant tank, called an autoclave. The method was invented in 1895 by the Italian oenologist Federico Martinotti, and perfected in France by Jean-Eugène Charmat around 1907. This method of producing effervescence, assuming almost no or brief maturation periods (usually up to a few weeks), is generally preferred when the winemaker wants to preserve a wine's underlying fresh fruit flavours and considers autolytic characteristics in the resultant wine as undesirable.

While the emphasis in the Charmat method (also called the Charmat process) is almost always placed on the final wine's freshness, some quality-focused producers slow down the secondary fermentation process by reducing the temperature, or allow the newly made sparkling wine to continue to rest in contact with lees for a few months for additional complexity. Some winemakers in the Prosecco zone, such as Fabio Zardetto, who use the Glera grapes, adopt the long Charmat method, locally known as Charmat *lungo*. It involves an extended maturation period following the second fermentation, lasting around six months (or longer). This method has also been adopted by Brazilian winemakers Lucas Foppa and Ricardo Ambrosi, who set up the wine-producing firm Tenuta Foppa & Ambrosi in Garibaldi (south Brazil) in 2019. Their Premier Cuvée 'Viticcio' Espumante Natural Brut Bianco, released in 2022, has been produced by fermenting base wine crafted from the grapes of Chardonnay, Pinot Noir and Riesling Italico, with the wine resting in contact with lees for six months.

Prosecco is a sparkling wine crafted by the Charmat method in Italy's five northeastern wine-producing provinces of the Veneto region (Belluno, Padua, Venezia, Vicenza, Treviso) and four provinces of the Friuli-Venezia Giulia region (Gorizia, Pordenone, Trieste, Udine) using at least 85% Glera grapes. Among other varieties favoured by the producers of prosecco for the final blend are Bianchetta Trevigiana, Chardonnay, Perera, Pinot Bianco (Pinot Blanc), Pinot Grigio (Pinot Gris), Pinot Nero (Pinot Noir) and Verdiso.

Some makers employ the long Charmat method to impart distinct yeasty characteristics to their wines, leaving the wine in contact with lees for considerably longer than the standard period of around six months. A notable example is provided by the winemakers at the family-run wine estate Eugenio Collavini, located in Corno di Rosazzo (in the Italian region of Friuli-Venezia Giulia). After the slow process of secondary fermentation of base wine made from the Ribolla Gialla grapes in large horizontal pressure tanks at 12°C, the makers of the wine Spumante 'Ribolla Gialla' Dosaggio Zero–2015 kept their wine in contact with lees for around 30 months prior to bottling.

The long Charmat method is sometimes associated with the use of special tanks equipped with screw agitators for mechanical stirring of lees during the prolonged contact of sparkling wine with the yeast sediment. Applied in the maturation process, which lasts for at least 12 months, these agitators bring the fermentation sediments back into suspension, increasing the effect of the lees contact on the complexity of the wine's flavour. This variation of the Charmat process bears the name of the Italian oenologist and inventor Nèreo Cavazzani, whose work on the development of this method began in the late 1970s. Winemakers at Astoria (Vini) applied this method for their Kálibro range of vintage sparkling wines. The bottling of Astoria Vini's Cuvée Brut 'Kálibro'–Metodo Cavazzani (Charmat Lungo)–Millesimato 2015 was preceded by storage of the wine on its lees for a year at a temperature of 6–8°C.

📖 Bakker and Clarke [2004] 2012: 23; Campanale and Stein 2022: 42; Cisilotto *et al.* 2023; Goldstein 2023; Morris 2019; Skelton 2021: 151; Sullivan 1998: 62; Sussman 2021: 12–14; TFA 2022; Vittek 2023; WSET 2020: 11–12, 61

collavini.it/en
kuban-vino.ru
it.zardettoprosecco.com/en
www.tenutafa.com.br
www.astoria.it/en
www.prosecco.it/en
www.prosecco.wine/en

Charmat-method sparkling red wines

A category of sparkling red wines made using the double-fermentation method with the second fermentation taking place in a sealed fermentation tank. Among sparkling red wines produced using the tank method that have recently attracted the interest of wine critics are Venturini Baldini 'Rubino del Cerro'–Secco (made from the grapes Lambrusco Montericco, Lambrusco Salamino and Lambrusco Grasparossa, grown in the Reggio Emilia province in the Emilia-Romagna region [in northern Italy]) and Hollick Sparkling Merlot–Coonawarra 2021, a monovarietal vintage wine from South Australia's Coonawarra region.

Edwards, P. 2023; Speller and Currin [1994] 2015: 470–1; WA 2022: 3–4, 18; WSET 2020: 54, 72

hollick.com
venturinibaldini.it/en
www.woodparkwines.com.au

Charmat-method sparkling rosé wines

A category of sparkling rosé wines made using the double-fermentation method with the second fermentation carried out in a sealed fermentation tank. An interesting example has been provided by the Italian producer Venturini Baldini. The firm's sparkling wine Venturini Baldini 'Cadelvento' Rosato Spumante–Secco has been crafted from the grapes of Lambrusco Sorbara and Lambrusco Grasparossa using the long Charmat method. An illustrative example is also offered by producer Domaine Bousquet, located in Argentina's wine-producing province of Mendoza. The company's salmon-hued wine Domaine Bousquet Charmat Organic Brut Rosé has been crafted from 75% Pinot Noir and 25% Chardonnay. Yet another wine made using the Charmat method is in the portfolio of the English producer Flint Vineyard, set up by Ben Witchell in the village of Earsham (Norfolk) in 2016. The vintage-dated wine Flint Vineyard Charmat Rosé 2021 (representing the company's range of dated Charmat-method rosé wines initiated in 2018) is a unique blend of wines made from the grapes of Pinot Noir, Rondo, Solaris, Reichensteiner, Bacchus and Cabernet Cortis.

Baker 2023; James 2013: 117; Jukes 2018; Kirton 2023; Mobbs 2022

domainebousquet.com/en
flintvineyard.com
venturinibaldini.it/en

Charmat-method sparkling white wines

A category of tank-fermented sparkling white wines, these wines are obtained from base wine subjected to a second fermentation in pressure-resistant tanks, a process triggered by the addition of tirage liqueur. Wines made using the standard Charmat method, with the wine maturation period following the secondary fermentation being brief, tend to exhibit primary aromas of the grapes from which the cuvée has been prepared. Flavour notes of these wines, frequently described as fresh, fruity and floral, can be expected of both a standard DOC prosecco, produced in one of the provinces of the Prosecco zone, and a prosecco-style wine of decent quality crafted in other winemaking areas. Charmat-method wines with increased complexity are obtained by keeping the sparkling wine on lees for a period of between a few weeks and a few months before the wine is bottled. Finally, sparkling white wines made using the long Charmat method — another derivation of the Charmat process, sometimes also referred to as the Cavazzani method — are wines with flavours enriched by extended lees contact (typically lasting for a year or longer), with the lees — in the case of modern wineries — stirred mechanically by the screw agitators or paddles with which some pressure tanks have been equipped.

The full spectrum of Charmat-method white wines can be found in the portfolio of Astoria (Vini), a well-respected producer of prosecco and other tank-fermented sparkling wines. A notable example of sparkling wine made using the method of secondary fermentation followed by almost no maturation on lees is Astoria Vini's Astoria 'Yu Sushi' Spumante–Extra Brut, crafted from a variety of grapes from the province of Treviso. By contrast, Astoria 'Casa Vittorino' Valdobbiadene Prosecco Superiore, Rive di Refrontolo–Brut, one of Astoria Vini's premium proseccos, exhibits fruity flavours (typical of prosecco) with added complexity, which may have resulted from the several-week period in which the wine rested in contact with lees prior to bottling. The flavour of the company's wine Astoria Cuvée Brut 'Kálibro'–Metodo Cavazzani (Charmat Lungo)–Millesimato 2015, crafted using 80% Chardonnay and 20% Pinot Bianco (also called Pinot Blanc), is described as showing notes of ripe fruit, acacia honey and vanilla, an effect produced by the wine's intense contact with lees before bottling for the 12-month period.

Grainger 2009: 29–32; Lindgren and Leahy 2023: 276–8; McKay *et al.* 2011: 341; Pomarici *et al.* 2019: 421–3; Speller [1994] 2015c: 647–8; TFA 2022; Walkey 2019

www.astoria.it/en
www.lemanzane.com/en
www.prosecco.it/en
www.prosecco.wine/en

chilling of Asti-method sparkling wines

Rapid reduction of temperature of the Asti-method wines during their fermentation in closed tanks. The temperature is lowered to reach the water-freezing point in order to stop the fermentation process when the desired concentrations of residual sugar in the fermenting liquids, corresponding to the desired pressures and alcohol contents, are obtained. For the semi-sparkling wines known by the name Moscato d'Asti, fermentation is halted when the alcohol content reaches between 4.5% abv and 6.5% abv, with typical sugar contents in these wines being around 130 g/l. By contrast, for fully sparkling wines, commonly known as Asti Spumante, fermentation is stopped when the alcohol content has reached between 6% abv and around 8% abv for sweet styles and above 10% abv for dry styles. This operation of

Photo 15. A vineyard in the hamlet of Formeniga, in the province of Treviso (northeastern Italy). This vineyard is an important source of the Glera grapes intended for the highest-quality proseccos in the portfolio of the wine producer Le Manzane, notably Prosecco Le Manzane 'Springo Blue' Rive de Formeniga DOCG.
Photo: MartaRęgowska, November 2023

refrigerating the liquid is practised by the producers of Asti wines in the region of Piedmont and their followers in the rest of the world. The operation is followed by the removal of yeasty sediment.

📖 Bastianich and Lynch [2002] 2005: 158; Robinson [2000] 2005: 222; Sacchetto 2023; Slinkard 2013: 202; Speller [1994] 2015b: 541–2; WSET 2020: 65–8; Wu 2015

💻 www.astidocg.it/en

choice of final closures

A winemaker's decision-making task concerning the appropriate closure (wine bottle stopper) for finished sparkling wines. While cork closures are mandatory in Champagne for finished sparkling wines, some alternatives for these wines have also been developed and are currently used in other sparkling winemaking regions. The most commonly applied include crown closures and screw caps.

📖 Crouvisier-Urion *et al.* 2021; Grainger 2021: 417–19; Liem 2017: 295–6; Phillips 2019

💻 www.champagne.fr/en

choice of temporary closures

A winemaker's decision-making task concerning the appropriate closure (wine bottle stopper) for their cuvée, a base wine or a blend of base wines destined for secondary fermentation in the bottle. Usually, the winemaker chooses between two alternatives: temporary (crown) caps and temporary corks.

📖 Fleet [1985] 1998: 233; Karlsson 2016; SL 2018/2019: 43; Varlamov and Aslamazov [2001] 2019: 250; Wilson [1998] 1999: 72

💻 www.champagne.fr/en
torello.com/en

choice of yeast strain and nutrients for red and rosé base wines

A decision-based winemaking practice taking place following a period of

Photo 16. Bottles of sparkling wine closed by temporary corks and secured with clasps called *agrafes*, maturing on their lees in the dark cellars of the corpinnat producer Torelló. These are bottles of Gran Torelló Corpinnat Brut Nature 2016, produced using grapes from the firm's Can Martí estate: 48% Xarello, 47% Macabeo and 5% Parellada, and released after over 72 months of rest in contact with lees.

Photo: MartaRęgowska, April 2023

pre-fermentation maceration of red berry skins which involves the selection of either slower-acting wild yeasts, which are capable of adding complexity to the resultant base wine, or cultured yeasts, which are much more predictable. Using the natural microflora, present both on the grapes and in the winery, requires close monitoring of the fermentation process, as the activity of the native yeast strains, dominated by non-Saccharomyces yeasts, can lead to significant depletion of yeast nutrients, notably nitrogen and thiamine, which must be supplemented. Prominent among the growing number of winemakers committed to applying indigenous yeasts for crafting base wines destined for rosé sparkling wines are Delphine Boulard, head of Champagne Francis Boulard et Fille, a family-owned champagne-producing business founded in 2009, and the Larmandier family, running the champagne-making firm Champagne

Larmandier-Bernier. Behind the employment of native yeast strains for primary fermentation is the conviction that these strains are integral to the expression of an individual terroir.

By contrast, cultured yeasts tend to be selected for their reliability and predictable contributions to the production of flavour compounds, such as esters and terpenes, which, depending on the strain used, can differ not only in quality but also in quantity. Inoculation of coloured musts with a selected yeast strain is now common. Musts with cultured yeasts also need monitoring, especially for the content of yeast nutrients, notably nitrogen, which impacts the sugar transport kinetics during fermentation. The content of nitrogen in the must is variable, ranging between 50 mg/l and 500 mg/l, with the minimum yeast assimilable nitrogen (YAN) requirement for low-risk fermentations being around 100 mg/l for red musts and 150 mg/l for lightly coloured musts. Musts with lower concentrations are usually considered as nitrogen-deficient.

Halliday and Johnson [1992] 2013: 218; Holzapfel 2014/2015: 21; Jackson [1994] 2008: 303–4, 541; Julien *et al.* 2014: 6–7; Makarov and Lutkov 2021; Zoecklein 2002: 3, 9–10

larmandier.fr/en
www.awri.com.au

choice of yeast strain and nutrients for white base wine

A decision-based winemaking practice taking place after the clarification of white juice which involves selection of either wild yeasts, which are credited with adding complexity and lending a distinct character to the resultant base wine, or cultured yeasts, which are appreciated for their predictability. Native yeasts are present both on the grapes and in the winery, but their use requires close monitoring of the fermentation process. The activity of some of the wild yeast strains, in most cases dominated by non-Saccharomyces yeasts, can lead to depletion of yeast nutrients, such as nitrogen and thiamine. If this is the case, these nutrients must be supplemented. A growing number of winemakers now vinify all or some of their gapes using indigenous yeasts. Prominent among them are Emmanuel Lassaigne of the champagne producer Jacques Lassaigne, who joined the family business in 1999; Vincent Laval, running the champagne-making family firm Georges Laval ever since 1996; and Jennifer Doyle, the head winemaker of the wine house Jansz Tasmania, part of the Hill-Smith Family Estates since 1997. The use of yeast strains native to a wine's place of origin is increasingly viewed as crucial for the full expression of its terroir.

Cultured yeasts tend to be selected primarily for their predictable contributions to the production of flavour compounds, such as esters and terpenes. Primary fermentation carried out by selected yeasts needs monitoring on account of changes in the concentrations of yeast nutrients, notably nitrogen. The content of nitrogen in the white must ranges between 50 mg/l and 500 mg/l. The minimum yeast assimilable nitrogen (YAN) requirement for low-risk fermentations of white musts is around 150 mg/l. Musts with lower concentrations are often regarded as nitrogen-deficient. Notably, YAN is typically found in relatively greater concentrations in the *taille* fractions than in the cuvée. In order to avoid stuck fermentations, nitrogen additions can be made using inorganic nitrogen, often in the form of diammonium phosphate, or organic nitrogen in the form of alpha amino acids (except proline).

Croguennec 2016: 292; Halliday and Johnson [1992] 2013: 218; Jackson [1994] 2008: 303; Julien *et al.* 2014: 6–7, 12–13, 28–9; Liem 2017: 249–50; OIV 2022: II.4.1–8; Ribéreau-Gayon *et al.* 2006: 460; SL 2018/2019: 23; White 2018; Willcox 2023; Zoecklein 2002: 3, 9–10

maisons-champagne.com/en
www.awri.com.au
www.hsfe.com
www.jansz.com.au
www.montgueux.com

clarification of Charmat-method sparkling wines

A winemaking operation of removing the yeast sediment and contaminants from chilled Charmat-method sparkling wine. The operation takes place either directly after the secondary fermentation of the wine or following a period of its maturation. The removal of yeast sediments from Charmat-method wines is usually achieved by a combination of centrifugation and filtration under pressure. A series of filtrations may be considered necessary if the wine is found to contain fat particles released from yeast cells as a result of extended lees contact and regular stirring of the lees in the maturation process, which may be the case with the long Charmat method, and especially with the Cavazzani method, as the latter requires regular agitation of the sediment.

Jackson [1994] 2008: 547; Nagórka 2013: 39; SL 2018/2019: 10; Rose 2021: 26–7; Sussman 2021: 13–14

clarification of coloured musts

A winemaking operation applied in the production of red and rosé base wines and partially fermented musts whereby impurities (including fragments of grape skin, pips, stalks, dust, etc.) are eliminated from the coloured must. This is usually achieved by natural sedimentation at low temperatures or using mechanical devices. As with white juice, very thorough clarification of red- and pink-coloured musts is not recommended, as it can lead to a sluggish fermentations and a reduced complexity of flavours in the finished wine.

Barros et al. 2022: 3,247; Dwyer 2018; Forbes [1967] 1989: 299; Moreno and Peinado [2010] 2012: 190–1; Ribéreau-Gayon *et al.* 2006: 448, 460, 468

clarification of continuous-method sparkling wine

A winemaking activity performed in the final stage of the continuous method of transforming the cuvée into sparkling wine with the aim of stabilizing the wine prior to bottling. The wine is racked into vats while leaving behind the sediment consisting of (dead) yeast cells immobilized on wood shavings. The sparkling wine is then cooled and held under pressure at temperatures close to (and sometimes even slightly below) the water-freezing point, which facilitates the precipitation of tartrates. The clear sparkling wine can then be transferred to another (pressure-resistant) tank, where it awaits dosage and (isobaric) bottling.

Campbell 2021; Carrascosa *et al.* 2011: 35; Epstein 2011: 65; Nagórka 2013: 39; OIV 2020a: 20; Smith 2022

clarification of transfer-method sparkling wine

A set of winemaking operations in the transfer method of producing sparkling wines that takes place after some period of contact between the wine resting in bottles and the lees remaining from the second fermentation in cellar bottles, with the maturation period being (in most cases) rather short, when compared with the usual lees contact time in the traditional method of transforming base wine into effervescent wine. The aim is to clear the effervescent wine before the winemaking operations that follow (which include the transfer of sparkling wine to pressure tanks with expedition liqueur and rebottling). The bottles of sparkling wine, stored neck down, are chilled to below 0°C, opened using a transfer machine, and the wine is slowly decanted under pressure into pressure-resistant receiving tanks. The sparkling wine is then subjected to filtration using pressure filters to preserve its carbonation.

Grainger and Tattersall 2005: 98; Grainger and Tattersall [2005] 2016: 137; Howe 2003: 149; Jackson [1994] 2008: 547; Skelton 2021: 152

clarification of white must

A winemaking operation used in the making of base wines and partially fermented musts intended for sparkling white wines whereby impurities (such as fragments of grape skin, pips, stalks, dust, etc.) are removed from the juice. Winemakers usually purge the juice by natural clarification at low temperatures, and — sometimes — also by centrifugation or fining, or a combination of these methods. The operation is applied to musts intended for all types of sparkling wine (fully and semi-sparkling wines included). Notably, the must destined for wines made using the Asti method also needs to be clarified before it is allowed to rest in a refrigerated form until required. Some makers of sparkling wine argue that clarification of the juice before fermentation, which is known to improve the overall quality of white base wines, ought not to be too thorough. Excessive clarification could result in sluggish fermentations and a significant reduction in the complexity of flavours in the finished wine.

Forbes [1967] 1989: 299; Halliday and Johnson [1992] 2013: 114; Nagórka 2013: 39; Rand 2022: 18; Ribéreau-Gayon *et al.* 2006: 422–3; WSET 2020: 65–6; Zoecklein 2002: 9

www.astidocg.it/en
www.champagne.fr/en

coeur de cuvée

The highest-quality juice obtained from the middle part of the first pressing of selected grapes for white base wine, excluding the first and the last portions of the juice. In practice, in the Champagne region, *coeur de cuvée* represents between 1,400 and 1,800 litres of (quality) juice from the first 2,050 litres of juice obtained as a result of pressing 4,000 kg of selected grapes. The use of juice from the heart of the first pressing to create base wines is often seen as a sign of the quality of the final sparkling wine, and the relevant information about this fact present on the label serves marketing purposes. Notable examples of vintage-dated sparkling wines whose names indicate the use of *coeur de cuvée* for their production are: Vilmart & Cie.'s Coeur de Cuvée 2015, an old-vine champagne made from Chardonnay and Pinot Noir grapes, and Chapel Down's Kit's Coty Coeur de Cuvée 2016, an English sparkling wine made from Chardonnay grapes.

Liem 2017: 47, 138, 281–2, 287, 294; Liem [2019] 2022; MacNeil [2002] 2015: 1,956–7; Pitcairn 2022

www.champagnevilmart.fr/en
chapeldown.com
www.gobelsburg.at/en

cold storage of juice

Conservation of grape juice in low temperatures in order to prevent bacterial activity and preclude the onset of fermentation, with the juice kept at temperatures equal or close to water-freezing point. When crafting base wine destined for double-fermentation sparkling wines, the storage of the juice

Photo 17. Vineyards around Schloss Gobelsburg, located in the Kamptal region of Lower Austria. Schloss Gobelsburg Langenloiser Sekt Extra Brut 2012, one of the producer's flagship sparkling wines, has been made exclusively from the juice obtained from the heart of the first pressing.
Photo: courtesy of Schloss Gobelsburg (Michael Moosbrugger), October 2020

at low temperatures is continued until the process of its clarification can take place. This juice storage method takes on particular importance in the production of wines using the Asti method, where the process of cold storage can — and often does — extend beyond the point of clarification. In the Asti method, the juice that is not immediately required is clarified and kept in a refrigerated form to preserve freshness until there is demand. The juice can remain in these conditions for over a year without losing its fresh and fruity aromatics.

Day *et al.* 2019: 186–7; Hubble 2023; Halliday and Johnson [1992] 2013: 114; Liem 2017: 17–33; WSET 2020: 65–6; Zoecklein 2002: 9

www.astidocg.it/en

collage

≈ fining of base wines and partially fermented musts

≈ fining of coloured musts

≈ fining of white musts

complete fermentation of coloured musts

A controlled biochemical process involving conversion of sugars contained in the red- or pink-coloured must — obtained from dark-skinned grapes and destined for red or rosé base wines — into ethyl alcohol, carbon dioxide and secondary products, such as esters, as a result of anaerobic metabolic activity of yeast. Lightly coloured musts are fermented in temperatures similar to those applied to the juice obtained from white grapes, which are in the range of around 12°C to around 20°C. When producing base wine destined for traditional-method rosé, Mike Whitney, owner of Adoria Vineyards (southwestern Poland), ferments pink-coloured musts from the Pinot Noir grapes at temperatures in the range 16–20°C, thereby trying to preserve volatile aromatics.

Temperatures for moderately or heavily pigmented musts tend to be higher than in the case of white juice, with a typical range being from around 16°C to around 25°C. In very rare cases, winemakers allow the fermentation temperature to reach around 30°C for a short period, while remaining cautious not to overheat the wine. Those of them who ferment red musts to obtain base wines destined for sparkling red wines (as in the production of sparkling shiraz) will usually aim at intensifying the extraction of pigments, tannins and flavour compounds from the skins of black grapes. Normally, the vinification process is allowed to continue until all or most of the sugar has been converted into alcohol. By that time, the alcohol content in the resultant red or rosé base wine will have reached between 9.5% and 11.5% abv.

Today, winemakers ferment their musts mostly in stainless steel, but some producers, especially those in the EU, experiment with different vessels. For example, the champagne house Laherte Frères has produced non-vintage wine Laherte Frères 'Rosé de Meunier' Extra Brut by fermenting musts in both stainless-steel vats and wood, using both *foudres* and barrels. A number of sparkling wine producers in Champagne, including Champagne Henri Giraud, are also in the process of evaluating merits of amphorae made from clay-based materials and sandstone.

Barros *et al.* 2022: 3,247; Broggi 2023; CIVC [2014] 2022: 9; Hnt [2015] 2022; Lascelles 2021; Liger-Belair and Cilindre 2021: 23; Robinson 2019b; Whitney 2022.

maisons-champagne.com/en
www.champagne.fr/en
www.champagne-laherte.com

complete fermentation of white must

A controlled biochemical process involving conversion of sugars contained in the white juice — obtained from light- or dark-skinned grapes and destined for white base wine — into ethyl alcohol, carbon dioxide and secondary products, such as esters, as a result of the anaerobic metabolic activity of yeast. Fermentation temperatures, typically monitored daily, are usually maintained at around 12°C to around 20°C, with temperatures at the low end of this range preferred for base wines with floral and fruity characteristics and those at the high end for obtaining base wines with low floral intensity and notes of ripe fruit. This process of vinification, allowed to continue until the last grams of sugar have been converted into alcohol, is typical in producing base wines destined for fully sparkling wines, such as Charmat-method sparkling wines and traditional-method sparkling wines, the latter including (among others): cap classique, cava, corpinnat, *crémant* and champagne. It is generally expected that, by the end of the fermentation process, the base wine will have reached an alcohol content of between 9.5% and 11.5% abv. The process is usually carried out in stainless steel. However, some producers, notably in Europe, sometimes also use wooden casks, barrels and *foudres* to ferment all or part of the juice. Moreover, a number of sparkling wine producers in Champagne are currently in the process of evaluating the merits of fermenters made from clay-based materials. The end of the fermentation process can be determined by a number of signals, including a natural clearing of the liquid, its vinous taste and aroma, and a drop in temperature. Nonetheless, this still needs to be confirmed by measuring the amount of residual sugar in the fermented must.

Broggi 2023; CIVC [2014] 2022: 9; Considine and Considine 1982: 2154; Croguennec 2016: 292; GBC 2018–2022: 1; Lascelles 2021; Liger-Belair and Cilindre 2021: 23; Pozo-Bayón *et al.* 2009: 290; Rand [2018] 2022: 226–33; Robinson 2019a; Rose 2021: 31; Whitney 2022; WSET 2020: 5–6; Zoecklein 2002: 9–10

maisons-champagne.com/en
www.boizel.com/en
www.champagne.fr/en

continuous fermentation

The process of secondary fermentation taking place in a system of sealed tanks set up in parallel, making the wine effervescent. Central to the continuous method of sparkling wine production, this process increases the alcohol content in the wine while producing carbon dioxide, which elevates pressure in the reticulated tanks. Yeasts are added continuously to the wine together with nutrients to sustain the fermentation process, with temperatures being rarely higher than 15°C. The main factor identified as particularly inhibitory to the desirable activity of yeast cells in the wine is the growing pressure in the fermentation tanks. Wood shavings are

Photo 18. The idea for launching the 'Sous Bois' range was born around 1990 as a result of Evelyne Roques-Boizel and her husband Christophe's experiments in vinifying part of their musts in old oak. One example of a vintage-dated champagne from this range is Champagne Boizel Cuvée 'Sous Bois' 2009, made from 40% Pinot Noir and 60% Chardonnay. The must intended for this wine was vinified and matured exclusively in old casks for around nine months.
Photo: author, July 2024

(often) introduced to the tanks containing the fermenting wine to increase the total surface area on which the yeast cells can accumulate, so that the lees can impart desirable flavour features to the resulting wine. The whole process usually takes from three to four weeks.

Divies [1993] 2002: 467; Goldstein 2006: 34; Gordon 2016; Jackson [1994] 2008: 548, 619; Kunkee and Ough 1966: 643–7; Nordestgaard [2019] 2020: 54; OIV 2020a: 20; Pambianchi 2006; Rose 2021: 26–7; Schiessl 2017b; Skelton 2021: 151; Spaziani 2019: 13; Walkey 2023

continuous method

A method of sparkling wine production, developed in the Soviet Union as an elaborate variant of the Charmat process, using a series of interconnected pressure tanks. The invention of this method of producing effervescence in the wine, which dates back to the second quarter of the 20th century, is attributed to Anton Frolov-Bagreev, the then head winemaker at the Abrau-Durso winery, now part of the Abrau-Durso Group. Production of sparkling wine by this method boomed in the Soviet Union until the mid-1960s. Currently, quality-oriented producers, Abrau-Durso included, tend to use the classical method for their best cuvées, and the Charmat method for medium- and high-quality wines. Nonetheless, the continuous method is still used by some producers, notably in Germany and some of the post-Soviet states, to make large volumes of inexpensive effervescent wines.

This method involves passage of the prepared cuvée, supplemented (in a continuous manner) with *tirage* liqueur, through a system of pressure-resistant tanks in which yeast cultures are retained. The first of these tanks is designated for initiating the process of secondary fermentation. Channelled through the other two or (in some cases) three (or more) sealed pressure tanks, the fermenting wine becomes effervescent. While there, the wine is held on (dead) yeast cells caught on wood shavings submerged in the liquid for varying periods, usually up to four weeks. The remaining two pressure-resistant tanks serve as containers for maturation of the resultant (sparkling) wine and its effective clarification before (isobaric) bottling.

The continuous method of sparkling winemaking is becoming increasingly rare, even among large producers who once adopted it to reap the benefits of the economies of scale. Not least among the reasons is that the equipment needed for this method requires high capital outlays, and a decent return on investment can only be expected if the demand for continuous-method sparkling wines remains high for a sufficiently long period, which does not allow firms to easily redefine their target groups, needed in the event of a significant upgrade in preferences among customers. Producers that have invested in an expensive and elaborate system of pressure tanks geared towards obtaining continuous-method wines, commonly perceived as mid-market products, may find it difficult to respond to increasing expectations in the market segment they have targeted, with a growing percentage of customers switching from mid-range wines to high-end products crafted using the classical method. From this viewpoint, the continuous method may be considered as suitable mainly for producers supplying sparkling wines to large mid-market segments that are particularly stable.

Bittner 2021: 145; Bogdanov 2013: 235–8; Buglass 2022: 238; Epstein 2011: 65; Fraley 2021; Genisheva *et al.* 2014: 36–9; Goldstein 2006: 34; Gordon 2016; Intardonato 2009; Nordestgaard [2019] 2020: 54; OIV 2020a: 20; Pambianchi 2006; Rose 2021: 26–7; Schiessl 2017b; Skelton 2021: 151; Spaziani 2019: 13; Walkey 2023

bacchus.kz/en
en.abrau.ru

continuous tirage

An operation used in the continuous method of sparkling winemaking. It involves introducing into a set of connected tanks base wine that has been biologically de-oxygenated by addition of a selected yeast culture and protected from oxygen for a short period, usually between one and two days. Importantly, the base wine, which has been blended with tirage liqueur, is introduced to the system of closed tanks under constant pressure and in a continuous manner. The activity is carried out to initiate the second fermentation in the prepared cuvée, which results in the production of effervescence.

Campbell 2021; Dupont 2023c; MacKay 2021; Nagórka 2013: 43–4; OIV 2022: II.4.1–1, II.4.4–3; Rose 2021: 26–7; Schiessl 2017b; Walkey 2023

continuous-method sparkling white wines

A subcategory of tank-fermented sparkling white wines, continuous-method wines are obtained from base wine subjected to a second fermentation in reticulated tanks, to which yeast and nutrient are continuously added in an attempt to sustain the process of secondary fermentation and to give the wine a slightly yeasty flavour. The production of these wines, with its history reaching back to the second quarter of the 20th century, boomed in the Soviet Union until the mid-1960s. Categorized as mid-market commodities, these wines were also produced in large quantities in Portugal in the 1980s. Currently, their presence in the market for sparkling wines is increasingly rare, as the main producers are switching to the Charmat method, requiring a single tank for the production of mid-range wines, or to the classical method, used for the production of quality sparkling wines. An example of this trend is provided by the Portuguese wine-producing company José Maria da Fonseca. The company's rich portfolio includes a series of fully sparkling wines launched since 1984 as an extension of the Lancers range, created towards the end of World War II (mainly for the American market). Until recently, sparkling wines sold under the label José Maria da Fonseca 'Lancers' Vinho Espumante–Brut were produced by the continuous method. From the new tech sheets prepared for these wines, the wine connoisseur can learn that they are now crafted using the classical method (also known as the traditional method or *méthode traditionnelle*).

Chroman 1986; JMF 2017; Mayson [1994] 2015: 471; Rose 2021: 26–7; Schiessl 2017b; Seidanova 2022

en.abrau.ru
www.jmf.pt

cooling of wine in autoclaves

A winemaking operation used in the manufacture of Charmat-method sparkling wines that involves arresting the secondary fermentation in wine placed in sealed tanks, called autoclaves, by reducing the temperature in the pressure tank in which the wine is stored. Target temperatures are usually in the range of 2–4°C, with some producers using temperatures that are either close to or slightly below water-freezing point. The resultant sparkling wine is usually kept at very low temperatures for several days for cold stabilization.

📖 Jackson [1994] 2008: 547; Prichard 2020; Ribéreau-Gayon *et al.* 2006: 468–9; SL 2018/2019: 10; WSET 2020: 12

corking and wirehooding

≈ sealing of bottles of sparkling wine with corks

≈ wirehooding

cuve close method

≈ tank method

cuvée [juice]

The first 2,050 litres of juice extracted from a single *marc*, a unit of measurement used in Champagne, corresponding to 4,000 kg of grapes placed in the press. The operation is conducted as part of the pressing of whole clusters of grapes destined for white base wine. Quality-conscious producers minimize the contact between the juice being obtained and the skins of grapes, ensuring that the colouring matter and tannins contained in the skins of grapes, most notably dark-sinned grapes, such as Pinot Noir or Meunier, do not taint the resultant juice, as these substances are usually considered undesirable in the making of quality sparkling white wines.

Being extracted by means of pressing, cuvée therefore does not include free-run juice; the initial 100–150 litres of juice per *marc* flows freely from the weight of grapes as they are loaded into the press, with the latter usually discarded by makers of quality wines, as the free-run juice tends to contain dust and other impurities found on the skins of grapes. Obtained from the first stage of pressing, and released to a great extent from the flesh of the berries' intermediate areas, the fraction of juice called cuvée is relatively rich in sugar and acids, and tends to be (much) more balanced and fragrant than *taille* — the juice obtained from the second stage of pressing.

📖 CIVC 2023b; Croguennec 2016: 292; Forbes [1967] 1989: 295–9; Liem 2017: 46–7, [2019] 2022; Pessey 2002: 48, 76–7; Ribéreau-Gayon *et al.* 2006: 460

💻 maisons-champagne.com/en
www.champagne.fr/en

Photo 19. Late autumn at Adoria Vineyards (a wine-producing estate in southwestern Poland). If the aim is to obtain white juice for traditional-method sparkling wine, the winemaker tends to minimize the contact of the juice with the grape skins in order to avoid excessive colouring of the juice.
Photo: author, November 2013

cuvée [wine]

An intentionally crafted blend of base wines (or partially fermented musts). A final blend can be made up of wines (or musts) from either a single year or different years. A blend of wines (or musts) from a single year can be composed of wines (or musts) from different grapes, grape-growing sites, or even regions (unless otherwise stipulated by the regulations in a particular winemaking area). Blends of wines from different years are usually composed of reserve wines (or musts) and base wine (or must) from grapes collected during the most recent harvest. Prestige cuvées are unique blends, reputed for their superior quality.

📖 Briscoe 2023; Freson 2021; Liem 2017: 33–4, 52, 61; Pessey 2002: 48; RV 2023; SL 2018/2019: 7; WSET 2020: 6; Zoecklein 2002: 13

💻 maisons-champagne.com/en
www.champagne.fr/en

cuvée filtration

An optional winemaking activity performed by many winemakers after blending, the aim being to reduce the risk of malolactic conversion occurring spontaneously at one of the subsequent stages of the sparkling winemaking process and to render the blended wine (or partially fermented must) generally stable. Malolactic conversion occurring in the traditional-method fermentation, which takes place in the bottle, may lead to an uncontrolled increase in the concentration of bacteria, an event associated with riddling difficulty and possible loss of the wine's palatability.

The choice of filters is largely dependent on the purity of the blend, with cross-flow filters being increasingly preferred to earth filters, as the former have the advantage of helping the producer reduce winery waste. Notably, the filtration of blends before the production of effervescence in base wines will sometimes be regarded as unnecessary, especially when malolactic conversion has already taken place, which may have been the case during (or immediately after) the primary fermentation.

📖 Halliday and Johnson [1992] 2013: 115; Nagórka 2013: 46; RV 2023: 14; SL 2018/2019: 7; Whitney 2022; WSET 2020: 6, 10; Zoecklein 2002: 13

D

débourbage

≈ natural clarification of coloured musts

≈ natural clarification of white must

dégorgement

≈ disgorgement

dégorgement à la glace

≈ mechanical disgorgement

dégorgement à la volée

≈ manual disgorgement

dépointage

≈ stacking *sur pointes*

disgorgement

Removal of sediment including deposits of dead yeast cells from the neck of each bottle of sparkling wine made using the traditional method. The act of ejecting the sediment follows the traditional-method fermentation and extended maturation on lees. It is facilitated by moving the sediment to the neck of the bottle through riddling, a winemaking practice that involves rotating bottles while gradually tilting them upside down. The sediment can be removed from the necks of inverted bottles by machine or by hand.

Importantly, the date of disgorgement marks the first contact of the wine with air in months or years, depending on the length of extended maturation on lees. Moreover, following the process of disgorgement, the yeasty sediment, also known as lees, no longer protects the wine from oxygen, which from then on can enter through the cork. Notably, the cork stoppers currently used by producers vary in the rate of oxygen transfer (OTR), which has an impact on the differences in the rate of evolution of the wine's bouquet.

The category of recently disgorged (or late-release) sparkling wines includes traditional-method wines that have experienced prolonged lees contact and have been disgorged a short time before their release. These wines are considered suitable for immediate consumption, because — despite their considerable age — they tend to be characterized not only by depth and mature flavours but also by remarkable freshness. Notable examples of this style are vintage-dated champagnes: Dom Pérignon Vintage 2006 Pléntitude 2 (disgorged following around 15 years on lees) and Champagne Bollinger R.D. 2008 (disgorged on 25 October 2023).

Information about the date of disgorgement, which is placed by some producers on labels, can be viewed as an indicator of the wine's flavour profile. Among the pioneers in Champagne in declaring the disgorgement date on their bottles are champagne houses Bruno Paillard and Philipponnat, with the former declaring the date of removal of the yeast deposit on every bottle since 1983, beginning this practice only two years after the house was founded.

📖 Avellan 2022b; Buxaderas and López-Tamames 2012: 20; CIVC 2023b; Dublin 2016; Gibb 2012; Halliday and Johnson [1992] 2013: 117; Jestin 2023; Karlsson 2016; Kirkland 2016; Krebiehl 2024a; Liem 2017: 294, 307, [2019] 2022; MacNeil [2002] 2015: 2,201; Nagórka 2013: 51; Pessey 2002: 48–9; Rand 2022: 20; Skelton 2021: 151–2; Stelzer 2014; Stevenson [2014] 2022: 245, 246–7; Stevenson [2015] 2022: 85–6; Stimpfig 2018; TFB 2016, 2018; Whitney 2022; Wilson 2018; WSET 2020: 9–10, 11, 14, 16, 17, 25, 46, 51, 61, 72, 85, 94, 104, 110, 111

💻 champagnebrunopaillard.com/en
maisons-champagne.com/en
www.cava.wine/en
www.champagne.fr/en
www.champagne-bollinger.com/en
www.domperignon.com
www.philipponnat.com
www.recaredo.com/en

dosage

≈ dosage in Charmat-method sparkling wines
≈ dosage in continuous-method sparkling wines
≈ dosage in traditional-method sparkling wines

Photo 20. These crown-capped bottles of sparkling wine placed in A-shaped racks (called *pupitres*) in the dark cellars of the cava producer Vilarnau (located in Valls d'Anoia-Foix, part of Comtats de Barcelona), are in the final stage of riddling, almost ready for disgorgement.
Photo: Marta Ręgowska, April 2023

dosage in Charmat-method sparkling wines

The practice of introducing a preparation known as expedition liqueur or dosage liqueur (sometimes also called shipping dosage, and often referred to by the French terms *liqueur d'expédition* and *liqueur de dosage*) into an air-free pressure tank and blending it with the Charmat-method wine that has been clarified and chilled to between −2°C and 0°C. The aim of this practice is to impart to the sparkling wine harmony and a distinctive character, determining its flavour and its final level of sweetness (see Table 1 under *sparkling wine* later in this section). The liqueur's main ingredient may be wine, grape must, or a blend of wine and grape must, where the must can be either unfermented, or partially fermented, or concentrated, or concentrated and rectified. The expedition liqueur added to Charmat-method sparkling wines usually contains sucrose, ascorbic acid and sulphur dioxide. Some makers also tend to add citric acid at this stage. In the production of sparkling wines using the Charmat method, the addition of *liqueur de dosage* is (usually) followed by (final) filtration and bottling.

📖 AO 2022: 5, 40; Cisilotto *et al.* 2023; Hornsey 2007: 4.3.2; OIV 2022: II.4.1–4, II.4.3–3; SL 2018/2019: 10; WSET 2020: 12

💻 www.prosecco.it/en

dosage in continuous-method sparkling wines

An optional winemaking practice of introducing a preparation known as dosage liqueur into an air-free pressure tank and blending it with the continuous-method wine that has been cleared of sediment and chilled. As in the case of Charmat-method wines, the aim of this practice, if followed, is to impart to the sparkling wine harmony and distinctive sensory characteristics while determining its final sweetness (see Table 1). Following this discretionary practice of adding the dosage liqueur, the wine is considered ready for (isobaric) bottling.

📖 Baldy [1993] 2009: 189; MacKay 2021; OIV 2020a: 20, 2022: II.4.3–3; Rose 2021: 26–7

dosage in traditional-method sparkling wines

The practice of adding up to several centilitres of preparation, known as expedition liqueur or dosage liqueur (often also referred to *liqueur d'expédition* or *liqueur de dosage*) to a bottle of traditional-method sparkling wine following disgorgement. This compensates for the slight loss of sparkling wine during the disgorgement process and determines the final level of sweetness in the wine (see Table 1).

> Some producers in Australia replenish their traditional-method sparkling red wines following disgorgement using dosage liqueur based on fortified (port-style) wine. The aim is to add richness and complexity to the final product. Illustrative examples of quality sparkling red wines to which small amounts of port-style wine were added include Campbells of Rutherglen 'Sparkling Shiraz' NV, for which the fruit was sourced from the Rutherglen region (in the state of Victoria), and Majella 'Sparkling Shiraz' Vintage 2022, made from the grapes from the Coonawarra region (in South Australia).

Most significantly, the addition of *liqueur de dosage* serves to adjust the sensory profile of the wine. The dosage liqueur typically comprises wine, grape must, or a blend of wine and grape must, where the must can be either unfermented, or partially fermented, or concentrated, or concentrated and rectified. Notably, *liqueur de dosage* may contain sugar. As recommended by the OIV (Organisation Internationale de la Vigne et du Vin) in the International Code of Oenological Practices, a document issued in 2022, the sugar employed in the production of expedition liqueur can be either sucrose extracted from sugar cane or sugar beet, or sugar obtained from grapes. These regulations require that the addition of the liqueur increases the actual alcohol content of the sparkling wine by no more than 0.5% abv. Among the additives that may be added to the dosage liqueur are: sulphur dioxide, ascorbic acid, metatartaric acid, sorbic acid and citric acid. The wide range of ingredients allowed for the creation of a dosage enables the winemaker to give their sparkling wine a distinctive finish.

In Champagne, the dosage liqueur is usually prepared several months in advance to prevent the development of turbidity, often using both the same cuvée reserved for this purpose and reserve wines. Winemakers at the family-run firm Winnica Turnau, located in the Polish part of Pomerania, create dosage liqueur (also called shipping dosage) for their Classique Brut using ice wine. Some of the most innovative winemakers are at Bacigalupi Vineyards, located on Westside Road in the northern part of the Russian River Valley, the heart of Sonoma County in northwestern California. The company's winemakers created a dosage liqueur for their wine Brillante Blanc de Noir Sparkling–2017 using two still wines: Bacigalupi Vineyards Pinot Noir–2019 and Bacigalupi Vineyards Chardonnay–2019, and honey sourced from a local beekeeper. When the dosage liqueur is considered ready, it is typically injected into bottles with an automatic or semi-automatic dosage machine, known in French as a *doseuse*. Nowadays, the most efficient of these machines are capable of handling more than 18,000 bottles an hour. It is a common practice to prepare sweeter liqueurs for younger sparkling wines, in which acidity will be relatively high, and less sweet liqueurs for older wines, as acidity in sparkling wines is known to round out as these wines age. Importantly, dosage liqueur could slightly modify the wine's flavour, as in cases where the winemaker uses oak-aged wine as the liqueur's main component. Interestingly, in the production of rosé sparkling wines, expedition liqueur is sometimes also used to correct colour differences.

Photo 21. Most sparkling wine producers present information about their wine's sugar content on the bottle label using categories of sweetness, from *brut zéro* (or *brut nature*) to sweet. Some producers provide more exact information, expressing the wine's sweetness level in grams of sugar per litre. Champagne de Castellane Demi-Sec is a classic blend of three grape varieties (40% Chardonnay, 30% Pinot Noir and 30% Meunier), released with 38 g/l of sugar.
Photo: author, July 2024

Bacigalupi 2023; Buxaderas and López-Tamames 2012: 20–2; Carrascosa *et al.* 2011: 37; Cisilotto *et* al. 2023; Edwards 2009: 316; Jackson [1994] 2008: 546; Kemp *et al.* 2017; Liem 2017: 61–5; Lynn 2020; Miller 2023; OIV 2022: II.4.1–4; Panesar *et al.* 2017: 503; SL 2018/2019: 46; Stimpfig 2018; WA 2022: 11; WSET 2020: 25

bacigalupivineyards.com
maisons-champagne.com/en
russianrivervalley.org
www.campbellswines.com.au
www.castellane.com/en
www.majellawines.com.au

double-fermentation method

≈ secondary fermentation method

double-fermentation sparkling red wines

Effervescent red wines crafted using the secondary fermentation method, with the second fermentation taking place either in the bottle or in a sealed tank. Hence, this category of sparkling wines encompasses both traditional-method sparkling red wines and tank-fermented sparkling red wines. Among the former one can find numerous examples of sparkling red wines crafted by quality-conscious makers in Australia, notably: Clare Dry, who — after joining Seppelt as a new lead winemaker in November 2020 — was responsible for the finishing of Seppelt 'Show Sparkling' Limited Release Shiraz 2012, a vintage-dated wine marketed in June 2023; and Stephen George, a long-time winemaker at Ashton Hills (located in Adelaide Hills, South Australia) with around four decades of experience and author of the age-worthy sparkling red wine Ashton Hills Clare Valley Sparkling Shiraz 2010. Sparkling red wines produced by the tank method are relatively rare. Quality tank-fermented sparkling red wines can be found in the Italian region Emilia-Romagna and in winemaking regions of South Australia.

Speller and Currin [1994] 2015: 470–1; Stelzer 2018: 61; WA 2022: 3–4, 18; Wheeler 2023; WSET 2020: 54, 72

seppeltgreatwestern.com.au
www.ashtonhills.com.au
www.seppelt.com.au

double-fermentation sparkling rosé wines

Effervescent rosé wines made using the secondary fermentation method, with the second fermentation taking place either in the bottle or in a sealed tank. This category of sparkling wines includes traditional-method sparkling rosé wines and tank-fermented sparkling rosé wines. Notable rosé sparkling wines made using the tank method include the Hungarian producer Törley's wines: sweet Törley 'Charmant' Rosé (55 g/l sugar) and dry Törley 'Excellence' Pinot Noir Rosé (20 g/l sugar). Primary examples of traditional-method sparkling rosé wines are provided by many champagne houses, including (among others) Champagne Bollinger, Champagne Veuve Doussot, Pierre Gimonnet et Fils, (Maison) Ruinart and Veuve Fourny et Fils.

Bell 2024; CIVC 2023b; Edwards 2009: 115; Hyland 2019; Köhler 2022; Liem 2017: 55, 279–80, 302

torleypezsgo.hu/en
www.champagne.fr/en
www.champagne-bollinger.com
www.champagne-veuve-fourny.com/en
www.champagneveuvedoussot.com
www.corpinnat.com/en
torello.com/en

double-fermentation sparkling white wines

Effervescent white wines from white- or black-skinned grapes produced using the secondary fermentation method, with the second fermentation taking place either

Photo 22. The sparkling wine Torelló Rosé Corpinnat Brut Reserva 2022, rested in bottles in contact with lees for more than 18 months before release. This wine has been crafted at Torelló Viticultors from the Grenache grapes sourced from the producer's Can Martí estate.
Photo: Marta Ręgowska, April 2023

in the bottle or in a sealed tank. This broad category of sparkling white wines encompasses bottle-fermented sparkling white wines (a subcategory for both traditional-method white wines — such as cap classique wines, champagnes and franciacortas — and transfer-method sparkling white wines, for instance those from the Tokaji range offered by the Hungarian producer Törley) and tank-fermented sparkling white wines (a subcategory including Charmat-method sparkling white wines, such as those produced in the Prosecco zone, and sparkling white wines made using the continuous method).

DeJean 2005: 152; Grainger 2009: 29–32; JMF 2017; Kemp *et al.* [2018] 2019; Lindgren and Leahy 2023: 276–8; Mayson [1994] 2015: 471; McKay *et al.* 2011: 341; Rose 2021: 26–7; Pomarici *et al.* 2019: 421–3

franciacorta.wine/en
torleypezsgo.hu/en
capclassique.co.za
www.champagne.fr/en
www.prosecco.it/en

E

early pressing for rosé must

A winemaking procedure that involves applying light pressure to black-skinned grapes (sometimes mixed with white grapes) to produce lightly coloured must destined for sparkling rosé wine. The operation is performed more slowly than in cases where the winemaker presses dark-skinned grapes to obtain white juice. Taking place prior to the onset of the primary fermentation, the earliest pressing of the grapes makes for slightly tinted musts destined for wines known (on account of their light hues) by the name *oeil de perdrix*. When musts for wines of this type are required, the winemaker normally presses grapes after only a few (usually two or three) hours of skin contact. A longer period of skin contact (usually up to three days) is needed for musts from red grapes destined for fully macerated rosé or musts from red grapes intended for wines crafted using the *saignée* method.

CIVC 2023b; Eads 2014; Hyland 2024; Liem 2017: 269–72; Napjus 2014; Nichols 2018; Parker [1995] 2008: 450

maisons-champagne.com/en
www.champagne.fr/en
www.champagnejeanvesselle.com/en

enrichment of juice from sparkling wine grapes

A winemaking operation in the making of base wine for sparkling wines which involves elevating the concentration of fermentable sugars contained in grape juice. This practice is viewed as necessary in order to compensate for natural sugar deficiencies in grapes destined for base wine in cases where the grapes are considered under-ripe, either because they have been grown in relatively cool climates — for instance in Canada, Champagne, England, New Zealand or Oregon — or following a particularly cool growing season. Increased levels of sugar concentration in the grape juice can be achieved using additive methods or subtractive (richness) methods. Additive methods involve supplementing grapes or the (fermenting) juice either with sugars of exogenous origin, a procedure commonly referred to as chaptalization, or by grape-based sweeteners, such as grape juice, grape concentrates and grape sugar. Subtractive methods involve concentrating the juice for base wine by removing water. They include: cryo-concentration (partial freezing of the juice followed by elimination of the ice that has formed) and reverse osmosis, whereby some of the water in the juice is eliminated through membranes under the action of pressures greater than the osmotic pressure of the juice.

Photo 23. The underground cellars of Moët & Chandon in Épernay house not only old wines but also artifacts prompting narratives about historical events. This is an oak *foudre* from the Algarve region given by Napoleon I to the then head of the firm Jean-Rémy Moët during the emperor's visit to Épernay in 1810.
Photo: author, July 2024

Additive methods of enriching grape juice in the making of base wines destined for sparkling wine take their origin from experiments that took place across France's wine-producing estates and laboratories in the second half of the 18th century, especially those conducted by the innovative chemist Pierre-Joseph Macquer. The use of sweetening agents was promulgated in the early years of the 19th century through the work of the French scientist Jean-Antoine Chaptal, Minister of the Interior under Napoleon Bonaparte in the period 1800–1804, notably through his treatise *L'Art de faire le vin*, published in 1807. Bearing his name, the winemaking procedure of chaptalization, customarily involving the addition of sucrose (sugar extracted from sugar cane or sugar beet) to grapes or fermenting juice, is strictly regulated, and even prohibited in many wine-producing countries and regions, especially those characterized by relatively warm climates such as Napa Valley, Italy, South Africa and Spain. Nowadays, to preserve the authenticity and identity of sparkling wines, international codes of oenological practices recommend the use of grape-derived agents, such as grape concentrates, grape juice and grape sugar.

Busby 1825: 156–67; Campbell [1994] 2015: 570; Eales 2019; GBC 2018–2022; Gough 1998: 74–80; Hall [1994] 2015: 585–6; Kladstrup and Kladstrup [2005] 2006: 63–4; OIV 2022: II.4.1–2; Platter and Fridjhon [1994] 2015: 752–3; Skelton [1994] 2015: 324–5; Unwin [1994] 2015: 221

www.oregonwine.org
www.nzwine.com
maisons-champagne.com/en

extended ageing of sparkling wine in cellars

Long-term storage of finished sparkling wine in cellars, lasting from over a year to even several decades. The ageing potential of sparkling wines depends not only on their individual qualities but also on the capacities of the bottles in which these wines rest, with wines stored in sealed magnums (1.5 litre bottles) and jéroboams (3 litre bottles) retaining dissolved carbon dioxide more efficiently during long-term storage than wines elaborated in sealed standard 0.75 litre bottles. While stoppers used for sealing bottles of sparkling wine provide effective protection against wine leakage, they are not fully impermeable to gas transfers. The slow diffusion of gaseous carbon dioxide (present in the headspace) into ambient air in the course of prolonged ageing entails (in accordance with Henry's law) a proportional decrease in carbon dioxide dissolved in the wine, where the dissolved CO_2 is considered essential for the formation of bubbles in a tasting glass. Results of the most recent studies on the ageing potential of champagnes, linked with the progressive decay of dissolved CO_2 during the process of maturation, and ageing suggest that the loss of effervescence in sparkling wine may be up to three times slower when it ages in a jéroboam than when it is stored in a standard bottle.

Prolonged storage of sparkling wine, whether in the cellar of a producer or of a connoisseur, requires a cool place away from bright light, ensuring constant temperatures, preferably no lower than 7°C and no higher than 11°C. Moreover, the bottles of sparkling wine sealed with corks are usually stored on their sides, the aim being to reduce the risk of the cork drying out. Fine sparkling wines that are considered likely to benefit from long-term storage, especially some of the finest traditional-method wines, rest in bottles first to recover from the shock of disgorgement. This marks a change from the storage of sparkling wine in a highly reductive environment offered by bottles in which the wine has been maturing in the presence of lees to the storage of sparkling wine in the presence of small amounts of oxygen that will have entered the head-space at disgorgement. At the house of Bruno Paillard the convalescence period usually increases with the time the wine has matured on lees, lasting between five months (as in the case of Bruno Paillard Rosé–Première Cuvée, which matured in contact with lees for three years before it was disgorged) and even around 18 months and beyond in the case of wines from the company's 'Nec Plus Ultra' range, such as the champagne Bruno Paillard Nec Plus Ultra–2008, which had been allowed to mature *sur lie* for around 10 years. Following the time they need to regain their balance, the wines will continue to evolve, with changes observable in both pressures and flavour features, where the latter are influenced, among other things, by the presence of small amounts of oxygen.

Tasters at the champagne house Bruno Paillard have observed that, following disgorgement, their wines' bouquets evolve with age, exhibiting increasingly complex flavours. They see this (oxidative) evolution as consisting of five stages. Their wines' flavours are described as evolving from fruit-dominated — perceptible soon after disgorgement — to floral, found in wines aged for around two to three years. Sparkling wine made using the traditional method has been found to take between around four and seven years to acquire spicy characteristics, and between eight and close to 10 years to gain toasted flavours. Fine sparkling wines subjected to post-disgorgement ageing for around 10 years and beyond can develop very rich flavours, including notes of gingerbread, candied fruit, honey and roasted almonds. Quality-conscious buyers of fine age-worthy traditional-method sparkling wines are often advised to be guided by the date of disgorgement when deciding when to open their bottles.

Avellan 2022b; Crawford 2023; Douglas 2020; Edwards 2009: 204; Edwards [2015] 2021; Heiligenthal 2012; Kirkland 2016; Liem 2017: 259–60, 300–1, [2019] 2022; Liger-Belair *et al.* 2013: 78–80; Liger-Belair *et al.* 2023: 22,844–52; Rountree 2023; Stevenson [2015] 2022: 85–7

champagnebrunopaillard.com/en

extended maturation on lees

Prolonged storage of bottles of sparkling wine in contact with the sediment containing dead yeast cells, taking place directly after traditional-method fermentation. Sparkling wines are stored in cellars, protected from light and at a constant temperature, typically within a range of 8°C to 12°C. Storage times differ from one wine region to another. In Champagne, the minimum time required for post-fermentation storage of bottles of sparkling wine is 15 months for non-vintage champagnes (a period that in practice is often extended to at least two years) and three years for vintage-dated champagnes (with the minimum timeframe often extended to between four and 10 years). Storage of sparkling wine on the lees is regulated differently in Spain's cava-producing region, with the baseline threshold for cava wines (used for the youngest cavas, representing the Cava de Guarda segment) being nine months. The minimum maturation periods for sparkling wines in the Cava de Guarda Superior segment are as follows: 18 months for cava reserva, 30 months for cava gran reserva, and 3 years for cava de paraje calificado. Producers of corpinnats are obliged to subject all their bottle-fermented sparkling wines to post-fermentation maturation on lees for a minimum period of 18 months. In practice, around 80% of corpinnats that reached the market in 2021 had matured in contact with lees for more than 30 months.

The length of maturation of sparkling wines on lees has its natural constraints. Not least of them is the slow diffusion of gaseous carbon dioxide into the ambient air through crown caps or temporary corks (as they are not 100% impermeable to gas transfers), entailing a gradual loss of dissolved carbon dioxide, a phenomenon explained by Henry's law. Notably, the period of maturation on lees must be completed well before the wine's dissolved CO_2 content begins to approach the critical concentration needed to produce bubbles once the wine has been served in a glass. Oenological studies conducted in the early 2020s by the research group Equipe Effervescence, Champagne et Applications, led by Gérard Liger-Belair of Université de Reims Champagne-Ardenne, confirm that increasing the size of the bottle in which sparkling wine matures on lees results in the wine's increased capacity to preserve its dissolved CO_2. The results of these studies show that the loss of effervescence in sparkling wines could be around three times slower in the case of wines maturing in a jéroboam (3 litre bottle) than in the case of wines maturing in a standard-sized (0.75 litre) bottle.

In the traditional method of sparkling winemaking, following the secondary fermentation, sparkling wines mature on lees in bottles closed using either crown caps or corks. Labour-intensive as it is, the traditional use of corks as bottle closures before the onset of sparkling fermentation and the maturation period that follows is now rare, but is regarded as having an impact on the development of desirable characteristics in sparkling wines, especially those that are destined for very long maturation periods, including not only the complexity of the wine's flavour profile but also the persistence of its bubbles. Among those subjecting their sparkling wines to the traditional on-lees maturation under cork are producers of champagnes, such as the champagne house Ruinart, having in its portfolio Dom Ruinart Blanc de Blanc Extra Brut–2010, maturing on lees for nearly 10 years, and the family-owned house Bérêche et Fils, producer of Bérêche et Fils Aÿ Grand Cru–2014, a wine that rested on its lees under cork for around six-and-a-half years until it was disgorged in November 2021. Maturation of sparkling wine on lees under cork is also practised by sparkling winemakers at the wine-producing company Le Lude, located in the Coastal Region's Franschhoek district, in Western Cape (South Africa). Before being disgorged and released, Le Lude's Brut Rosé 'Agrafé'–2013, created by the winemaker Paul Gerber, rested on lees for several years in bottles closed by temporary corks, with every cork secured by a clasp called an *agrafe*.

CIVC [2014] 2022: 14, 19, 29, 35; Eedes 2019; Gramona and Mata 2021; Jackson [1994] 2014: 710–12; Jolly *et al.* 2021: 136–52; Liem 2017: 55–6, [2019] 2022; Liger-Belair *et al.* 2018: 40–5; Liger-Belair *et al.* 2023: 22,845–51; Mercer 2021; Schmitt 2020; Torresi *et al.* 2011: 1,232–3

www.bereche.com/en
www.cava.wine/en
www.champagne.fr/en
www.ruinart.com/en-gb
www.lelude.co.za
maisons-champagne.com/en

extraction of skin components for red must

A winemaking process accompanying the maceration of red berry skins during the vinification of grapes intended for red base wines or musts. As a result of this process, desirable compounds contained in the skins (necessary for the development of the wine's colour, flavour profile and tannin) are leached into the must. The process is usually encouraged by the use of pumping-over systems and (manual) plunging of the skins of red grapes. Some makers of sparkling red wines also choose to heat their dark-skinned grapes (gently) before the fermentation begins.

Angelosante 2020; Jackson [1994] 2008: 347, 597, 716; Nordestgaard 2017: 54–60; OIV 2022: II.1.1–8; Ribéreau-Gayon *et al.* 2006: 383–5, 450–5; Stelzer 2018: 28–30; WA 2022: 17–19

Photo 24. Vintage-dated cuvées maturing in bottles on lees in the dark cellars of Maison Ruinart in Reims. The average maturation period for vintage champagnes made by Ruinart's *chef de cave* Frédéric Panaïotis (such as, for example, Dom Riuinart Blanc de Blancs 2009 and Dom Ruinart Rosé 2009) is 10 years.
Photo: author, July 2024

F

fermentation continued in bottles

A key stage in the production of semi-sparkling wines using the ancestral method, during which the winemaker restores conditions that are conducive to fermentation, with the result that the yeasts (often wild yeasts) present in the partially fermented must are able to resume the fermentation process, previously interrupted by lowering the temperature to or slightly below water-freezing point. This winemaking practice consists of bottling the partially fermented must and sealing the bottles, which results in the production of effervescence in the fermenting must. The method is used by producers of pétillant naturel wines, commonly referred to as pét-nats, and by wine producers in regions with long traditions of crafting ancestral-method wines. The latter include, among others: producers of semi-sparkling white wines known as Clairette de Die Méthode Ancestrale, made in the Diois region, an area in the central part of the Rhône Valley (southeastern France); and producers of semi-sparkling rosé wines crafted in the area of Bugey (central-eastern France), called Bugey Cerdon wines. In the Diois region, the resumed in-bottle fermentations of half-fermented musts take upwards of four months at a constant temperature of around 12°C. Winemakers in the Bugey area ferment their half-fermented musts in bottles for a minimum of two months at a temperature in the range of 10–12°C but, in most cases, the fermentation process is allowed to continue for a longer period before the sediment is removed (by cold filtration).

An increasing number of producers of pétillant naturel wines around the world are replicating the age-old concept of continued fermentation, while adapting it to local conditions, many of them following traditions cultivated in the area of Limoux (southern France), where producers leave the yeast deposit in the bottle, containing wine from the Mauzac grapes, as part of the wine's style. When adaptations to the general concept are considered necessary or useful — for instance, as a result of the use of local grapes and yeasts native to a given geographical area, as well as differences in target alcohol contents in the resultant wines — they concern mainly such factors as fermentation temperatures and the length of the fermentation process. Average fermentation temperatures for pétillant naturel wines crafted in non-traditional wine regions range between 13°C and 16°C, with the resumed fermentation process usually lasting several weeks. Occasionally, makers of ancestral-method wines keep them on lees before releasing them (with or without the yeast sediment). An interesting, albeit unique, example is the Classic Penedès single-fermentation vintage wine Albet i Noya 'Efecte: Ancestral' Reserva Brut Natural 2020. The wine, which boasts an alcohol content of 12% abv and an excess pressure of around five atmospheres — parameters rather unusual for ancestral-method wines, with their alcohol contents typically below 10% abv and excess pressures usually lower than four atmospheres — had matured in contact with lees for over 15 months before the yeast deposit was removed.

Brazil 2023; Gardner 2015b; Gerling *et al.* 2023; Kojder 2022; Sohm and Muhlke 2019: 36; Teclemariam 2023; Zohar 2020

albetinoya.cat/en
clairette-de-die.com/en/
vinsdubugey.net/en
www.chapoutier.com/en
www.domaineladentelle.com
www.limoux-aoc.com/en

Photo 25. The winemaker Richard Woodfine tending the vines at Newbarn Farm in Buckinghamshire, UK (around 30 miles northwest of central London). He crafts ancestral-method lightly sparkling wines made from the grapes of Chardonnay and Pinot Noir.
Photo: courtesy of Woodfine Wine (Richard Woodfine), August 2022

www.newbarnfarm.org
www.vinetrail.co.uk
www.woodfinewine.com

fermentation continued in closed tanks

A key stage in the production of semi- and fully sparkling wines using the Asti method, which consists of closing a valve in the temperature-controlled pressure-resistant fermentation tank where the fermenting must is stored so that the carbon dioxide produced as a result of the ongoing fermentation process is retained, which results in the production of effervescence. The fermenting must is closely monitored (primarily for the amount of sugar remaining in the must, which gradually decreases), as is the fermentation temperature (which is rarely allowed to exceed 18°C). The valve in the tank is closed when the amount of residual sugar required to give the wine its desired pressure has been reached. For Moscato d'Asti wines, fermentation is arrested when the alcohol content in the fermenting liquid reaches around 5% abv, the maximum alcoholic strength recommended for these semi-sparkling wines being 6.5% abv, and the typical sugar content around 130 g/l. Fermentations usually take longer for fully sparkling Asti wines, also known by the name Asti Spumante. The alcoholic strength required for these wines typically ranges between 6% abv and around 8% abv for sweet styles (such as Abbazia di San Gaudenzio's Asti Spumante-Dolce from the company's Fortuna range, which has an abv of 7.5% and a sugar content in the range of 85–95 g/l) and above 10% abv for dry styles (such as Tosti 1820's Asti Secco, a wine with an abv of 11% and a sugar content of 17 g/l).

Fielden [1994] 1998: 297; Gardner 2015b; Speller [1994] 2015b: 541–2; WSET 2020: 64–7; Wu 2015

www.abbazia.it/en
www.astidocg.it/en
www.tosti.it/en

fermentation in autoclaves

A key step in the Charmat process — a cost-effective method of making base wine effervescent — involving a secondary fermentation in the base wine induced using *tirage* liqueur, a mixture of a small amount of wine, additional sugar, yeasts and yeast nutrients (with the possible addition of clarifying agents) blended with the base wine in a sealed pressure tank. When fully sparkling wine is to be obtained, the process of secondary fermentation is allowed to continue until the over-pressure in the closed pressure tanks, called autoclaves, begins to exceed five atmospheres. The fermentation process taking place in these tanks is controlled by regulating temperature. In the case of Charmat-method sparkling wines made in Italy's northeastern regions of Veneto and Friuli-Venezia Giulia, the temperature range during the secondary fermentation conducted in autoclaves is 12–18°C for typical prosecco and prosecco-style sparkling wines, with the fermentation process in these wines normally allowed to continue for about 30 days, with *sur lie* maturation times usually not exceeding several weeks, as the winemakers want to limit the duration of the autolytic processes to preserve the fruity character of their wines. The typical range of fermentation temperatures for high-quality tank-fermented sparkling wines, especially those made using the long Charmat method or the Cavazzani method, tends to be lower, oscillating between 12°C and 16°C, with the fermentation process allowed to continue for up to 50 days. This period is usually followed by an extended *sur lie* maturation, which in the case of the Cavazzani method can last even one year or longer, with the maturation temperature often reduced to 6–8°C.

Grainger 2009: 29–32; Lindgren and Leahy 2023: 276–8; MacNeil [2002] 2015: 815–16; Skelton 2021: 150–1; Sussman 2021: 13–14; WSET 2020: 61

collavini.it
proseccobaldi.it
www.adamiprosecco.it
www.astoria.it/en
www.bepindeeto.it
www.prosecco.it/en
www.prosecco.wine/en

fermentation of coloured musts

A controlled biochemical process involving conversion of sugars contained in the pink-coloured juice or red must destined for red and rosé base wines into ethyl alcohol, carbon dioxide and secondary products, such as esters, as a result of anaerobic metabolic activity of yeast. Temperature ranges for primary alcoholic fermentation can differ considerably. Lightly coloured musts are fermented in temperatures similar to those applied to the juice obtained from white grapes, usually between around 12°C and around 20°C, the aim being to preserve volatile aromatics. By contrast, a typical temperature range for moderately or heavily pigmented musts is between around 16°C and around 25°C, the main aim being to further intensify the extraction of colourants, tannins and flavour compounds. The fermentation process can be either partial (as in the case of wines made using the ancestral method) or complete (as in the case of preparing base wines destined for standard Charmat-method sparkling wines or traditional-method wines, such as cap classique, cava or champagne). Today, most winemakers run primary fermentations using stainless steel; others conduct the fermentation process in wood; still others are experimenters with fermentation vessels made of less commonly used materials, such as sandstone or materials derived from clay.

Barros *et al.* 2022: 3,247; Broggi 2023; CIVC [2014] 2022: 9; Hunt [2015] 2022; Jackson [1994] 2014: 708; Liem 2017: 49; Liger-Belair and Cilindre 2021: 23; Smith 2020; Whitney 2022; Wislocki 2022

fermentation of white must

A controlled biochemical process involving conversion of sugars contained in the grape juice destined for white base wine into ethyl alcohol, carbon dioxide and secondary products, such as esters, as a result of anaerobic metabolic activity of yeast. Temperatures are usually maintained within a range between around 12°C and around 20°C. Temperatures at the low end of this range are often preferred when the winemaker is aiming for base wines exhibiting delicate floral and fruity aromas. Conversely, when the goal is base wines with low floral intensity, temperatures at the high end of the range are preferred. Importantly, the fermentation process can be either partial (as in the case of wines made using the ancestral method) or complete (as in the case of preparing base wines destined for traditional-method sparkling wines, such as cap classique, cava or champagne). Today, the process is usually carried out using stainless-steel tanks, with some winemakers preferring to ferment the juice in wood. There are also a number of experimenters using an array of fermentation vessels made of alternative materials, including clay-based vessels.

CIVC 2013: 25, [2014] 2022: 9; Jackson [1994] 2014: 708; Liem 2017: 49; Liger-Belair and Cilindre 2021: 23; Whitney 2022; Wislocki 2022; WSET 2020: 5–6; Zoecklein 2002: 9–10

ficelage

The age-old practice of securing the cork in the neck of the bottle by means of string. This was the usual method of securing corks in Champagne after the edict of 25 May 1728 issued by King Louis XV granted sparkling wine producers and merchants the right to transport their wines in bottles. Gradually supplanted by wirehooding, a technique that began to be used on a large scale in the last quarter of the 19th century, and has gained general acceptance since the end of the Great War (1914–1918), the use of string to secure cork stoppers is currently practised by several champagne producers who consider this method of securing corks a valuable reference to their heritage. A notable example is Plantevigne Dubosquet, a company offering the Louis Dubosquet

Photo 26. A bottle of Champagne A.D. Coutelas 'Cuvée 1809', sealed with a cork secured with string.
Photo: author, September 2023

Champagnes range, including bottles of grand-cru champagnes sealed with corks tied down with crossed string. Other notable champagne producers with bottles of sparkling wine in their portfolios that are closed with corks secured using the traditional method are the family businesses A.D. Coutelas and A.R. Lenoble.

Edwards 2009: 12; Gilby 2020; Holthausen 2020; Stevenson 2022

champagne-adcoutelas.com
lenoble-champagne.com/en
maisons-champagne.com/en
m.plantevigne-dubosquet.com/en

filtration of base wines and partially fermented musts

A winemaking procedure that involves separating undesirable substances from base wine or partially fermented must using filters. Commonly applied for microbial stability, the method of sterile filtration reduces the foaming capacity of base wines by removing positively charged colloids, and for this reason chitosan treatments are increasingly considered a beneficial alternative. Base wines and partially fermented musts are often subjected to coarse filtration to reduce the quantity of lees while not eliminating them completely, retaining colloids regarded as beneficial to the development of the wine at subsequent stages of its production.

Colangelo *et al.* 2018: 301; Enartis 2019: 10; Halliday and Johnson [1992] 2013: 115; WSET 2020: 9–12

final pressing of whole clusters

Extraction of the remaining fraction of grape juice, considered too coarse for making base wine for quality sparkling wines. In champagne production, the extraction of the last remaining juice, referred to as *rebêche*, is conducted after the first 2,550 litres of juice destined for base wine have been obtained. To save time, which is needed most for the production of base wine, the operation resulting in obtaining *rebêche* is sometimes performed using much more powerful presses than those used to obtain quality juice fractions: cuvée and *taille*.

Fielden [1994] 1998: 286–7; Forbes [1967] 1989: 297–8; Krebiehl 2023; Liem 2017: 46–7, 301; Skelton 2016: 33; Tardi 2016: 15–17

maisons-champagne.com/en

fining of base wines and partially fermented musts

A stabilizing treatment of base wines intended for sparkling wines involving the use of clarifying agents. The purpose of adding these substances to the base wine is to expedite the binding of colloidal particles present in the wine and their subsequent precipitation, a process facilitating their removal. Regarded as an aid to natural clarification, this treatment is generally recommended for use only to the extent necessary. While fining agents show numerous advantages, research institutions, such as the Australian Wine Research Institute (AWRI) and Organisation Internationale de la Vigne et du Vin (OIV), recommend that their use in the cellar ought to be preceded by laboratory trials, with a control sample being prepared where no clarifying agents have been used. The aim is to enable the winemaker to assess the effects of the addition of clarifying agents to the remaining samples. Importantly, with all base wines and partially fermented musts showing some differences in their components, the results of preliminary tests also assist in determining the exact doses of the clarifying substances to be added.

Fining agents that winemakers usually apply at this stage include bentonite, casein, chitosan, egg white, gelatin and isinglass. Bentonite clay is used mostly for white base wines, its use for red and rosé wines or partially fermented musts being limited on account of its ability to reduce colour by adsorption of anthocyanins. When added to base wine or partially fermented must in moderate doses, bentonite clay precipitates positively charged proteins, facilitating the removal of protein-polysaccharide complexes, which promotes protein stabilization. Casein, the principal protein in milk, is applied in fining mainly white wines or partially fermented musts. Its use assists in reducing the concentration of phenolics associated with bitterness and browning. Chitosan as a fining agent is regarded as a useful alternative to bentonite, with the adsorption of proteins by the latter being rather unspecific, and hence likely to affect base wines' individual aroma profiles. The use of chitosan, with a maximum recommended dose of 100 g/hl, has been reported as contributing to the microbial stability of base wines (or partially fermented musts), while maintaining their foaming capacity and leaving their aromas largely unaffected.

At this stage, egg whites are sometimes applied to red base wines or partially fermented musts to remove phenolics associated with harsh astringency. Gelatin and isinglass are protein-based fining agents which are often used in conjunction with small doses of oenological tannins (typically 2–4 g/hl), the use of the latter associated with increasing the effect that these agents have on the base wine. Gelatin, with a typical dose ranging between 4 g/hl and 7 g/hl, tends to be added to white base wines in order to reduce bitter aftertaste, and to red and rosé base wines to soften the wine when it is viewed as too harsh or tannic. Isinglass, also known as fish glue, obtained from the swim bladders, gills and ears of certain fish, such as sturgeon and cod, is typically applied as a fining agent in white base wines or partially fermented musts. Fish glue, with a usual dose of 1.5–2.5 g/hl, is less active towards condensed tannins than gelatine or casein. After using this agent, the lees produced tend to be light and fluffy. For this reason care must be taken not to disturb the lees when racking or filtering.

Colangelo *et al.* 2018: 301–8; Enartis 2019: 10; Forbes [1967] 1989: 311; Halliday and Johnson [1992] 2013: 115; Jackson [1994] 2008: 435–8; Lambri *et al.* 2016: 388–9, 392; Marín *et al.* 2020; OIV 2022: II.3.2–10, II.3.2–18; Ribéreau-Gayon *et al.* 2006: 461–2; SL 2018/2019: 31; Zoecklein 2002: 10, 13

www.awri.com.au

fining of coloured musts

A method of clarifying coloured musts, especially rosé musts, with the use of clarifying agents, typically bentonite, casein and gelatin. This winemaking practice is sometimes used with the aim of refining the aromas of the resultant rosé base wines. The use of bentonite for pink-coloured musts, with the usual doses rarely exceeding 50 g/hl, can result in a slight colour decrease by the adsorption of anthocyanins, while rendering the coloured juice brighter and less susceptible to oxidation. As in the clarification of white musts, preference is given to spontaneous (natural) settling, with fining agents being used only in necessary cases. Both wine researchers and winemakers advise delaying the use of bentonite clay at least by several hours if the must has been treated with pectolytic enzymes (usually added at doses lower than 2 g/hl), which facilitate natural settling. As in the case of white juice, particular care should be taken to avoid excessive clarification of coloured musts, as turbidity levels below 50 NTU (nephelometric turbidity units) tend to result in sluggish fermentation. Excessive removal of proteins from coloured musts by the use of fining agents could also lead to an insufficient presence of particles in the corresponding sparkling wine that provide a base for the formation of bubbles, which can lower the intensity of effervescence and reduce foam stability of the resultant wine.

Barros *et al.* 2022: 3,247; Dwyer 2018; Jackson [1994] 2008: 541–2; Moreno and Peinado [2010] 2012: 190–1; Ribéreau-Gayon *et al.* 2006: 448, 460, 468

www.awri.com.au

fining of white must

A method of clarifying white musts — obtained from white or dark-skinned grapes — with the use of clarifying agents, typically gelatin, bentonite and casein. Gelatin is sometimes added to white juice, particularly pressings, to reduce the levels of phenolic compounds associated with astringency, bitterness and browning. Likewise, bentonite clay or potassium caseinate, or mixtures of bentonite and casein, can be added to the *taille* fraction of white juice to assist in the precipitation of tannins, and to facilitate natural settling and to ensure subsequent protein stability. Bentonite, with the usual doses not exceeding 50 g/hl, is commonly added to white musts following an initial period of natural sedimentation to enhance this agent's efficacy. Notably, bentonites are known to stimulate fermentation by offering physical support for the yeast. While fining agents show numerous advantages, research institutions, such as the Australian Wine Research Institute (AWRI), recommend that clarifying agents should be used with utmost care and only when necessary, with spontaneous (natural) clarification being preferred. Importantly, careless use of fining agents at this early stage may result in excessive protein removal, negatively affecting the foaming potential of base wines used for the production of sparkling wines.

Jackson [1994] 2008: 541–2; Joshi *et al.* 2017; Lira *et al.* 2013: 222; McKay *et al.* 2011: 330; Moreno and Peinado [2010] 2012: 352; Ribéreau-Gayon *et al.* 2006: 428, 466–8; Vernhet 2022: 88

www.awri.com.au

finishing of traditional-method sparkling wines

A set of winemaking operations in the traditional method of creating sparkling wines involving one of two alternative steps following the disgorgement process, namely: dosage in traditional-method sparkling wines in bottles in which these sparkling wines have been made; or the complex operation known as *transversage*, whereby the wine receives dosage liqueur (also called *liqueur d'expédition*) and is then transferred to non-standard-sized bottles with a capacity of less or more than the standard 75 cl.

ATF 2017: G.24.190; Hay 2022; Jeandet and Charters 2012: 20; Liem [2019] 2022; TFB 2016; UGA 2019

Photo 27. At the final stages of production, the pressure in a bottle of sparkling wine made using the traditional method reaches around 6 bars. Pressure of this magnitude is found in this bottle of champagne at the champagne house Champagne Charles de Cazanove, at Place de la République in Reims.
Photo: author, June 2023

first pressing of grapes for white must

Extraction of the initial fraction of juice from grapes destined for white base wine or partially fermented must using a wine press. In the region of Champagne, the first pressing is strictly regulated. Sparkling wine producers in this wine region must comply with the regulation on the maximum permitted amount of juice from this pressing, which is 2,050 litres of juice for every 4,000 kg of freshly picked grapes loaded into the press. This (initial) volume of juice is commonly referred to as cuvée.

Fielden [1994] 1998: 286–7; Forbes [1967] 1989: 297–8; Liem 2017: 46–7; Pessey 2002: 77; Skelton 2016: 33; WSET 2020: 4–5, 53

www.champagne.fr/en
maisons-champagne.com/en

fractional pressing

≈ whole cluster pressing

G

grape juice press fractioning

≈ whole cluster pressing

H

handling of white must

A decision-based procedure concerning the most appropriate method for the preparation of a given juice fraction for the fermentation process, with different pressing fractions handled separately. The choice between juice processing methods is based on the winemaker's decision about the desired degree of protection of the juice from its exposure to oxygen prior to the onset of the fermentation process. The general trend in handling juices destined for white base wines, especially those made from white grapes such as Chardonnay, is to limit oxygen exposure during post-press operations. Although, depending on the grape varieties used, controlled aeration of white musts can result in varying degrees of colour stabilization of white base wines, results of oenological research confirm that protective juice handling, which includes minimizing the impact of passive oxygenation that occurs during winemaking processes leading to fermentation, minimizes the yellowing and browning of the juice and has a positive effect on the preservation and development of flavour compounds responsible for floral and confectionery notes, detected during sensory analyses of the resulting (base) wine.

📖 Day *et al.* 2019: 198–9; Forbes [1967] 1989: 300; Nagórka 2013: 64; Ribéreau-Gayon *et al.* 2006: 417–18; Webb [1994] 2015a: 648; WSET 2020: 65–6; Zoecklein 2002: 9

💻 www.champagne.fr/en
maisons-champagne.com

horizontal blending

A winemaking procedure of combining two or more base wines of the same vintage made from different grape varieties or created from grapes sourced from different grape-growing sites. Australian producers are known for their remarkable ability to blend wines across grape varieties, vineyards, subregions and regions. The main focus at St Huberts is on blending wines from various grape varieties. A traditional-method sparkling wine made by Greg Jarratt from grapes grown solely in the Yarra Valley region, St Huberts Blanc de Noirs 2010 has been crafted using 87% Pinot Noir and 13% Chardonnay. Another notable example of wine made from different grapes (51% Pinot Noir, 42% Chardonnay and 7% Meunier) harvested in the same year is Taltarni's Cuvée Rosé 2015. This traditional-method sparkling wine, crafted by Robert Heywood, is also a blend of wines from grapes grown in Australia's three different states: Victoria, South Australia and Tasmania.

📖 Burton [2014] 2016; CIVC 2023b; Sanderson 2008; Stelzer 2018: 44; WA 2018: 4.26, 2023: 35–9

💻 maisons-champagne.com
www.sthuberts.com.au
taltarni.com.au
vinselcep.com

Photo 28. Base wine from the Macabeo grapes from old vines picked during the 2017 harvest is a component of the vintage-dated cuvée crafted at the cava-producing estate Vins El Cep, located in Comtats de Barcelona (in the northeast of Spain). The wine Vins El Cep 'Clos Gelida' Exclusive Brut Reserva 2017 has been produced from 40% Xarello, 25% Macabeo, 20% Chardonnay and 15% Parellada.
Photo: Marta Ręgowska, April 2023

I

isobaric bottling

≈ isobaric bottling of Charmat-method sparkling wine

≈ isobaric bottling of continuous-method sparkling wine

isobaric bottling of Charmat-method sparkling wine

The winemaking operation of transferring sparkling wine made using the Charmat method (also known as the Martinotti method) from a pressure-resistant tank in which the wine has been stored into bottles. The operation is performed under constant pressure, the aim being to avoid (any) loss of carbon dioxide that has accumulated in the secondary fermentation process taking place in autoclaves. Before the wine is transferred to bottles, it is usually chilled in order to stabilize and reduce effervescence. The bottling operation is performed using a counter-pressure filler.

📖 Cowderoy 2021: 36–41; Intardonato 2009; OIV 2020a: 20, 2022: II.4.3–6; Prichard 2020; SL 2018/2019: 10; WSET 2020: 12

isobaric bottling of continuous-method sparkling wine

The winemaking operation of transferring sparkling wine made using the continuous method from pressure-resistant tanks in which the wine has been stored into bottles. As with Charmat-method wines, the operation is performed under constant pressure. The aim is to avoid (any) loss of carbon dioxide that has accumulated in the process of continuous fermentation. Before bottling, the wine is usually chilled. The operation of bottling is performed using a counter-pressure filling system, in which the bottle is filled under pressure with carbon dioxide, which is then replaced with chilled sparkling wine from the pressure tanks.

Intardonato 2009; OIV 2020a: 20, 2022: II.4.3–6; SL 2018/2019; WSET 2020: 12

Italian method

≈ Charmat method

J

jetting

A sparkling winemaking technique used by some producers of traditional-method sparkling wines prior to the sealing of bottles before their release. The aim is to minimize and control the amount of air in the bottle, especially in view of the possibility of oxygen ingress during disgorgement. A minute droplet of (sulphited) water or wine is injected into the bottles of sparkling wine at high speed to induce foaming before these bottles are stoppered, with the liquid volume introduced typically ranging between 50 μl (0.05 ml) and 100 μl (0.1 ml) per bottle. As a consequence, carbon dioxide bubbles rise from the surface of the sparkling wine to the lip of the bottle, filling the headspace and thus expelling the air.

Winemakers favouring this winemaking technique, a concept borrowed from the brewing industry — which is known to have been using jetting technology since the early 1960s — see it as an effective method of controlling the amount of oxygen in the headspace of the sealed bottle of sparkling wine, oxygen being considered a major contributor to the quality of the wine, having an impact on its flavour. The method finds favour among an increasing number of producers. As acknowledged by Séverine Frerson, former cellar master at the champagne house Piper-Heidsieck, and — since October 2020 — Perrier-Jouët's eighth *chef de cave* in the house's long history (and the first woman to hold the position), this method serves to increase the chances that bottlings (most notably those destined for extended post-disgorgement ageing) will evolve in a consistent manner.

Frerson 2018; Hemming 2010; IOC 2018: 3; Kemp *et al.* 2015: 30–1; Liem 2017: 65; Rand [2018] 2022: 230; Riley 2018; Robillard 2012: 85, 87–9; Stevenson 2015b: 62; TFB 2016

ioc.eu.com
www.perrier-jouet.com/en-ww

L

late pressing for red base wine or partially fermented must

A winemaking procedure taking place following the initial fermentation of coloured musts which involves applying pressure to fully or partly fermented black-skinned grapes with the aim of obtaining coloured base wine (or base must) intended for sparkling red wine. Typical grape varieties used for base wines (obtained from the late pressing) intended for sparkling red wines in Australia, a country in which the production of these wines has a particularly long tradition, include: Shiraz, Pinot Noir, Cabernet Sauvignon, Merlot, Durif and Chambourcin. At Hammerling Wines, a winemaking firm set up in 2018 in Berkeley (California) by the winemaker Josh Hammerling, red base wines are obtained as a result of pressing fermented Gamay grapes. By contrast, Carboniste 'Rouge Brute', a low-pressure sparkling red wine crafted in California by Dan and Jacqueline Person, has been produced from base wines obtained as a result of pressing the grapes of Merlot and Albariño.

Dhillon 2021; Person and Person 2024; Stelzer 2018: 60–2; WA 2022: 18–19; WSET 2020: 111

carboniste.com
www.castleglenaustralia.com.au
www.hammerlingwines.co

M

maceration of red berry skins

The contact between the skins of black-skinned grapes — such as Pinot Noir, Meunier and Syrah — and grape must, treated as the first stage in the vinification of rosé and red base wines. This process is usually conducted in cold temperatures, typically around 8°C, for periods ranging between several hours (for light-coloured musts) and several days (for red musts). The skin contact results in the colouring of the must and the extraction of skin components into musts intended for rosé or red base wines. Recent oenological studies show that extended cold pre-fermentative maceration of Syrah grapes cultivated in Brazil's Northeast Region leads to increased concentrations of flavonoids in final red sparkling wines subjected to maturation on lees, notably kaempferol-3-O-glucoside, quercetin-3-β-d-glucoside, isorhamnetin-3-O-glucoside and petunidin-3-O-glucoside, thus enhancing the antioxidant capacity of these wines.

The maceration method is sometimes applied by sparkling winemakers in Champagne as part of the process of crafting rosé champagne, including not only macerated rosé (*rosé de macération*) but also blended rosé (*rosé d'assemblage*). Laherte Frères 'Rosé de Meunier' Extra Brut, a dark-coloured, well-structured non-vintage wine, classified as *rosé d'assemblage*, has been made from 100% Meunier grapes vinified in three different ways. This blended rosé has been composed using 30% macerated rosé wine, 10% red base wine and 60% white base wine (including 40% wood-aged reserve wine).

Barros *et al.* 2022: 3,254; CIVC 2023b; Liem 2017: 301, [2019] 2022; Moore 2022: 58–9; Ribéreau-Gayon *et al.* 2006: 448–9

www.champagne-laherte.com

Photo 29. Maceration of the juice from the Pinot Noir grapes with their skins has been one of the key steps in the production of Grempler Sekt–Pinot Noir 2021. This traditional-method sparkling red wine has been made from the estate-grown fruit by the winemaker Krzysztof Fedorowicz at Winnica Miłosz, a boutique sparkling wine producer in the village of Łaz, around eight miles east of the city of Zielona Góra (western Poland).
Photo: author, August 2024

malolactic conversion in red and rosé base wines and partially fermented musts

A biochemical process commonly used in the making of red and rosé base wines, whereby (strong) malic acid is converted to (milder) lactic acid and carbon dioxide through the action of lactic acid bacteria. The conversion process, which may last up to six weeks, is known to be facilitated by several factors, including: a low concentration of sulphur dioxide in the base wine (preferably below 30 mg/l); temperatures between around 18°C (but no lower than 16°C) to around 22°C (but rarely higher than 25°C); and the addition of lactic acid bacteria, the strain of which ought to be compatible with the fermentation yeast used. This winemaking process is typically carried out at the end of the primary fermentation or shortly after the fermentation is completed, producing a wine with desirable sensory attributes and increased microbiological stability. Some of the many examples of sparkling red and rosé wines made using malolactic conversion at the stage of creating base wine include: Champagne Philippe Martin's 'Rosé De Saignée', a blend made from the grapes of Pinot Noir and Meunier, picked during the 2018 harvest in Cumières, disgorged in 2022; Chandon Australia's Sparkling 'Pinot Shiraz', a non-vintage traditional-method sparkling wine created from approximately 70% Pinot Noir and 30% Shiraz grapes sourced from a diverse range of cool-climate sites within the state of Victoria; and Mount Avoca's 'Jack Barry' Sparkling Shiraz, a non-vintage sparkling red wine made in the Australian wine-producing region of Pyrenees (Victoria). Untypically for red wines, malolactic conversion will usually be blocked in some well-structured Lambrusco wines, notably those created from Lambrusco Grasparossa grapes, the aim being to preserve the wine's acidity.

AWRI [2011] 2020, 2020a, 2020b; Bartowsky 2022: 261–7; CIVC 2023b; Du Toit 2020; Lerm 2020; WA 2023; WSET 2020: 70–2

www.champagne.fr/en
champagne-philippe-martin.com/en
www.chandon.com.au
maisons-champagne.com
mountavoca.com

malolactic conversion in white base wines and partially fermented musts

A biochemical process in white base wines whereby (strong) malic acid is reduced to (milder) lactic acid and carbon dioxide, which leads to a reduction in the acidity of the wine. The process is typically considered beneficial, as it facilitates the development of desirable flavour compounds in the wine and ensures its microbiological stability. The conversion is encouraged — *inter alia* — by: ensuring a low concentration of sulphur dioxide in the base wine (preferably below 30 mg/l); maintaining temperatures in the range of 18–22°C; and adding lactic acid bacteria, the strain of which must be compatible with the fermentation yeast used. The process takes between four and six weeks and is usually carried out towards the end of the primary fermentation, or shortly afterwards, producing a softer-tasting base wine, showing fragrant notes of brioche and butter. The base wine is then gradually cooled to 10–12°C.

Nonetheless, wines that retain their malic acid tend to preserve the freshness of the grapes from which they have been produced, a feature believed to contribute to a greater longevity of the resultant wines. Among the producers known for avoiding the malolactic conversion process in the production of their wines are the two champagne houses Lanson and A.R. Lenoble. All the same, while generations of Lanson cellar masters have been preserving the principle of blocking malolactic conversion, Hervé Dantan, who joined the champagne house as *chef de cave* in 2013, combines the house's traditional style (which accentuates freshness) with a small proportion of wines that have undergone malolactic conversion when the aim is to bring to the final blend softness or additional flavours. Champagne Lanson 'Le Black'–Création 258 contains no more than 25% of wines with completed malolactic conversion.

AWRI [2011] 2020, 2020; Bartowsky 2022: 261–7; CIVC [2014] 2022: 8, 35, 2023b; Du Toit 2020; Hall 2021; Lerm 2020; OIV 2022: II.3.1.2.3; Schmitt 2016; WA 2022

lenoble-champagne.com/en
www.champagne.fr/en
lanson.com
maisons-champagne.com

malolactic fermentation in sparkling base wines and partially fermented musts

≈ malolactic conversion in red and rosé base wines and partially fermented musts

Photo 30. Maison Ruinart has in its portfolio a number of champagnes that have undergone malolactic conversion. These include (among others) Ruinart Blanc de Blancs, Dom Ruinart Blanc de Blancs 2010 and Ruinart Brut.
Photo: author, July 2024

≈ malolactic conversion in white base wines and partially fermented musts

manual disgorgement

A method of clearing traditional-method sparkling wines of sediment (which contains the deposit of dead yeast cells) using no mechanical devices. Requiring exceptional dexterity developed with experience, this winemaking activity is increasingly rare. This operation, which is sometimes used for old vintage wines, typically involves holding the bottle upside down in a vertical position or at a slight angle against the forearm, with the base of the bottle level with the crook of the arm. The disgorger (or *dégorgeur*) removes the temporary stopper — usually a crown cap, or in rare cases a cork — with a bottle opener (or a pair of pincers) while bringing the bottle back to the upright position, and allowing the sediment collected in the neck of the bottle (in most cases in the *bidule*, a plastic cylinder underneath the cap) to be driven out by the pressure. Then the disgorger sniffs the champagne to make sure that the wine has no unpleasant odours and puts the thumb over the lip of the bottle to prevent the foam from spilling, before placing the bottle on the dosage machine. For experienced disgorgers, the disgorgement rate averages around four hundred bottles a day.

Cole 2021: 24–5; Forbes [1967] 1989: 327–8; Grainger and Tattersall [2005] 2016: 141; Liem 2017: 60–1; TFB 2016, TFB 2018; Trento DOC 2022

www.champagne.fr/en
maisons-champagne.com

manual riddling

A time-honoured practice in the traditional method of transforming cuvée into effervescent wine which involves rotating bottles of sparkling wine placed in A-shaped wooden racks called *pupitres*, each with a holding capacity of 120 bottles. The aim is to dislodge the yeast deposit formed in the bottle during the second fermentation and extended maturation. Bottles of sparkling wine are turned repetitively, each time slightly towards an inverted vertical position, each bottle given a gentle shake before it is turned. This winemaking operation, traditionally performed by a *remueur*, can last from several weeks to a few months. At the end of the process, the yeast sediment is transferred to the bottle neck, from where it can be more easily removed.

This method of clearing bottles of sparkling wine of sediment dates back to the second decade of the 19th century. The introduction of the first riddling racks, around 1816, has been ascribed to Antoine Müller, Veuve Clicquot's cellar master at the time. Labour-intensive and time-consuming, manual riddling has now been largely replaced by an automated process using (computer-controlled) devices, acclaimed for their efficiency. Nonetheless, many sparkling wine-producing houses, including those in the Champagne region, still riddle by hand, often for small batches of wine or only for very special cuvées. Taking place inside underground cellars, this practice is maintained for various reasons, including: non-standard bottle shapes, tradition or exclusiveness.

Burgess 2022; CIVC [2014] 2022: 35; Forbes [1967] 1989: 155; Halliday and Johnson [1992] 2013: 116; Kunc 2012: 142; Livingstone-Learmonth 2005: 639; McCoy 2022: 32; Rand 2022: 19–20; SL 2018/2019: 40, 54

maisons-champagne.com/en

Martinotti method

≈ Charmat method

Martinotti-method sparkling red wines

≈ Charmat-method sparkling red wines

maturation of base wines and musts

A winemaking process in which new base wine or partially fermented must rests in conditions deemed appropriate for shaping its features, optimizing acid and alcohol proportions, as well as the proportions of esters formed as a result of their interactions. Maturation of partially fermented musts intended for single-fermentation wines, including (most) Asti-method wines, for instance Moscato d'Asti or Brachetto d'Acqui, is generally considered undesirable, as is the process of malolactic conversion, as — in most cases — the objective is to preserve in these wines their distinct acidity.

New base wines, which are to be made sparkling, including those destined for quality tank-fermented wines, and those intended for premium bottle-fermented wines, such as cap classique, cava, champagne, etc., are sometimes allowed to rest in conditions regarded as beneficial to their further development. Base wine used for the production of Adriano Adami's wine Vigneto Giardino–Rive di Colbertaldo–Asciutto from the Valdobbiadene DOCG zone rested on its lees in a steel tank for more than six months. Made by Hervé Jestin, the single-vineyard champagne Leclerc Briant 'La Croisette'–Brut has been crafted using base wine from 100% Chardonnay grapes. The wine underwent malolactic conversion and was allowed to rest on the lees for 10 months in oak barrels and terracotta eggs.

Halliday and Johnson [1992] 2013: 114; Jestin 2023; Leahy 2021: 38; WSET 2020: 66; Zoecklein 2002: 10

www.gancia.com/en
www.leclercbriant.fr/en
www.adamiprosecco.it
www.alasiavini.it
www.brachettodacqui.com/en

mechanical clarification of coloured musts

A winemaking operation conducted in the production of coloured musts, especially musts intended for rosé base wines or partially fermented musts, which involves separating the must from suspended matter by filtration or centrifugation. This clarification method, sometimes referred to as the dynamic settling method, is applied less frequently than the natural sedimentation method. The use of mechanical devices, such as filters and centrifuges, requires significant financial investment. Moreover, excessive clarification using machines can inhibit the ability of the must to ferment. Both filtration and centrifugation are applied mostly for particularly turbid rosé musts.

Joshi *et al.* 2017: 378; OIV 2022: II.2.1–34; Ribéreau-Gayon *et al.* 2006: 92–3, 108; WSET 2020: 65–6

mechanical clarification of white must

A winemaking operation used in the production of white base wines or partially fermented white musts which involves separating white juice obtained from sparkling wine grapes from solid matter suspended in the grape must by filtration or centrifugation. This method of clarification, sometimes referred to as the dynamic settling method, is used less frequently than the static cold settling (or natural sedimentation) method, not least as the use of mechanical devices requires significant financial investment, and because excessive clarification can inhibit the ability of the must to ferment and reduces the flavour complexity of the finished base wines. The use of filters, which can cause loss of aroma intensity, or centrifuges, which have been reported as being likely to lead to excessive contact of the must with oxygen, is usually reserved for cases of particularly turbid white musts.

Joshi *et al.* 2017: 378; McKay *et al.* 2011: 329; OIV 2022: II.2.1–34; Ribéreau-Gayon *et al.* 2006: 92–3, 108, 428; WSET 2020: 65–6

mechanical disgorgement

A method of clearing traditional-method sparkling wines of sediment (which includes the deposit of dead yeast cells) using bottle neck freezers. The method involves placing bottles of sparkling wine neck down in the openings of a disc-shaped tray, thus immersing their necks for several minutes in a container filled with a solution chilled to around −25°C. As a result, the yeasty sediment is transformed into a solid pellet of ice. The bottles are then turned upright and the stoppers are removed, so that the pressure built up inside the bottle can expel the frozen pellet, thus preparing the wine for finishing operations, which include, among other things, topping up the wine with expedition liqueur, also known by the French term *liqueur d'expédition*. Nowadays, depending on the equipment being used, the disgorgement rate using mechanical devices varies from 2,000 to around 18,000 bottles of wine per hour.

Buxaderas and López-Tamames 2012: 20; Cole 2021: 24–5; Grainger and Tattersall [2005] 2016: 141; Radil [2016] 2018; TFB 2016, 2018; Trento DOC 2022

www.champagne.fr/en
maisons-champagne.com

Photo 31. Michael Whitney at Adoria Vineyards in the village of Zachowice (around 20 miles southwest of the city of Wrocław) is using a bottle neck freezer. Necks of the bottles of traditional-method sparkling wine are exposed to very cold temperatures to solidify the sediment and thus ease its subsequent removal.

Photo: courtesy of Adoria Vineyards (Michael Whitney), July 2017

méthode ancestrale

≈ ancestral method

méthode artisanale

≈ ancestral method

méthode champenoise

≈ traditional method

méthode classique

≈ traditional method

méthode dioise

≈ ancestral method

méthode gaillacoise

≈ ancestral method

méthode rurale

≈ ancestral method

méthode traditionnelle

≈ traditional method

metodo ancestrale

≈ ancestral method

metodo classico

≈ traditional method

metodo Martinotti

≈ Charmat method

metodo tradizionale

≈ traditional method

modified Charmat method

≈ Asti method

multi-regional blending

A winemaking procedure that consists of combining base wines originating in different vine-growing regions. This is an extreme case of blending of wines from various grape-growing sites, particularly popular in Australia. Unfettered by strict appellation laws, Australia's winemakers are known for their creativity, which manifests itself in the art of blending, the usual aim being to create balanced and consistent wines by taking advantage of regional strengths. One example of sparkling wine created by blending across wine regions is Coates' 'La Petite Cuvée' NV, with the grapes used (mainly Shiraz with a small amount of Cabernet Sauvignon) sourced from South Australia's two wine regions: McLaren Vale and Adelaide Hills. Another example of a multi-regional blend is Taltarni's 'Taché' Sparkling Rosé 2015, a traditional-method sparkling wine made by Robert Heywood and Peter Warr. The wine has been created using 52% Chardonnay, 39% Pinot Noir and 9% Meunier, these grapes grown in a number of wine-growing regions, located in the states of Victoria, South Australia and Tasmania.

Martin *et al.* 2012: 1,082; Stelzer 2018: 34; WA 2018: 4.26, 2023: 35–9; Wheeler 2019

winetasmania.com.au
www.coates-wines.com
taltarni.com.au

Under Wine Australia Regulations 2018, a single geographical indication (GI) is allowed on the label of a bottle of wine provided that at least 85% of the wine has been obtained from the region indicated. With Australia's producers known for blending wines from grapes originating in various regions, the document enables the producer to indicate up to three registered regions or localities on a single label. A multiple GI claim may be made if at least 95% of the wine has been obtained from grapes grown in the regions indicated, with the contribution of each region to the final volume of this wine required to be at least 5%, and with the GIs required to be listed in descending order of the proportions of the relevant grapes in the final blend. In practice, if 83% of the resultant wine has been created from the McLaren Vale fruit (South Australia), 13% from grapes grown in the Yarra Valley (Victoria) and the remaining 4% from the Tumbarumba region (New South Wales), then the wine would need to be labelled 'McLaren Vale/Yarra Valley'. In such a case, the designation 'Tumbarumba' will not be allowed on the label.

multi-vintage blending

≈ vertical blending

muselage

≈ wirehooding

N

natural clarification of base wines and partially fermented musts

A stabilizing treatment of base wines or partially fermented musts involving the storage of the base wine or must in low temperatures, carried out to: inhibit malolactic conversion in the wine (or must) if the aim is to make high-acid wines; provoke the precipitation of tartrates; and facilitate the settling of lees. In the case of base wines prepared for the secondary fermentation, this operation is typically followed by racking the wine (while leaving the lees behind), or separating undesirable substances from the base wine using filters. In the case of partly fermented musts intended for single-fermentation sparkling wines, lowering the temperature at this stage serves to slow down or temporarily arrest the fermentation process in order to retain the desired concentration of sugars in the must before the valve in the pressure-resistant tank in which that must resides is closed and the temperature is raised again to complete the conversion of sugars into ethyl alcohol, carbon dioxide (and secondary products, such as esters), a step necessary for the production of effervescence.

📖 Jackson [1994] 2008: 542, 547, 720; Mercer 2021; Ribéreau-Gayon *et al.* 2006: 468–9; SL 2018/2019: 27; WSET 2020: 66

💻 www.astidocg.it/en

natural clarification of coloured musts

A winemaking operation used in the production of red and rosé base wines or partially fermented musts that involves separating the must obtained from sparkling wine grapes from sediments without disturbing the must. This action,

Photo 32. Clos Lanson, a one hectare plot of Chardonnay located in Reims. Natural clarification of musts, both coloured and white, is regarded as the safest clarification method, as it preserves a sufficient amount of substances needed for a wine's development at later stages of its production. From this vantage point, careful handling of the must at an early stage of base winemaking may have been among the key (albeit indirect) factors that have contributed to the great maturation and ageing potential displayed by Champagne Lanson 'Le Vintage' 2012, disgorged in June 2023. Judicious treatment of the must may also serve to explain the sustained formation of fine bubble chains in this wine once it is in the glass.

Photo: author, June 2023

carried out at low temperatures, consists of two consecutive stages. Settling of sediments in the grape must, sometimes called *débourbage*, is followed by racking. The initial stage involves resting the juice destined for red or rosé base wine or partially fermented must for a period that usually ranges between several and — in some cases — 48 hours. The juice remains (gently) chilled throughout the process. The low temperatures, usually around 5–8°C, stave off the occurrence of spontaneous fermentation while enabling the sediments to fall to the bottom of a tank in which the must has been placed: the *cuve de débourbage*. The clean must can then be drawn off through a hole in the wall of the tank, located above where the sediment has settled.

Barros *et al.* 2022: 3,246–7; CIVC 2023b; Forbes [1967] 1989: 299–300; Halliday and Johnson [1992] 2013: 114; Joshi *et al.* 2017: 378; Kojder 2022; TFB 2016; Waterhouse *et al.* 2016: 166; WSET 2020: 5, 65; Zoecklein 2002: 9

jaillance.com/en
www.champagne.fr/en

natural clarification of white must

A winemaking operation used in the production of white base wines or partially fermented white musts that involves separating white juice from sediments without disturbing the juice. This action, carried out at low temperatures, consists of two stages: static (cold) settling of juice from sparkling wine grapes, a winemaking process also known by the French word *débourbage*, and racking of must, an activity sometimes also referred to in French as *soutirage*. The initial stage of natural clarification of fresh grape juice involves resting the juice destined for white base wine or partially fermented must for a period that usually ranges between several hours and around 30 hours. The presence of sulphur dioxide, usually added at an earlier stage, can shorten the natural sedimentation to several hours. The juice remains (gently) chilled throughout the process. Low temperatures, usually around 5°C, stave off the onset of alcoholic fermentation while enabling the sediments to fall to the bottom of a tank in which the must has been placed: the *cuve de débourbage*. The clean must is then drawn off through a hole in the wall of the tank, located just above where the sediment has settled.

CIVC 2023b; Croguennec 2016: 292; Forbes [1967] 1989: 299–300; Halliday and Johnson [1992] 2013: 114; Joshi *et al.* 2017: 378; Kojder 2022; TFB 2016; Waterhouse *et al.* 2016: 166; WSET 2020: 5, 65; Zoecklein 2002: 9

jaillance.com/en
www.champagne.fr/en

P

partial fermentation of coloured musts

A controlled biochemical process that involves conversion of some of the sugar present in the red- or rosé-coloured must (obtained from dark-skinned grapes) into ethyl alcohol, carbon dioxide and secondary products, as a result of anaerobic metabolic activity of yeast. The must is allowed to ferment until a desired amount of sugar has been reached, with typical alcohol contents for partially fermented musts intended for single-fermentation sparkling red or rosé wines being in the range of 4.5–7% abv. Partial fermentations of red and rosé musts, often carried out at low or very low temperatures to preserve natural fruit flavours, usually take place in stainless-steel vats. Some producers use fermenters made from other materials. One example is provided by winemakers at Lightning Rock Winery, a small-scale wine-producing company in the Okanagan Valley, part of the Canadian wine region of British Columbia, who conducted partial fermentation of Pinot Noir grapes destined for their vintage (semi-sparkling) wine Lightning Rock 'Elysia Vineyard' Rosé Pét-Nat 2022 in fermentation vessels made of oak. By contrast, the partial fermentation of must from the Pinot Noir grapes intended for the (recently released) vintage (semi-sparkling) wine Lightning Rock 'Canyonview Vineyard' Pinot Noir Rosé Pét-Nat 2023 was carried out in an egg-shaped fermenter made of concrete.

Fermentation temperatures vary from producer to producer depending on the type of (semi-sparkling) wine they want to obtain. While the fermentation temperatures for obtaining partially fermented

Photo 33. A classic blend of Chardonnay and Pinot Noir, Woodfine Sparkling Chardonnay and Pinot is an English ancestral-method sparkling rosé from Buckinghamshire crafted by the winemaker Richard Woodfine. The first part of the alcoholic fermentation was conducted in oak and stainless steel. The winemaker is a proponent of a minimal waste strategy. Skins of grapes needed for the production of this sparkling wine have been subsequently reused in the production of cider, another product in the winemaker's Interconnections range.
Photo: Courtesy of Woodfine Wine (Richard Woodfine), August 2022

red or rosé musts (intended for pétillant naturel wines) are usually upwards of 13°C, in some wine regions the must is vinified at much lower temperatures. Producers of low-alcohol semi-sparkling rosé wines in the area of Bugey (central-eastern France), called Bugey Cerdon wines, carry out the initial phase of the fermentation process at very low temperatures, ranging from around 5°C to around 10°C. When the alcohol content in the fermenting must reaches the benchmark of around 6% abv, the fermentation is interrupted by refrigerating the vats containing the half-fermented must, with temperature reduced to (or several degrees below) the freezing point of water.

📖 Lorch [1994] 2015: 175–6; Makarov and Lutkov 2021: 290, 299–300; Schiessl 2023; Sohm and Muhlke 2019: 36

💻 www.gancia.com/en
vinsdubugey.net/en
www.brachettodacqui.com/en
www.domaineladentelle.com
lightningrockwinery.com
www.renardat-fache.com

partial fermentation of white must

A controlled biochemical process that involves conversion of some of the sugar contained in the white juice — obtained from light- or dark-skinned grapes and destined for white base wine — into ethyl alcohol, carbon dioxide and secondary products, as a result of anaerobic metabolic activity of yeast. The clarified juice is allowed to ferment until a desired amount of sugar has been reached, with typical alcohol contents for partially fermented musts intended for single-fermentation sparkling white wines being in the range of 4.5–7% abv. Partial fermentations are usually carried out at low temperatures using stainless-steel tanks, not least as these vessels enable the winemaker to control fermentation temperatures, but some producers, notably in Europe, sometimes choose to ferment all or part of the juice in wood. One example is the barrel-fermented wine Weszeli Pét Nat 'Blanc de Blanc', made from Grüner Veltliner grapes sourced from the vineyards located around the Austrian town of Langenlois.

Fermentation temperatures vary from producer to producer depending on the type of wine they want to obtain. While typical fermentation temperatures for obtaining partially fermented white juice (intended for pétillant naturel wines) are between 12°C and 16°C, the juice in some wine-producing regions is vinified at much lower temperatures. This is the case with producers in the Diois region, an area in the central part of the Rhône Valley (southeastern France), where the juice intended for wines known as Clairette de Die Tradition or Clairette de Die Méthode Ancestrale is allowed to ferment at very low temperatures, ranging between 5°C and 10°C, until the alcohol content reaches around 4.5% abv. Low temperatures slow down the partial fermentation process, which — under these conditions — can take up to several months.

Adams 2018; Buglass 2022: 227; Goode 2023; Joshi *et al.* 2021: 566–7; WSET 2020: 66

clairette-de-die.com/en/
jaillance.com/en
www.clairette-mongegranon.com
www.weszeli.at/en

perpetual blend

≈ perpetual reserve

perpetual blending

An economical method of combining vintage-dated wines used increasingly by small-scale producers, mainly in the Champagne region, which allows them to benefit from the complexity of a blend without the need to store many wines from different harvests separately. The method involves storing base wine in a single vessel, typically made of wood, stainless steel or concrete, while refreshing the wine regularly (usually every year) with a small amount of new base wine (made from grapes collected during the most recent harvest). The amounts of wine to be removed from and added to the batch each year are at the discretion of the cellar master. They can be fixed or dependent on the quality of the harvest. This blending method is gaining traction among an increasing number of small-scale producers, mainly in Champagne, one reason for its adoption being that many among them — especially grower estates — lack the space and/or equipment necessary to store large amounts of various (vintage-dated) wines individually.

Many winemakers view this method of storage and blending of their wines across vintages not only as an optimal way of expressing the full nature (and character) of a grape-growing site, but also as a stylistic departure from the conventional manner of storage and blending (which involves allowing wines to mature separately according to vintage year, grape variety and place of origin). The most acclaimed producers making use of the method of keeping and blending wines from grapes coming from different harvests, which results in the continuous evolution of the original reserve wine, fuelled each year by a fresh batch, are Bruno Paillard, Champagne Barons de Rothschild, Champagne Dehours et Fils, Champagne Huré-Frères, Champagne Jacques Selosse, Champagne Pierre Péters, and also Louis Roederer. Some producers, including the house of Bruno Paillard and Champagne Dehours et Fils, keep several different perpetual reserves, each linked with a different grape-growing site.

Claffee 2013; Edwards 2009: 201; Krebiehl [2017] 2022: 234, 236, 238–43; Liem 2017: 300, [2019] 2022; Mowery 2023a; Palmer 2024

champagnebrunopaillard.com
champagne-dehours.fr/en
champagne-bdr.com/en
www.champagne-hure-freres.com
www.champagne-peters.com/en
www.louis-roederer.com/en
www.selosse-lesavises.com/en

perpetual cuvée

≈ perpetual reserve

perpetual reserve

Wine that has been refreshed regularly — typically every year — with new base wine(s) crafted using grapes coming from the most recent harvest. Multi-vintage base wine of this type is usually produced by adding various amounts of wine from each year's vintage to a chosen storage vessel. Vessels in which producers keep this wine, also known in French as *réserve perpétuelle*, range from concrete or stainless-steel tanks (of differing sizes) to oak barrels or large wooden vats called *foudres*. Constantly refreshed, each *réserve perpétuelle* gradually becomes an increasingly complex blend composed of diminishing amounts of wines from (an ever-growing number of) different harvests. The wine is normally built up over many years before the cellar master decides that its composition has reached an acceptable level of complexity and balance, and is ready to be used as a final blend or as part of another cuvée destined for high-quality sparkling wine.

The perpetual reserve initiated in 1990 by Joseph Henriot has given rise to the sixth edition of Champagne Henriot 'Cuve 38'–La Réserve Perpétuelle, Blanc de Blancs, a champagne based entirely on perpetual cuvée, composed of 100% Chardonnay wines from every successive harvest from 1990 to 2012, which is a collective expression of the terroirs of the four Côte des Blancs villages, namely Avize, Le Mesnil-Sur-Oger, Oger and Chouilly. A different approach has been adopted at Champagne Dehours et Fils, where only around 30% of the company's *réserve perpétuelle*, which dates back to 1998, is added to the final blend to make Champagne Dehours & Fils 'Grande Réserve'–Brut.

Castaing 2015; Claffee 2013; Fallowfield 2022; Liem 2017: 52, 174, 220–1, 225, 242, 264, 266, 300, 302–3; Mowery 2023a; Palmer 2024

www.champagne-henriot.com/en
champagne-dehours.fr/en

plunging of red grape skins

A winemaking operation conducted with the aim of increasing the extraction of skin components and compounds from areas adjacent to the berry skin (such as natural colourants, flavour compounds and tannins) into the must intended for sparkling red wine. The method involves manual submerging of the skins of red grapes to intensify their contact with the must. Carried out at regular intervals, this activity also contributes to the micro-aeration of the must, a process facilitating colour stabilization. When applicable, the submersion of the red berry skins, used widely among the makers of sparkling red wines in Australia, is (usually) conducted two or three times daily.

Angelosante 2020; Bryan 2019; Ribéreau-Gayon *et al.* 2006: 350; WA 2022: 17–19; WSET 2020: 111

pfeifferwinesrutherglen.com.au
www.balgownie.com
taltarni.com.au
www.ashtonhills.com.au

poignettage

≈ *poignettage* during sparkling wine maturation
≈ *poignettage* for distribution of expedition liqueur

poignettage during sparkling wine maturation

An optional winemaking activity conducted by some producers during the extended maturation of sparkling wine on lees, part of the traditional method of producing effervescence in wines. During this activity, bottles containing sparkling wine, stacked horizontally, undergo shaking and restacking, the aim being to prevent the yeasty sediment that forms in the bottle during the maturation process from sticking to the glass. Bottles with wines designed for long-term maturation are usually agitated once a year.

📖 Burton and Flewellen 2013; Grainger and Tattersall [2005] 2016: 139; Jackson [2002] 2009: 162; Liem [2019] 2022; McKay *et al.* 2011: 334; Sandler and Pinder 2003

💻 torello.com

poignettage for distribution of expedition liqueur

A winemaking activity conducted at the end of the process of making sparkling wines using the traditional method. This activity consists of shaking the bottles containing traditional-method sparkling wine, such as cap classique, cava, champagne, etc. after expedition liqueur (also commonly referred to as *liqueur d'expédition*) has been added to the wine as part of the dosage operation, and the cork has been secured by the wire muzzle. The aim of shaking the bottle at this stage is to facilitate the process of blending the sparkling wine in the bottle with the expedition liqueur. The winemaking stage is typically accompanied, or followed immediately, by careful inspection of each bottle in order to check the limpidity of the wine, a process known in Champagne as *mirage*.

📖 Forbes [1967] 1989: 332; Liem [2019] 2022; SL 2018/2019: 46; Vizetelly 1882: 165–6

💻 www.champagne.fr/en
maisons-champagne.com

Photo 34. A winemaker at the sparkling wine-producing company Torelló holding one of the bottles of quality sparkling wine maturing in the cellar. In the case of Torelló's vintage cuvées, *poignettage* is done regularly over many years. Bottles of the corpinnat Gran Torelló Grandes Añadas Brut Nature 2013 undergo shaking and restacking once every year, a practice believed to add to this sparkling wine's palette of flavours.
Photo: Marta Ręgowska, April 2023

post-blending stabilization

A winemaking practice applied in the making of base wines or partially fermented musts after blending, the aim being to stabilize the cuvée and conserve its desirable characteristics. This winemaking practice may be particularly necessary when numerous base wines of different chemical properties have been included in the final blend, as such a practice often encourages the presence of unwanted matter. Usual options for obtaining a pure and stable blend include tartrate stabilization (followed by racking), the use of filters, or the application of both these methods in sequence.

📖 Halliday and Johnson [1992] 2013: 115; LA 2020: 3–4; SL 2018/2019: 6–7, 9–10, 33; WSET 2020: 6, 10; Zoecklein 2002: 10

post-disgorgement ageing

≈ ageing of sparkling wine

post-fermentation treatment of coloured base wines and partially fermented musts

An important stage in the production of sparkling red (and rosé) wines which takes place following the initial fermentation process. In the case of pink-hued base wines or partially fermented musts (obtained from early pressing), the liquids are usually kept in cold storage to preserve their characteristics, especially those responsible for natural aromas and colour. In the case of red-hued liquids, during this stage the winemaker may decide to press the dark-skinned grapes immediately to obtain red base wine or partially fermented must, or to carry out a brief post-fermentation maceration, the aim of the latter being to enhance the integration of fine tannins. Post-fermentation maceration was carried out under the watchful eye of Balgownie Estate's chief winemaker Tony Winspear during the vinification of Shiraz grapes when crafting the firm's Balgownie Estate 'Black Label' Sparkling Shiraz NV.

Garrido-Bañuelo *et al.* 2021; McCubbing 2024; Pickett 2023; Stávek *et al.* 2012: 1,133–45; WA 2022: 18

www.balgownie.com

pre-fermentation cold treatment of white must

One of the key stages in the vinification of grapes destined for white base wine or partially fermented must, reliant on exposing the roughly clarified juice to low temperatures, usually close to water-freezing point, to preserve the fresh and fruity aromatics of the grapes and to prevent any bacterial activity while not allowing the fermentation process to begin. Fermentation is delayed because the winemaker may need time for activities such as the inoculation of the juice with a selected yeast strain and/or enrichment of the juice. Cold treatment — enabled by placing the juice in temperature-controlled stainless-steel tanks — is vital for the preparation of Asti-method wines, in which case the juice may remain chilled for over a year until there is demand.

Burgess 2016; MacLean 2024; Nagórka 2013: 85–6; Sacchetto 2023; WSET 2020: 66

www.astidocg.it/en

pre-fermentation heating of red grapes

A winemaking method in the production of sparkling red wine that involves increasing temperature of the selected grapes to above 30°C shortly prior to the fermentation process. After the crushed grapes have been heated, with a high temperature maintained for a specified time, the unfermented (coloured) must is normally cooled to a temperature considered optimal for the fermentation process, which in the case of red musts ranges between 16°C and 25°C. The result of heat treatment is the relatively quick release of desirable compounds contained in the skins of red grapes, notably colourants (with a shorter treatment required for very high temperatures and a longer treatment required for moderate temperatures). If the aim is to speed up the extraction of colourants while protecting subtle varietal aromatics of delicate grapes, such as Pinot Noir, heating may take place at a temperature as low as 32°C, in which case that temperature is usually maintained for around a dozen hours. Significantly, it has also been found that heating could attenuate (negative) effects of high concentrations of methoxypyrazines, flavourants responsible for the green-bell-pepper characteristics (undesirable in red wines), which are leached into the must from the skins of Cabernet Sauvignon grapes. Importantly, at this stage, heat treatment for obtaining must intended for red sparkling wine is used less frequently than other methods supporting extraction, notably the pumping-over method and the manual plunging of the skins.

Jackson [1994] 2008: 347, 597, 716; Nordestgaard 2017: 54–59; OIV 2022: II.1.1–8; Ribéreau-Gayon *et al.* 2006: 383–5

pre-fermentation skin contact for rosé must

A method of treatment of red grapes destined for sparkling rosé wine — commonly referred to as *rosé de macération* — that involves tinting or colouring the must by allowing various periods of skin contact between the juice and the red berry skins. When the aim is to obtain slightly coloured must for wines traditionally called *oeil de perdrix* for their pale salmon and copper hues, the winemaker tints the juice of red grapes, such as Pinot Noir, by allowing minimal contact between the juice and the skins of the red berries. The contact time between the grape must and the red berry skins is much shorter when preparing must for *oeil de perdrix* wines (usually lasting no more than a few hours) than for a fully macerated rosé or *rosé de saignée* (in which cases maceration times usually range from several hours to around three days). The technique based on minimal skin contact has been applied at Champagne Jean Vesselle for the production of Jean Vesselle Oeil de Perdrix–Brut, made using 100% Pinot Noir grapes. The distinct colour of wines made by this method is obtained partly prior to and partly as a result of slow pressing.

Photo 35. A cork in wire cage topped with a pink cap, used for stoppering a bottle of Ridgeview Fitzrovia Rosé NV, a fruit-driven Chardonnay-dominated blend with 17% Pinot Noir and 15% Meunier. A vintage-dated sibling of this wine from the 'Fitzrovia' range — Ridgeview Fitzrovia Rosé 'Cuvée Merret' 2004 — was served on 24 May 2011 during a state banquet held at Buckingham Palace by Queen Elizabeth II to honour President Barack Obama of the United States.
Photo: author, March 2024

Berrill 2011; CIVC 2023b; Eads 2014; Hyland 2024; Liem 2017: 230; 278–9; Napjus 2014; Nichols 2018; Parker [1995] 2008: 450

www.champagne.fr/en
www.champagnejeanvesselle.com/en

pre-fermentation treatment of coloured musts

An important stage in the making of coloured base wines and base (partially fermented) musts destined for sparkling wines. This winemaking stage usually assumes the coloured must is kept in cold storage for a short period, during which the winemaker can perform background pre-fermentation activities, such as clarifying coloured musts and selection of yeast strains. When producing red base wines or base (i.e. partially fermented) musts, this stage tends to be considered as an extension of the cold maceration process. When the aim is to produce pink-hued base wines or (partially fermented) musts intended for sparkling wine, it is during this stage that the winemaker presses grapes for rosé must needed before the fermentation process is allowed to begin.

Halliday and Johnson [1992] 2013: 138; TFB 2017a; WSET 2020: 18; Zoecklein 2002: 8, 22

preparation of base wines and partially fermented musts

The first of the two stages in the making of sparkling wines, involving a sequence of operations from the rigorous selection of grapes to the stabilization of base wines and/or fermenting musts. The preparation of base wines (with an alcohol content of no less than 7% abv) or partially fermented musts (alcohol content usually in the range of 4.5–7% abv) intended for sparkling wines requires the concentration of all the skills needed for producing still wines, whether white, red or rosé. Regardless of the type of sparkling wine intended, the key operations revolve around protective juice handling, especially with juices or musts destined for white and rosé base wines, and the management of fermentation, which can be partial or complete.

The resultant base wines and partially fermented musts, destined for semi- or fully sparkling wine, tend to have low (notably in the case of partially fermented musts destined for single-fermentation wines) to moderate alcohol contents, relatively high acid concentrations, and low to moderate pH values. When crafting double-fermentation sparkling wines, notably traditional-method wines, parameters that are (generally) regarded as requisite for base wines and their blends include acid concentrations higher than 5.5 g/l, pH levels below 3.3 and alcohol contents between 9% and 11.5%. In order to ensure that the base wines retain their ability to referment, their free sulphur dioxide contents must be kept low (ideally below 20 mg/l).

Cisilotto *et al.* 2023; Di Gianvito *et al.* 2019: 352; IOC 2018: 2; Joshi *et al.* 2017: 378; Just-Borràs *et al.* 2023; OIV 2022: I.1.3–1, II.4.1–1; WSET 2020: 23; Zoecklein 2002: 11–12

maisons-champagne.com/en
www.champagne.fr/en

preparation of cuvée

≈ blending of base wines or partially fermented musts

preparation of sparkling wines

A sequence of winemaking activities conducted to obtain special wines characterized on contact with fresh air by the production of effervescence, which results from the release of natural (endogenous) carbon dioxide. The main activities are

Photo 36. Some beekeeping producers combine sparkling wine production with a small-scale production of honey. Notable among them are: the champagne houses Lanson (the firm's wine and honey sourced from the estate of Clos Lanson in Reims) and AR Lenoble (whose products take their origins from the village of Chouilly). The cap classique producer Anthonij Rupert Wyne offers sparkling wines and honey sourced from L'Ormarins, the firm's wine estate located in the Coastal Region's district Franschhoek.
Photo: author, July 2023

blending of base wines, stabilization of the resultant blend, choice of an adequate sparkling method, selection of bottle closures, and post-disgorgement ageing.

Bacigalupi 2023; Hall 2021; Hogendoorn *et al.* 2015; Howe 2003: 143–51; OIV 2022: I.1.4–4; Willcox 2020

bacigalupivineyards.com
lenoble-champagne.com/en
champagnelelarge-pugeot.com/en
leftcoastwine.com
lormarinswines.com

pressing of grapes for white must

A sparkling winemaking operation in the making of white base wine which involves exerting physical pressure on selected grapes. The operation involves gentle pressing of whole grape berries or a mash of lightly crushed grapes. In the case of making quality sparkling wines, the activity is usually carried out immediately after the harvest, with the possibility of contact between grape juice and skins — typically regarded as undesirable in the making of white base wines — being reduced to a minimum, unless the head winemaker is aiming to produce a rare skin-contact sparkling white wine, as is sometimes the case in the creation of certain pétillant naturel wines, commonly referred to as pét-nats. Quality-conscious producers

Beekeeping in the winegrowing area, though still rare, is slowly becoming a trend among environmentally conscious sparkling wine producers, who not only recognize the bees' considerable contribution to the ecological balance but also see opportunities to use the fruits of their labour in key stages of sparkling wine production. The family-owned (and -operated) wine producer Left Coast Estate, located in Oregon, west of Salem, is home to numerous beehives. The producer uses honey as a source of sugar for the yeast to support the fermentation of musts from Pinot Noir grapes destined for the company's range of pétillant naturel wines called Queen Bee Bubbly. Another example comes from Bacigalupi Vineyards, a wine producer from Sonoma County's Russian River Valley, whose managers have drawn inspiration from the winemaking philosophy and experiments of beekeeping champagne producer Lelarge-Pugeot. In 2020, after three years of ageing on lees, Bacigalupi Vineyards' wine Brillante Blanc de Noir Sparkling-2017 was disgorged prior to being topped up with *liqueur d'expédition*, which contained honey sourced from a local beekeeper.

prefer a gradual extraction of juice by slowly increasing the pressure on the grapes placed in the press.

Several types of wine press are currently in use in sparkling winemaking. Traditional vertical (fixed-basket) presses used in Champagne are characterized by a relatively broader surface area compared to standard versions of the vertical press, a feature viewed as desirable for pressing whole grape clusters. Other wine presses used in the Champagne region are horizontal presses, which include: automated plate press, pneumatic press with central bladder, pneumatic press with lateral membrane, and hydraulic press with tilted plates.

Considine and Considine 1982: 2,150; Forbes [1967] 1989: 295–9; Jackson [1994] 2014: 440; Hornsey 2007: 4.2.4; Liem 2017: 46–7; McKay *et al.* 2011: 328; Pessey 2002: 77; Tardi 2016: 15–18

www.champagne.fr/en
maisons-champagne.com/en

pressing of lightly crushed grapes

A non-traditional and less commonly used method of pressing grapes for white base wines. While quality-conscious producers usually favour pressing whole bunches, this pressing method, which involves liberating the juice from grape berries that have previously been subjected to gentle crushing, is sometimes considered desirable for its efficiency. The extraction of juice from crushed grapes rather than from whole bunches tends to be viewed as appropriate only in cases where the grapes selected for the winemaking process are in good condition. The activity is especially important in the production of so-called sparkling orange wines, in which light-skinned grapes, for instance Malvasia di Candia Aromatica, are macerated for periods of varying lengths, typically between a few hours and several days, or — in very rare cases — from several days to periods of up to two months, before going to the press.

CIVC 2023b; Gardner 2015a; Halliday and Johnson [1992] 2013: 114–19; Nagórka 2013: 88; Walkey 2016; Zoecklein 2002: 5–6

en.poderepradarolo.com

pre-treatment of sparkling wine grapes

A sparkling winemaking procedure that involves preparing previously selected black-skinned and/or light-skinned grapes for vinification, initiated either by the operation of pressing of grapes (when white base wines are being created) or by maceration of red berry skins (when red and rosé base wines are being made). With the majority of sparkling wines, especially white sparkling wines, for which no (or very limited) contact between the juice and berry skins is intended, care is taken to transport freshly harvested grapes to the pressing centre undamaged and as quickly as possible. This is essential to avoid staining juices of dark-skinned grapes destined for white base wines (or partially fermented musts), bearing in mind that in Champagne around 70% of sparkling wines are made using black grapes. Grape clusters are typically carried in wicker baskets or perforated crates. The objective is to ensure that the berries remain cool and well aerated during transport. By contrast, the black grapes destined for red base wines (or partially fermented musts) are often destemmed and crushed before they are submitted to cold pre-fermentative maceration.

Barros *et al.* 2022: 3,246–7; CIVC [2014] 2022: 12; Forbes [1967] 1989: 295; Moore 2022: 56–7, 59; Williams 2022;

www.champagne.fr/en
francis-boulard.com
maisons-champagne.com/en

Photo 37. Freshly harvested Chardonnay grapes being transported to Spier Wine Farm's pressing centre in the Stellenbosch district (part of the Western Cape's Coastal Region, South Africa).

Photo: courtesy of Spier Wine Farm (Lize-Marié Nieman and Johan Jordaan), January 2024

primary fermentation

≈ fermentation of coloured musts

≈ fermentation of white must

prise de mousse

≈ continuous fermentation

≈ fermentation in autoclaves

≈ traditional-method fermentation

≈ transfer-method fermentation

Pumping-over of coloured musts

A method of encouraging the extraction of skin components from dark-skinned grapes, including colourants, flavour compounds and tannins. Taking place prior to (or at the beginning of) primary fermentation, the process involves submersion of the skins by transferring the must from the bottom to the top of the vessel in which the grapes and must are stored, so that the skins rising to the top of the tank remain in contact with the must. This action also promotes the micro-aeration of the must, thus assisting in colour stabilization. The method is used in many wineries in Australia and the United States.

The frequency of the transfer of the must between lower and upper parts of the tank varies from producer to producer, but — on average — the must held in the tank is usually pumped over the skins from three to four times daily. At the winemaking plant of the Australian producer Grant Burge Wines, musts from the grapes of Shiraz and Cabernet Sauvignon, sourced from two different sites in the Barossa Valley (South Australia) and intended for the bottle-fermented sparkling red wine Grant Burge 'Shiraz Cabernet' NV, are pumped over three times a day.

Angelosante 2020; Bryan 2019; González-Lázaro *et al.* 2019; Ribéreau-Gayon *et al.* 2006: 341, 349–53, 382; SL 2018/2019: 30; WA 2022: 17–19; WSET 2020: 111

grantburgewines.com.au
pfeifferwinesrutherglen.com.au
taltarni.com.au
www.korbel.com

R

racking of blended base wines and partially fermented musts

Transfer of the prepared blend of base wines or partially fermented musts to another container with the aim of separating precipitated tartrates from the wine or must. This operation is often followed by (rough) filtration, with much care being taken not to eliminate excessive amounts of proteins and polysaccharides, which may play a decisive role in the formation of bubbles, an indicator of a sparkling wine's quality.

Abdallah *et al.* 2007: 377, 381–3; Forbes [1967] 1989: 311; Halliday and Johnson [1992] 2013: 115; Liger-Belair *et al.* 2008: 2,493; Nagórka 2013: 93; Stelzer 2016

rebêche

Juice obtained as a result of the final pressing of whole clusters of grapes. The juice is generally considered too coarse for producing base wine destined for high-quality sparkling wines, such as champagnes or corpinnats. Rather than being used for the making of sparkling wines, this low-quality juice, characterized by considerable bitterness and lack of acidity, is generally intended for distillation.

📖 Forbes [1967] 1989: 298; Liem 2017: 46–7, 301; SL 2018/2019: 14; Tardi 2016: 17, 39, 59, 113, 179

💻 maisons-champagne.com/en

rebottling of ancestral-method sparkling wines

A winemaking practice in the making of sparkling wines using a single fermentation finishing in the bottle, used for some pétillant naturel wines cleared of sediment by means of filtration, such as those created in the Diois region, a winemaking area in the central part of the Rhône Valley (southeastern France). This practice, employed for wines known as Clairette de Die Tradition or Clairette de Die Méthode Ancestrale, involves pouring the clear lightly effervescent wine into new bottles before their release.

📖 Goode 2023; Livingstone-Learmonth 2005: 662–3; Waldin 2013

💻 clairette-de-die.com/en/
www.domaine-long.com/en/

remise en cercles

A rare and expensive winemaking technique used exclusively for very special sparkling wines. The technique, applied in the blending of base wines for premium sparkling wines, most notably champagne, involves opening bottles of mature sparkling wine (either those following disgorgement or those aging *sur lattes*) and de-gassing the content in order to blend it with the base wine(s), preparing a prestige cuvée. This winemaking practice has been used by Rodolphe Péters at Fleur de Miraval for the production of the 'Exclusivement Rosé' range, the first edition of which (ER1) was released in October 2020. This sparkling wine was created by opening bottles of seven-year-old *blanc de blancs* champagne and adding it to the final blend before initiating secondary fermentation using the traditional method.

The winemaking practice of blending base wine with wines that have been made effervescent by means of a secondary fermentation in the bottle, has also been used at Moët & Chandon by the company's long-time *chef de cave* Benoît Gouez to develop the multi-vintage champagne Moët & Chandon 'MCIII', bottled in 2004, disgorged in 2014 and released the following year. The roman numeral in the name refers to three types of ageing vessel used in the creation of the blend: metal, wood and glass. Notably, 25% of the cuvée destined for this special wine was made up of bottled sparkling wines from grapes collected during three great harvests: 1993, 1998 and 1999. Bottles containing these wines were opened and emptied into a vat, where the wines were blended with reserve wines aged in stainless-steel tanks and reserve wines aged in wood.

📖 Goode 2016; Gouez 2020; Green 2023b; Perrin and Péters 2022; Simon [2020] 2022; TFB 2017b

💻 fleurdemiraval.com/en
www.moet.com/en-int
www.vineyardbrands.com

removal of sediment from ancestral-method sparkling wines

Elimination of the yeast deposit formed in bottles containing effervescent wine obtained from a single fermentation. Once viewed as desirable for aesthetic reasons, while aiming to offer customers a stable commercial product, the elimination of lees from bottles containing sparkling wine is now usually considered unnecessary, with the deposit often regarded as part of the wine's style.

While the elimination of the yeast deposit from sparkling wines in the world's many winemaking regions is left at the discretion of the producer, this activity is mandatory in the French region Montlouis-sur-Loire, where ancestral-method sparkling wines, officially referred to as pétillant originel, are made from the grapes of Chenin Blanc. These wines are left to mature on lees for a minimum of nine months before the partially bottle-fermented lightly sparkling wines are cleared of sediment.

Ancestral-method sparkling wines released to the market without the yeast sediment are also offered by producers in the central part of the Rhône Valley (southeastern France). These wines are commonly known by the name Clairette de Die Méthode Ancestrale, sometimes also called Clairette de Die Tradition. Made from the grapes of Muscat Blanc à Petits Grains and Clairette Blanche, these lightly sparkling wines are clarified before release, typically by means of decanting the wine into vats followed by cold filtration.

A different technique of clarifying pétillant naturel wines made using the ancestral method is practised at Chandon Australia by the company's chief winemaker Dan Buckle. Bottles of Chandon Ancestrale Rosé 2019 were riddled before the wine was partially disgorged, a clarification technique somewhat similar to that practised in the making of traditional-method sparkling wines, such as champagne or cap classique.

Campanale and Stein 2022: 47; Hooke 2021; Livingstone-Learmonth 2005: 662–3; Sussman 2015; Vierra 2021; WSET 2020: 11

clairette-de-die.com/en/
www.clairette-de-die-lantheaume.fr
www.chandon.com.au
www.domaine-long.com/en/

removal of sediment from Asti-method sparkling wines

Elimination of the yeast deposit formed in pressure-resistant tanks containing effervescent wine obtained using a single fermentation. The yeasty sediment is most often removed from the wine by centrifugation or filtration. This operation is regarded as necessary for all Asti-method wines to prevent refermentation of the wine after bottling, which would (otherwise) lead to a change in the wine's parameters required for its category, including (among other factors) its (fruit) flavour profile, sugar concentration levels, alcohol content and over-pressure.

Fielden [1994] 1998: 297; Mowery 2023b; Sacchetto 2023; WSET 2020: 65–8; Wu 2015

www.astidocg.it/en
www.hubertsekt.sk/en

remuage

≈ riddling

réserve perpétuelle

≈ perpetual reserve

retention of sediment

A winemaking practice in the making of ancestral-method sparkling wines; it involves keeping the yeast sediment that has formed in the bottle containing effervescent wine obtained from a single fermentation. Once bottled, the wine is subjected neither to disgorgement nor to filtration. The deposit of yeast sediment will make the resultant wine slightly cloudy, with the decision whether to decant or not being left to the consumer.

There are numerous producers releasing to market ancestral method wines containing sediment, including wine houses as far apart as Benjamin Bridge from Gaspereau Valley in the Canadian province of Nova Scotia and the family-run winemaking firm Winnice Kojder, located in the Polish part of Pomerania, an emerging wine region on the southern shore of the Baltic Sea. Another example of bottles of ancestral-method effervescent wine released with fermentation sediment is provided by Chandon Australia. The wine Chandon Ancestrale Rosé 2019 has been only partly disgorged before release, which means that a small amount of lees remains in the bottle.

Hooke 2021; O'Keefe 2018, 2019; Rose 2021: 26–7; Teclemariam 2023; Waldin [1994] 2015: 557; WSET 2020: 11

benjaminbridge.com
www.chandon.com.au

Photo 38. Grapes ripening in the estate vineyard at Winnica Miłosz, a boutique sparkling winemaking house in the village of Łaz, around eight miles east of the city of Zielona Góra (western Poland). The house has in its portfolio single-fermentation wines with retained sediment. These include Winnica Miłosz Pét-Nat Zweigelt 2019 (a pink-hued pétillant naturel wine from the grapes of Zweigelt) and Winnica Miłosz 'Wolne Wino' BIO 2023 (a lightly sparkling rosé crafted using fully ripe Pinot Noir grapes).
Photo: author, August 2024

retrousse

A winemaking operation typically performed when pressing grapes destined for white base wine using vertical basket presses; it involves breaking up the cake of fruit and skins formed at the bottom of the press following each pressing. This activity is both laborious and time-consuming when performed manually, with winery workers traditionally using pitchforks to move the pile composed of grapes and their skins, which still contains a lot of juice, back to the centre of the press before it is subjected to another pressing. Once conducted exclusively by hand, the *retrousse* operation can now also be done mechanically, or automatically, using modern and ultra-modern presses, such as the cylindrical-shaped pneumatic press and the fully automatic PAI press (*pressoir automatique à plateau incliné*), with the operation's intensity increasingly controlled by advanced software.

CIVC 2013: 39, 2023b; Edwards 2009: 35–6; Jackson [1994] 2008: 540–1; Liem 2017: 46–7; Tardi 2016: 37–9

maisons-champagne.com

riddling

A winemaking operation conducted to prepare bottles containing traditional-method sparkling wine for the removal of the yeasty sediment, called lees, on which the wine has matured. This operation, which can be performed either by hand or by machine, involves dislodging the yeast deposit that has formed in the bottle during the second fermentation and extended maturation by repeated turning of the bottle progressively towards an inverted vertical position while shaking the lees into the neck of the increasingly inverted bottle, thus facilitating their subsequent removal. Depending on the method used, this operation can take between around a week — in the case of a fully mechanized process — and around eight weeks in cases where riddling is performed manually with utmost care.

📖 Buxaderas and López-Tamames 2012: 20; CIVC 2023b; Jackson [1994] 2014: 710–12; Liem 2017: 58–61, [2019] 2022; Rand 2022: 19–20

rigorous selection of sparkling wine grapes

A sparkling winemaking procedure which involves the choice of grapes and the choice of an optimal time for the harvest. The relevant decisions depend on the region in which the vines are grown. For categories that encompass top-quality sparkling wines, the selection of grapes is (in many cases) regulated by competent authorities. Some of these bodies may also be tasked with determining the exact dates for the beginning of harvest, which can vary depending on the year of harvest and the place in which the vines are grown.

Some cap classique producers in Western Cape's Coastal Region in the Paarl district select their grapes based on the age of the vines, with old vines usually considered relatively resilient to damage caused by extreme weather events, such as heatwaves. Albertus Louw and Natalie Kühne, winemakers at Perdeberg Group, an active member of Old Vine Project, source their Chenin Blanc grapes from a bushvine vineyard, planted in 1977. The grapes have been used to create two quality sparkling wines from the firm's Vineyard Collection range: Perdeberg Cellar's Old Vine Chenin Blanc Cap Classique 2019 and Perdeberg Cellar's Chenin Blanc Cap Classique 2020. For some producers of top-quality sparkling wines in Catalonia, the use of grapes traditional to the region is of great importance. At least 90% of the grapes for corpinnat — quality sparkling wine made in the heart of the Penedès region — must come from the region's autochthonous (historical) varieties. These varieties, authorized by Associació de Viticultors i Elaboradors Corpinnat (AVEC), include: Garnacha, Macabeo, Parellada, Monastrell, Subirat Parent, Sumoll and Xarello.

In many wine regions, grapes for most sparkling wines are harvested earlier than those intended for still wines, the aim being to achieve the high-acidity and low-alcohol profile usually desired for the majority of wines representing the sparkling wine category. One exception to this general rule is the fruit for sparkling red wines, whose attributes typically include richness and intensity of flavour. Hence, the grapes needed for sparkling red wines tend to be picked at roughly the same time as those destined for still wines.

The grapes authorized in the region of Champagne include: Pinot Noir (an early-ripening grape variety, accounting for around 38% of grapes grown in the region), Chardonnay and Meunier (each accounting for about 31% of the area under vine). Other grape varieties approved for the production of champagne are very rare, accounting for up to 0.3% of the grape-growing area. These are: Arbanne, Petit Meslier, Pinot Gris and Pinot Blanc. Every year, the date of the beginning of the harvest is fixed by a decree based on a provisional harvest calendar suggested by the region's viticultural association known as AVC (Association Viticole Champenoise).

Another example of restrictions on the use of grapes is South Africa's organization

Cap Classique Producers Association (also known as CCPA), which requires that, as from the 2023 harvest, sparkling wines destined for the special category Cap Classique with Extended Lees Ageing be created using solely one or more of the (grape) varieties most commonly used for champagne, namely: Chardonnay, Pinot Noir and/or Meunier.

By comparison, the primary varieties authorized in the cava-producing zone include: Macabeo, Xarello and Parellada. Other grapes authorized by Consejo Regulador de la Denominación de Origen Cava (also called Cava Regulatory Council) are: Chardonnay, Garnacha, Subirat Parent, Trepat, Pinot Noir and Monastrell. In order for a producer's cava wines to qualify for the Cava de Guarda Superior segment, the grapes used must come from vineyards that are at least 10 years old.

Champagne producer Leclerc Briant adheres to the principle that winemaking begins in the vineyard, using the energies inherent in nature to nourish the vines. Organic and biodynamic practices, rooted in the nonconformist history of Leclerc Briant, are now honed by the company's cellar master Hervé Jestin, with the vines sprayed with a combination of copper-based solutions, plants and essential oils. Plants that are typically used in biodynamics for spraying the vine include: nettle (which bolsters vines' resistance to mildew by increasing the rate of sap flow in the vines), wicker (known to stimulate the vine's natural defences), comfrey (for its healing and protective properties), yarrow (reducing the risk of powdery mildew on the vines owing to the plant's high sulphur content), and also chamomile (stimulating sap movement in unfavourable weather conditions, especially drought).

Ultimately, decisions on grape selection are made in the vineyard at harvest. These decisions rest on such factors as the health of grapes, their physiological maturity and their sugar content. There can be marked differences in sugar content between grapes from vines grown in different parts of a single vineyard. Grapes from clusters that receive the most sunshine will usually have the highest sugar content, often much higher than those growing under the foliage. Because of these factors, among others, hand-picking is the norm in the production of top-quality sparkling wines, such as champagnes, cavas de paraje calificado and franciacortas.

CRC 2022; Denig 2020; Field 2022a; Forbes [1967] 1989: 35; Hartley [1999] 2022; Johnson and Rand [1977] 2022: 58; Joshi *et al.* 2017: 378; Kemp *et al.* 2015: 19; LHV 2016; Liem 2017: 53–5, 75, 77–9; Nagórka 2013: 98; Speller [1994] 2015a: 357; Szabo 2022; WSET 2020: 3–4, 18, 22, 34–5, 38, 65, 93, 96, 100, 110, 115

www.cava.wine/en
www.champagne.fr/en
www.champagne-salon.fr
www.corpinnat.com/en
maisons-champagne.com
oldvineproject.co.za
www.spier.co.za
www.wosa.co.za
franciacorta.wine/en
perdeberg.co.za

rural method

≈ ancestral method

Russian method

≈ continuous method

Photo 39. Spier's cellar master Johan Jordaan during his routine tour of the vineyards in Western Cape (South Africa). Among the most acclaimed members of his winemaking team is Heidi Dietstein, a specialist in cap classiques, in the making of which she uses two grape varieties: Chardonnay and Pinot Noir, sourced mainly from the districts of Stellenbosch (Coastal Region) and Robertson (Breede River Valley).

Photo: courtesy of Spier Wine Farm (Lize-Marié Nieman and Johan Jordaan), January 2024

S

saignée method for rosé must

A method used in the production of macerated sparkling rosé wines, sometimes referred to as *rosé de saignée*. The method consists of draining a portion of red grape must (typically between 10% and 15%) — coloured as a result of brief maceration of red berry skins (usually lasting from several hours to around two days) — into a separate vessel. While this operation creates an opportunity for crafting well-structured (usually medium-bodied) rosé base wines, its use also increases the proportion of organic matter in the mash of red grapes relative to the volume of must remaining in the original vessel, thus helping the winemaker produce red must with increased concentration of flavours. This must can be used for making either still red wines or red base wines that could subsequently become part of a blend prepared for secondary fermentation. When making champagne or other quality sparkling wines, the operation of draining the coloured must is conducted when the winemaker decides that red grape skins have been in contact with the must for long enough to impart to that must the desired colour and flavour. The method is favoured by the makers of sparkling rosé at the champagne house Fleur de Miraval, where it is applied in obtaining delicate coloured juice from Pinot Noir grapes, which is then vinified and blended with wine crafted from the Chardonnay grapes.

📖 CIVC 2023b; Liem 2017: 55, 269, 302; Mercer 2020b; Moore 2022: 58–9; Pessey 2002: 96; Rose 2021: 60

💻 fleurdemiraval.com/en

sealing of bottles of cuvée with temporary corks

A method of closing bottles filled with wine prepared for secondary fermentation in the bottle using cork closures. While the use of cork at this stage of the sparkling wine production process remains relatively rare, not least because of its high cost, its application is increasingly justified by the latest discoveries in the field of oenology. The use of cork at bottling before the second fermentation has been found to lead to stylistic changes in the wines. Differences in style between cap classique wines obtained by secondary fermentation and maturation in the crown-capped bottle and cap classique wines that have fermented and matured in contact with cork have been (tentatively) confirmed, with the method of using cork closures resulting in sparkling wines characterized by greater foam stability and bubble retention times when served in the glass, with relatively smaller-sized bubbles and greater flavour complexity. Among the growing number of producers using corks for the second fermentation and maturation of their wines are the family-owned champagne-making companies Bérêche et Fils and R. Pouillon et Fils, and the cap classique producers Le Lude and Graham Beck.

📖 CIVC 2023b; Fleet [1985] 1998: 233; Jolly *et al.* 2021: 136–52; Liem 2017: 56; Minnaar *et al.* 2021: 19–23; OIV 2022: II.4.1–2, II.4.1–10; Schmitt 2020; Willcox 2023

💻 www.lelude.co.za
maisons-champagne.com/en
champagne-pouillon.com/en
www.bereche.com/en
www.spier.co.za

sealing of bottles of cuvée with temporary crown caps

A method of closing bottles of wine prepared for secondary fermentation in the bottle using crown cap closures. Crown caps are widely applied at this stage of the sparkling wine production, not least as these closures are less expensive than corks. The use of crown caps as bottle stoppers prior to the second fermentation when creating cap classique wines has been reported to result in sparkling wines with profiles different from those of wines that fermented and matured under cork. Sparkling wines created by the bottle method using crown caps as temporary closures have been reported as being characterized by visually larger bubbles and more pronounced autolytic character than sparkling wines created by the bottle method using corks.

📖 Fleet [1985] 1998: 233; Jolly *et al.* 2021: 136–52; OIV 2022: II.4.1–2, II.4.1–10; Reddy and Joshi 2021: 442; Schmitt 2020

Photo 40. Closed with temporary corks, bottles of Gran Torelló Corpinnat Brut Nature 2017 left to mature in contact with lees for a minimum of 72 months.
Photo: Marta Ręgowska, April 2023

sealing of bottles of sparkling wine with corks

Closing bottles with sparkling wine by means of special corks, commonly referred to as champagne corks, as they are mandatory in Champagne for finished wines. Made to particular specifications, these corks must be up to 31 mm in diameter and 48 mm in length. The cork is compressed by machines to a diameter of around 17 mm and driven into the neck of the bottle, resulting in a tight seal, thus keeping carbon dioxide in the bottle. Cork being a porous material, the significant difference in pressure between the headspace in the bottle and the ambient air in the cellar, where sparkling wines are typically aged, causes CO_2 molecules to diffuse slowly through the cork, resulting in the wine losing its CO_2 contents over time. The loss of effervescence during extended periods of post-disgorgement ageing will not be the case with bottles of sparkling wine aged underwater, a concept increasingly being explored in sparkling winemaking practice following discoveries of well-preserved bottles of champagne from shipwrecks in the late 1990s and the first decade of the 21st century.

A number of producers in the Champagne region ageing their wines in cellars have embraced Mytik Diam, a type of (technological) cork offered by the Oeneo group since 2005 and continuously improved thereafter to more effectively preserve CO_2 in the bottle. With contaminants detrimental to sparkling wine, in particular TCA (2,4,6-trichloroanisole) molecules, having been removed from the cork flour used for their manufacture (by treatment with supercritical carbon dioxide), corks from the Mytik Diam range have found favour with sparkling wine producers because of their effectiveness in combating cork taint, one of the most serious faults affecting flavour. Champagne producers using these technological closures include Champagne Billecart-Salmon, Charles de Cazanove and Moët & Chandon.

Crouvisier-Urion *et al.* 2021; Liem 2017: 65, 295; Liger-Belair 2016: 527–32; Phillips 2019; Rand 2022: 20; Rodríguez-Bencomo *et al.* [2004] 2012: 189; SL 2018/2019: 8, 11, 50; Vine [1981] 2012: 257–8

maisons-champagne.com/en
oeneo.com/en
www.champagne.fr/en
diam-cork.com

sealing of bottles of sparkling wine with crown caps

A non-traditional method of closing bottles of finished sparkling wines using metal capsules rather than corks. The method has been developed in an effort to avoid cork taint, a serious defect affecting wines' flavour, and as an attempt to solve the problem of variations in oxygen transmission levels found in natural cork stoppers owing to their structural heterogeneity. Crown closures are offered with a range of liners, varying in oxygen transfer rates (OTR), with the OTR of the liner having an impact on the outcome of the sparkling wine ageing process. While crown-capping of finished sparkling wines is (still) illegal in Champagne, the method finds favour with some producers in the New World, not least as it is viewed as a relatively inexpensive and efficient alternative to sealing bottles with corks. Crown caps are applied mostly for wines made using the ancestral method. Nonetheless, these closures are sometimes also applied for wines created using the traditional method. Traditional-method sparkling wines released under crown caps are offered by some producers in Canada, including: Flat Rock Cellars 'Riddled Sparkling' 2007, Grange of Prince Edward

'County Crémant' 2013 and Flat Rock Cellars 'Riddled Sparkling' 2017.

📖 Goode 2013; Liem 2017: 295; Phillips 2019; Taber [2007] 2009: 180; Vine [1981] 2012: 257–8

sealing of bottles of sparkling wine with screw caps

A method of closing bottles of finished sparkling wines using metal capsules that twist into grooves present in the necks of sparkling wine bottles, thus sealing them tightly. Equipped with liners, these closures typically provide a tighter seal than natural cork stoppers, with the latter, being structurally heterogenous, exhibiting significant differences in terms of oxygen transfer rate (OTR). Screw caps are used for sealing sparkling wines by De Bortoli Wines, a family-owned company founded in Australia by Vittorio de Bortoli — an Italian immigrant — in 1928. The firm's screw-capped bottles of tank-fermented sparkling wine De Bortoli Willowglen Brut Cuvée have been available on Australia's sparkling wine market since their debut in 2012.

📖 Gargett 2012; Hooke 2012; Pons and Loisel 2021; Sheather 2012; Stelzer 2012

💻 www.debortoli.com.au

second pressing of grapes for white must

Extraction of the second fraction of juice, using a wine press, from grapes destined for white base wine or partially fermented must. In the Champagne region, where methods continue to serve as a model of pressing grapes for the world's other sparkling wine-producing regions, the second pressing is strictly regulated. Champagne producers must comply with the regulation on the maximum permitted amount of juice from this pressing, which is 500 litres for every 4,000 kg of freshly picked grapes loaded into the press. The resultant juice is commonly known as *taille*.

📖 Fielden [1994] 1998: 286–7; Forbes [1967] 1989: 297–8; Liem 2017: 46–7; Skelton 2016: 33

💻 www.champagne.fr/en
maisons-champagne.com/en

secondary fermentation

≈ continuous fermentation

≈ fermentation in autoclaves

≈ traditional-method fermentation

≈ transfer-method fermentation

secondary fermentation method

A method of making effervescent wines using two fermentations, with the latter — induced by the addition of *tirage* liqueur (containing yeast and sugar) — taking place in sealed (pressure-resistant) bottles or tightly sealed pressure tanks. When the second fermentation takes place in the bottle, one can speak about the bottle method. When the second fermentation is induced and managed in base wines placed in sealed pressure tanks, the process is often referred to as the tank (or bulk) method.

📖 Amatuzzi 2013: 56–7; CIVC 2013: 29; Goldstein 2006: 32–7; Hornsey 2007: 4.3.2; Jackson [2002] 2009: 162; Pessey 2002: 52–3; Quaile 2023: 159; Rand 2022: 18–20; Welke *et al.* 2022: 5.2

💻 maisons-champagne.com/en
www.capclassique.co.za
www.cava.wine/en
www.champagne.fr/en

securing of final corks

A winemaking practice applied during the sealing of bottles containing finished sparkling wines with cork stoppers. It is essential to secure sparkling wine corks when shipping sparkling wines in bottles. It involves restraining the cork (using wire,

Photo 41. Bottles of champagne placed on riddling racks called *pupitres* in the cellar of Maison Lanson in Reims.
Photo: author, June 2023

metal clips, or string) in order to ensure that it remains firmly seated in the bottle neck, thus preventing wine leakage and uncontrolled loss of pressure inside the bottle.

Epstein 2011: 61; Liem 2017: 65, 299; Lloyd [2017] 2024; Vizetelly 1882: 165–6

maisons-champagne.com

securing of final corks with *agrafes*

The practice of securing (immobilizing) the cork placed in the neck of the bottle of finished sparkling wine by means of a metal clip, called an *agrafe*. Metal clips are applied primarily for the purpose of securing temporary corks before the onset of the second fermentation, in which case they serve as a more traditional alternative to crown caps. However, these clips are currently also used occasionally by some sparkling wine producers to prevent the final cork from being expelled by the pressure inside the bottle, a function usually performed by the widely used wire cages, known as *muselets*. Some of the most prominent among these rare producers are: Champagne Moutard Père & Fils, a family-run champagne house from Buxeuil, in Côte des Bar (the southern part of Champagne), and Le Lude, located in the Coastal Region's winemaking district Franschhoek, in Western Cape (South Africa).

CIVC 2023b; Jackson [1994] 2014: 710; Jolly *et al.* 2021: 136–7; Liem [2019] 2022; Lloyd [2017] 2024

www.lelude.co.za
maisons-champagne.com/en
www.famillemoutard.com/en

securing of temporary corks with *agrafes*

The rare practice of securing (immobilizing) the cork placed in the neck of the bottle containing base wine(s) prepared for the second fermentation by means of a metal clip, called an *agrafe*. The aim of this winemaking practice is to prevent the cork from being expelled by the pressure mounting inside the bottle as a result of the accumulation of carbon dioxide produced during the second fermentation process. The *agrafe* method enables a return to the age-old practices in which the second fermentation and maturation processes take place in the presence of natural cork. This winemaking practice has been used at the champagne house Ruinart by the house's cellar master Frédéric Panaïotis to create the wine Dom Ruinart Blanc de Blancs 2010. *Agrafes* have also recently been applied for the traditional-method fermentation and maturation by Le Lude, a sparkling wine producer in Western Cape's Coastal Region's district Franschhoek, in South Africa.

CIVC 2023b; Ferreira 2020; Forbes [1967] 1989: 316; Lloyd [2017] 2024; Mocke 2017; Sherwood 2018

www.lelude.co.za
www.ruinart.com/en-int
www.wosa.co.za

short ageing of sparkling wine in cellars

A resting period for some finished sparkling wines, especially those created using the traditional method, considered necessary by quality-conscious producers in connection with the recent removal of lees formed during the wine's maturation on lees. When creating fine wines using the traditional method, the removal of lees, commonly referred to as disgorgement, marks a transition from the storage of sparkling wine in a highly reductive environment offered by bottles in which the sparkling wine has matured in the presence of yeast lees (with yeast cells having used much or nearly all of the oxygen available in the headspace, and yeast derivatives having notable antioxidant properties) to the storage of sparkling wine in the presence of the small amounts of oxygen that entered the bottle during disgorgement. The convalescence periods guaranteed for disgorged wines to recover from the shock of disgorgement differ from producer to producer, usually increasing with the time the wine had matured on lees. Producers of top-quality champagnes tend to hold their sparkling wines for at least five months before release if these wines had been cellared in the presence of lees for two or three years before they were cleared of their sediment. The resting period will usually be over a year in the case of sparkling wines that were allowed to mature on lees for 10 years or longer.

📖 Lambert-Royo *et al.* 2022; Liem 2017: 300–1; Liem [2019] 2022; Paillard 2015; Stevenson [2014] 2022: 245; Wilson 2018

💻 champagnebrunopaillard.com/en

Disgorgement, which marks for bottle-fermented traditional-method sparkling wine a transition from its storage in a highly reductive environment to storage in the presence of small amounts of oxygen, is compared by some champagne producers to surgery, with the wine being viewed as a patient, who — after the operation — needs time to recover. Both managers and winemakers at the champagne house Bruno Paillard perpetuate this analogy, noting that older patients take longer to fully recover than younger ones. The makers of champagne Bruno Paillard 'Blanc de Blancs' Grand Cru cellar this wine between eight and 10 months between its disgorgement and release, given its four years on lees. By contrast, they have aged their champagne Bruno Paillard 'Nec Plus Ultra'-2008 for almost two years, given the fact that the wine was disgorged after having rested for 10 years in contact with lees.

short maturation on lees

Storage of young bottle-fermented sparkling wine in contact with lees for a relatively short time (in some cases as short as three months) before the wine is transferred to tanks for blending and the removal of the sediment. The short-term storage of sparkling wine on lees, an integral part of the (time-saving) transfer method, which has gained a particular following among producers in Australia and New Zealand, enables winemakers to achieve the minimum complexity expected of quality wines, not least as a result of the autolysis of yeast, which modifies the wine's organoleptic properties.

📖 Ellis 2012: 167; Epstein 2011: 64; Jackson [2002] 2009: 373; Walkey 2023; WSET 2020: 10, 85, 111, 115

Photo 42. The gate of Hexham Vineyard, located at the foot of Warramate Hills in Yarra Valley, Australia's state of Victoria. Hexham Vineyard provides fruit for sparkling wines of the producer Soumah. Having rested in contact with lees for a brief period, the sparkling wine Soumah Blanc de Blancs NV is released shortly after it has been cleared of sediment.

Photo: courtesy of Soumah of Yarra Valley (Georgia Mehrtens), February 2023

single-fermentation method

A method of making effervescent wines using one fermentation consisting of two stages, the final stage finishing either in sealed (pressure-resistant) bottles or in a sealed pressure tank. When the second stage of the fermentation process takes place in the bottle, one can speak about the ancestral method (also sometimes referred to as *méthode ancestrale*, *méthode artisanale*, *méthode rurale*, etc.). When the fermentation continues in tightly sealed pressure tanks, the process is referred to as the Asti method.

Barr 2019; Buglass 2022: 227; Caliari *et al.* 2015: 393–8; Clarke [2000] 2003: 20, 28, 41; Collin [2004] 2009: 207–8; Epstein 2011: 65; Fielden [1994] 1998: 294; Maker 2021; Welke *et al.* 2022: 5.2, 5.3.2; WSET 2020: 65–8; Wu 2015

single-fermentation sparkling red wines

Effervescent wines created using the one-fermentation method, the final phase of which finishes in the bottle or in a closed fermentation tank. This sparkling wine category, which is (mostly) made up of lightly sparkling red wines, consists mainly of wines made by a single fermentation continuing in bottles, including those obtained by bottling partly fermented red must, and those made by refermenting almost completely fermented red must, a winemaking process facilitated by adding to the base (red) wine a (relatively) small amount of (sweet) red must before the bottles containing this wine are sealed. Importantly, winemaking processes such as maceration of the skins of red grapes and stabulation — a form of pre-fermentative cold treatment — of red must are sometimes used before fermentation begins.

Lebanese wine-producing company Mersel (Wine) is known for Mersel LebNat PetNat 'Red'–2021, a lightly sparkling red wine made from Cinsault grapes, grown in the Bekaa Valley at Deir El Ahmar (northeastern Lebanon). This wine was created by the stabulation of red must in tanks at 8°C for five days, followed by the initial stage of fermentation at 14°C lasting three weeks. The wine was then transferred to bottles, where the fermentation was completed, and in which the wine rested on yeast lees for a year. Created by the winemaker Eddie Chami, this single-fermentation (ancestral-method) wine was subjected to disgorgement before it was released.

One example of semi-sparkling red wine created using the refermentation method is Podere Cipolla's Lambrusco dell'Emilia 'Ponente 270'–2020, made by Denny Bini from a blend of various Lambrusco grapes (Lambrusco Grasparossa, Lambrusco Sorbara, Lambrusco Salamino, Lambrusco Maestri, Lambrusco Montericco) and Malbo Gentile, grown on the Reggio Emilia hills in northern Italy. Undisgorged and unfiltered, this wine is characterized by noticeable turbidity, caused by yeast sediment left in the bottle.

Campanale and Stein 2022: 47; Goode 2020b; Gustafsson 2023; Speller and Currin [1994] 2015: 470–1; Sussman 2020a, 2020b; Teclemariam 2023; White 2020

www.merselwine.com

single-fermentation sparkling rosé wines

Effervescent wines created using the one-fermentation method, the final phase of which finishes in the bottle or in a closed fermentation tank. This sparkling wine category, encompassing (mostly) lightly sparkling wines, consists predominantly of wines made by a single fermentation continuing in bottles, including those obtained by bottling partly fermented lightly coloured juice from red grapes, and those made by refermenting almost completely fermented lightly coloured juice from red grapes, a process facilitated by adding to the base wine a (relatively) small amount of naturally sweet grape juice before the bottles containing this wine are sealed.

Notable examples are provided by crown-capped bottles of lightly sparkling wines in the portfolios of MontRubí, a producer from the Penedès region in Catalonia, and Medici Ermete & Figli, a family business from Reggio Emilia in the Emilia-Romagna region (in northern Italy). MontRubí 'L'Ancestral Rosé'–2021 has been made by partial fermentation in the bottle using indigenous yeast and natural sugars found in Sumoll grapes, the cultivation of which has been revived by the producer since the early 2000s. By contrast, Medici Ermete's 'Phermento'–2021 is a very dry sparkling wine (0.5 g/l sugar) with pronounced acidity (7.75 g/l), made by refermenting in the bottle the original base wine created from Lambrusco di Sorbara grapes.

Some single-fermentation semi-sparkling rosé wines have a cloudy appearance, a result of the presence of yeast residue left in the bottle. This is the case with the refermented wine from Reggio Emilia: Medici Ermete's 'Phermento'–2021. The practice of producing ancestral-method semi-sparkling rosé wines in the Bugey area (southeastern France) differs in this respect from the age-old winemaking practices in northern Italy, as the French winemakers remove dead yeast cells from their wines through filtration. One example is Cave Wojtkowiak's Bugey Cerdon Méthode Ancestrale (Demi-Sec)–2020, made from 80% Gamay grapes and 20% Poulsard grapes.

Campanale and Stein 2022: 47; Joe 2023; Lorch [1994] 2015: 175–6; Parzen 2018; Speller and Currin [1994] 2015: 470–1; Sussman 2020b

www.cerdon-wojtkowiak.fr/en
montrubi.com/en
www.enotecaemiliaromagna.it/en
www.medici.it/en

single-fermentation sparkling white wines

Effervescent white wines created using the one-fermentation method, the final phase of which finishes either in the bottle or in a closed fermentation tank. This category of sparkling white wines encompasses both ancestral-method sparkling white wines and wines crafted by a single fermentation continued in tanks, such as most Asti wines (made using modified Martinotti-Charmat method), or sparkling ice wines, a speciality of the winemakers in Canada. Wines representing either of these two groups range from lightly sparkling to fully sparkling.

Gerling *et al.* 2023; Phillips 2017: 75; Wu 2015; Zamuner 2021

www.astidocg.it/en

single-fermentation sparkling wines

≈ single-fermentation sparkling red wines
≈ single-fermentation sparkling rosé wines
≈ single-fermentation sparkling white wines

sparkling method

One of the key stages in the production of sparkling wines whereby base wines or partially fermented musts obtain effervescence. This stage involves the choice of a suitable fermentation method, with the winemaker having to decide between continuing the first fermentation in a sealed vessel, such as a bottle or pressure-resistant vat, and preparing their cuvée for a second fermentation, which can take place in a sealed bottle or closed pressure tank. This stage of production also includes sparkling wine maturation, a monitored process promoting the formation and shaping of desirable characteristics of the resultant wine.

CIVC 2013: 29; Jackson [2002] 2009: 162; MacNeil [2002] 2015: 384–5, 2,195, 2,219–20; Nagórka 2013: 103–4; OIV 2020a: 20; Schneider 2019: 54; Sussman 2021: 7–10; Welke *et al.* 2022: 110

www.astidocg.it/en
capclassique.co.za
www.cava.wine/en
www.champagne.fr/en
www.corpinnat.com
www.wosa.co.za

Photo 43. Many makers of effervescent wines use more than one method of production. One example is the family-run winemaking business Winnice Kojder, located in the Polish part of Pomerania. Winemakers at Winnice Kojder use Johanniter grapes to craft both traditional-method (i.e. double-fermentation) sparkling wines and pétillant naturel (i.e. single-fermentation) wines.
Photo: author, September 2024

sparkling red wine

A broad category of effervescent wines created (solely) by fermenting red grape must or red base wine in a closed container. These wines have red hues of various intensity, resulting primarily from contact between the grape must and red grape skins. The category encompasses sparkling wines made using a single fermentation and sparkling wines created using a secondary fermentation in the bottle. The production of red sparkling wines, though not geographically determined, continues to be associated mainly with two wine-producing countries, namely Italy and Australia. Key grape varieties for the Italian wines are Lambrusco grapes, and the most important varieties for the sparkling red wines made in Australia include Shiraz, Pinot Noir, Cabernet Sauvignon, Merlot and Durif.

Charpentier 2010: 176; Demmond 2023; Jeffery and Wilkinson 2014: 81; WA 2022: 17–18, 2023: 34–7

lambrusco.net/en
www.majellawines.com.au

sparkling rosé wine

A category of effervescent wines resulting from the fermentation of grape juice or still wine in a closed container, displaying a range of pink hues, an effect of the contact between grape must and the skins of red grapes or blending red and white base wines, as when creating most of the traditional-method sparkling rosé wines in the region of Champagne. The colour of rosé sparkling wine can range from soft pink (sometimes with pale onion-skin hues) to salmon pink, raspberry pink, strawberry pink to light brown or tawny (in the case of old and very old wines). Notably, sparkling rosé wines obtained by blending tend to be much paler, with their hues less intense, than those made using macerated fractions of must.

Fleur de Miraval, a business enterprise dedicated exclusively to rosé champagne, has a portfolio of sparkling rosé wines made using innovative methods. Conceived around 2015 at the joint initiative of Brad Pitt and the Perrin and Péters families, this buoyant small champagne producer combines the *saignée* method with the art of blending. The company's first wine: Champagne Fleur de Miraval 'Exclusivement Rosé', released in October 2020, initiated a range of rosé champagnes dedicated to high-end consumers with refined tastes. Known as ER1, the blend is composed of 25% *saignée* wines, made from the grapes of Pinot Noir from Vertus, and 75% wines created using Chardonnay grapes, sourced from various villages of the Côte des Blancs area. The Chardonnay component came from three sources: perpetual reserve (a blend of wine from the current vintage and reserve wines from previous years, the oldest dating back to 2007); (de-gassed) premium seven-year-old *blanc de blanc* champagnes from opened bottles emptied into a blending vat; and *vins clairs* produced using Chardonnay grapes collected in 2016. Aged on lees for three years in the house's cellars at Le Mesnil-sur-Oger, the wine was disgorged in June 2020 before 20,000 bottles of this limited-edition rosé champagne were released.

This category of sparkling wines includes wines that have been made effervescent utilizing the method of a single fermentation and those created using double-fermentation methods. Within either of these groups one can distinguish between sparkling rosé wines, the colours of which have been extracted by direct pressing of dark-skinned grapes, and sparkling rosé wines, the colours of which have been obtained by macerating the must of dark-skinned grape berries on their skins for desired periods. The latter, commonly referred to as 'macerated rosé' or *rosé de macération*, include wines known as *rosé de saignée*, for which maceration periods are usually rather short (typically no longer than around a dozen hours). In the region of Champagne, the majority of sparkling rosé wines are made by blending white and red base wines. For this reason a wine representing this subcategory is often referred to as 'blended rosé' or *rosé d'assemblage*.

📖 CIVC [2014] 2022: 23; Gabay 2018: 161–2; Green 2023b; Köhler 2022; O'Donnell 2020; Perrin and Péters 2022; WSET 2020: 7, 16, 17, 24–5, 30, 35, 75–6, 106, 112

💻 fleurdemiraval.com/en
www.cava.wine/en
www.champagne.fr/en
www.vineyardbrands.com

sparkling white wine

A broad category of effervescent wines resulting (solely) from the fermentation of white grape juice or still white wine in a closed container, their colours typically ranging from a pale green yellow to lemon and golden yellow hues, sometimes with orange highlights (which could serve as an indicator that the wine has been created mainly from black-skinned grapes or that the final blend contains a large proportion of reserve wines), and sometimes falling into shades of old gold (which is usually a sign of the wine's high complexity and great maturity).

Sparkling wines representing this category are divided by method of production into single-fermentation sparkling white wines and double-fermentation sparkling white wines. When created only from light-skinned grapes, such as Chardonnay, Chenin Blanc or Riesling, a sparkling wine from this category is often labelled *blanc de blancs*; by contrast, a sparkling wine made exclusively from black-skinned grapes, such as Pinot Noir or Meunier, will be referred to as *blanc de noirs*.

Atkins 2021; Capretti 2022; Hui [2004] 2012: 344, 349; Jeffery and Wilkinson 2014: 56; Rose 2021: 57–8; SL 2018/2019: 54; Stevenson 2011: 168–75; Zoecklein 2002: 25

www.champagne.fr/en

sparkling wine

Special wine obtained in the process of alcoholic fermentation of grape juice or wine taking place in a closed container, such as a bottle or tank, resulting in the production of natural (endogenous) carbon dioxide, which is released on contact with fresh air, causing the resultant wine to effervesce. Pressure requirements for the category of sparkling wines in the broad sense, related to their carbon dioxide contents, are set at above 2.5 bar (2.47 atm) when measured at a temperature of 20°C, with lightly sparkling wines having pressures in the 2.5–3.5 bar range. Typical pressures found in sparkling wines proper range from 3.6 bar to 6 bar or more.

Photo 44. Ridgeview Single Vineyard Oak Reserve is an English traditional-method sparkling white wine with an intense, well-integrated oak component. The base wines for this sparkling wine were made from Chardonnay grapes grown on the Ridgeview estate in Sussex using French oak barrels.
Photo: author, March 2024

These wines are characterized by a diversity similar to — if not greater than — that of still wines. They are made from grapes grown in climates ranging from those characterized by oceanic and continental trends, as in Champagne, to Mediterranean climates of various types, as in most of the cava zone Comtats de Barcelona, California's Napa Valley and South Africa's Coastal Region districts of the Western Cape, to the semi-arid climate of the Uco Valley, influenced by elevations reaching up to 1,650 metres, to the tropical climate of Nashik, India's main winegrowing region. The main grapes used in the production of sparkling wine include, but are not limited to, Chardonnay, Pinot Noir, Chenin Blanc, Meunier, Semillon, Riesling, Malbec and Shiraz. These wines can differ not only in fizziness, and the character of the foam, but also in colour, alcoholic strength and sweetness.

While the vast majority of the world's sparkling wines produced are white and rosé, red sparkling wines, traditionally made in Australia and western parts of Italy's region Emilia-Romagna, continue to find a loyal following, as do two recently emerging sparkling red wines made by Domaine Chandon: Chandon California's sparkling wine created from Pinot Noir and Zinfandel, and Chandon China's wine produced from Marselan grapes. Most sparkling wines in general, and second-fermentation wines in particular, have an abv of around 12% or more, while the alcoholic strength of single-fermentation sparkling wines, including those created using the ancestral method and those made by a modified Charmat method, typically ranges between 5.5% abv and around 10% abv. Sparkling wines are made in various degrees of sweetness (see Table 1), the smallest amounts of sugar found in *brut nature* wines (0–3 g/l of sugar), and the wines with the highest sugar content categorized as sweet wines (50+ g/l of sugar).

Champagnes, the archetypal sparkling wines created in northeastern France, have had their sugar contents tailored to preferences in (the main) export markets since at least the third quarter of the 19th century, when champagnes containing no or very little sugar were first marketed in Britain. Historically, at the beginning of the second decade of the 20th century, sweet sparkling wines were predominantly shipped to Russia, whereas Americans preferred a moderately dry style and the English chose mainly extra-dry and *brut* sparkling wines, chiefly champagnes. While these old taste habits may still have an impact on consumer demand in many traditional markets, even with the current general tendency among consumers to choose drier wines, these markets now appear to be much more diverse in terms of expectations, not least as a consequence of the globalization trends of the first quarter of the 21st century, especially those related to international migration, often entailing changes of residence and even nationality. A marked rise in sparkling wine consumption worldwide in the period 2002–2018, from slightly less than 12 million hectolitres to around 19 million hectolitres, is explained by a combination of two main factors: deseasonalization of the consumption of sparkling wines, and the availability of these wines in a consistently wider price range. Towards the close of the second decade of the 21st century, the largest markets for sparkling wine (by volume) were: Germany (3.2 million hectolitres), France (2.6 million hectolitres) and the US (2.6 million hectolitres), followed by Russia (1.6 million hectolitres), Italy (1.6 million hectolitres) and the UK (1.5 million hectolitres).

The sparkling wine industry has received regular assistance from natural scientists, especially since the 18th- and 19th-century discoveries of the nature of the fermentation of grape juices and wines, a process leading to changes in their characteristics. This progress can be attributed to some of the most prominent names in chemical and biological sciences, including Antoine Laurent Lavoisier, who in 1789 explained the nature of the carbonic gas formed during fermentations and causing effervescence in wines, and Louis Pasteur, who by the mid-1860s had proved that yeast cells are responsible for converting sugar in grape juice (and wine) into alcohol, carbon dioxide and secondary products, such as glycerol. Today, wine serves as an object of advanced oenological studies, with wine scientists in the 21st century making use of precise measurements of its contents. Differences in carbon isotopic compositions of CO_2 bubbles in sparkling wines are sometimes used to indicate their origin: cane sugar, the source of carbon in a significant proportion of Brazilian sparkling wines, is isotopically distinct from beet sugar, typically used for second fermentations in the production of sparkling wines in Europe.

Capretti 2022; Charpentier 2010: 176; CIVC [2014] 2022: 23; Demmond 2023; Forbes [1967] 1989: 153–5, 161–2; Gabay 2018: 161–2; Gerling *et al.* 2023; GBC 2018–2022: 17, 33; Howe 2003: 150;

Hui [2004] 2012: 344, 349; Jeffery and Wilkinson 2014: 56; Lavoisier [1789] 1806: 179–80; Martinelli *et al.* 2003: 2,625–30; OIV 2020a: 4, 9–11, 2021: 16, 2022: I.1.4–4; Pessey 2002: 112; Phillips 2017: 75; Vizetelly 1882: 164; Willcox 2022; WSET 2020: 7, 11, 13, 16, 17, 24–5, 30, 35, 75–6, 106

www.cava.wine/en/categories-types
capclassique.co.za
www.champagne.fr/en
www.chandon.com
www.chandon.com.au
www.corpinnat.com/en
maisons-champagne.com
www.moet.com/en-gb
www.napavalley.wine
napavintners.com
www.nzwine.com
www.wineaustralia.com
www.wosa.co.za

Table 1 Standard reference bands for sugar content in sparkling wines

Source: GBC 2018–2022: 17, 33; OIV 2022: I.1.4–4

Sweetness type	Minimum (g/l)	Maximum (g/l)
nature	–	3
extra-brut	–	6
brut	–	12 (+3)
extra-dry	12	17 (+3)
dry	17	32 (+3)
demi-sec	32	50
sweet	50+	–

sparkling wine producers

Companies making sparkling wine, with the choice of grapes used for its creation typically regulated by special bodies designated to ensure the quality of the wine and control the sparkling wine industry in the region in which the companies operate. These companies can differ markedly, not merely in terms of geographic location but also in sparkling methods used in the manufacturing process. In the diverse sparkling wine market, there is space for companies using traditional methods, such as producers in Champagne, e.g. the champagne house (i.e. *négociant manipulant*) Louis Roederer and the grower-producer (i.e. *récoltant manipulant*) Larmandier-Bernier; and for companies using single-fermentation methods, such as sparkling ice-wine producers in Ontario, Canada's second-largest province, most notably Magnotta Winery and Inniskillin, officially referred to as Inniskillin Niagara Estate Wines.

The oldest sparkling wine producers are champagne producers. Nicolas Ruinart founded a winemaking house dedicated to sparkling wines under his own name in 1729. Based in Reims, Maison Ruinart was acquired in 1963 by Moët & Chandon, and is now part of the luxury goods group LVMH. Following the establishment of Maison Ruinart, a number of champagne-making businesses emerged, notable among them: Chanoine, set up in 1730, referred to as Chanoine Frères since 1815, presently part of the business group Lanson-BCC; Forest-Fourneaux, built in 1734, known as Taittinger since 1931; and Moët & Chandon, established in 1743. In the latter half of the 18th century foundations were laid for the now-prominent champagne houses: Champagne Henri Abelé, with its origins dating back to Théodore Vanderveken's entrepreneurial initiative of 1757; Veuve Clicquot, whose foundations were established by Philippe Clicquot in 1772; and the champagne house bearing the name Heidsieck & Cie, established in 1785, the progenitor of today's three champagne-making firms, namely Heidsieck & Co Monopole, Piper-Heidsieck and Charles Heidsieck, the latter set up in 1851. Some of the now-renowned champagne houses created in the course of the 19th century are Joseph Perrier, set up in 1825; G.H. Mumm, 1827; Bollinger, 1829; De Venoge & Cie,

1837; Krug, 1843; Pol Roger (Sir Winston Churchill's favourite brand), 1849; and Mercier, founded in 1858.

Europe's historically significant sparkling wine producers operating outside of France include those in Germany, Italy and Spain. The oldest known sparkling wine house in Germany is G.C. Kessler, founded in 1826 in Esslingen am Neckar by Georg Christian von Kessler, formerly a valued employee at the champagne house Veuve Clicquot-Ponsardin. The creation of Piedmont's first sparkling wines in the image of champagnes is credited to the brothers Carlo and Edoardo Gancia. Their company, set up in 1850 and now officially known as Fratelli Gancia & C., began producing sparkling wines from Moscato grapes as early as the mid-1860s. In the early 1870s, head of the family firm Codorníu, Josep Raventós i Fatjó, marked the birth of Spain's sparkling wine industry by making sparkling wine using the traditional method, typical of Champagne, while using grapes typical of the Penedès region.

Photo 45. A vineyard at Avenay Val d'Or (Vallée de la Marne), a five-minute walk from the Avenay train station. These parcels constitute a source of grapes for two very old and renowned champagne producers: Maison Perrier-Jouët and G.H. Mumm, with their origins dating back to 1811 and 1827 respectively.
Photo: author July 2024

The wine producer Seppelt continues to honour the memory of the French winemaker Charles Pierlot, pioneering the production of *méthode traditionnelle* sparkling wines in Australia in the 1890s, launching traditional-method sparkling wines as part of the Pierlot by Seppelt range. Created to express features of the cool-climate terroir of the Henty region in the far southwest corner of Victoria, the range includes two non-vintage sparkling wines made by Seppelt's senior winemaker Adam Carnaby: Pierlot Henty Brut Rosé NV and Pierlot Henty Brut Cuvée NV.

Towards the close of the 19th century, the concept of creating sparkling wines through the fermentation of base wines in the bottle, which had been used for over 150 years in Champagne, was soon to be implemented in Australia and the United States. Founded in 1865 by Joseph Best, and purchased in 1888 by Hans William Henry Irvine, Great Western Vineyard, after hiring the French winemaker Charles Pierlot, began production of sparkling wines using the traditional method in the 1890s, gaining recognition as the first producer of sparkling wines of this type in Australia. In 1918 the company was acquired by Seppelt, currently one of Australia's most renowned sparkling wine producers. The history of sparkling wines in the state of California is inextricably linked to that of the wine company known as Korbel Champagne Cellars (F. Korbel & Bros.), founded in 1882 at Guerneville, in Sonoma County, by the Czech brothers Francis, Antone and Josef Korbel. The company's production of quality sparkling wines is attributed to the skill of its winemakers: Franz Hazek, employed by Francis Korbel in 1896, and sparkling-wine master Jan Hanuska, author of the bottle-fermented sparkling wine Korbel Sec, released by the Korbel family before the Prohibition era in the United States (1919–1933), and Korbel Brut, available in the US market since the early 1940s. The company was purchased by the Heck family in 1954. F. Korbel & Bros.'s wines have been served at many state occasions hosted by US presidents, as have sparkling wines marketed by California's other producers, notably Schramsberg.

F. Korbel & Bros.'s sparkling wines Korbel Natural' and Korbel Blanc de Noirs can be numbered among the favourite wines of US presidents of the latter half of the 20th century. Korbel Natural' has been served on numerous formal occasions, including: a dinner in Century Plaza, Los Angeles, held in honour of the Apollo 11 astronauts by Richard Nixon on 13 August 1969; a state dinner organized by Gerald R. Ford on 8 October 1974 for the First Secretary of the Central Committee of the Polish United Workers' Party, Edward Gierek; and a state dinner held by Jimmy Carter in honour of German Chancellor Helmut Schmidt on 13 July 1977. Korbel Blanc de Noirs was served at two state dinners hosted by Ronald Reagan: one for India's Prime Minister Indira Gandhi on 29 July 1982 and one for Sri Lanka's President Junius Richard Jayewardene on 18 June 1984.

Located in Napa Valley's Diamond Mountain District, with its origins traceable to 1862, the Schramsberg property was revitalized in 1965 by Jack and Jamie Davies, who in that year began producing quality sparkling wines using Pinot Noir and Chardonnay grapes. The company gained international renown on 25 February 1972 when its Blanc de Blancs 1969 was served by US president Richard Nixon at the Toast to Peace in Beijing during a reciprocal dinner organized for Chinese Premier Zhou Enlai. Ever since then, of the sparkling wines made in the US, Schramsberg's have been among the most frequently chosen for events, including state events, organized by the White House. Examples include: Schramsberg Crémant Demi-Sec 1979, featured on the menu for President Ronald Reagan's dinner honouring Queen Beatrix of the Netherlands, held on 19 April 1982; Schramsberg Crémant Demi-Sec 2011, served at the state dinner held by President Barack Obama for China's President Xi Jinping on 25 September 2015; and Schramsberg Crémant Demi-Sec 2014, served during President Donald Trump's state dinner held in honour of France's President Emmanuel Macron on 24 April 2018.

The period from the late 1950s to the present has witnessed the phenomenon of corporate growth on a scale without precedence in the history of the sparkling wine industry. This phenomenon's origin is attributed to the vision of Robert Jean de Vogüé, the former head of Moët & Chandon, who in 1959 took the decision to transplant the company's centuries-old know-how to areas outside the European continent that had by then remained relatively untamed by either grape-growers or winemakers. By the end of 2014, the company Moët-Hennessy, part of the luxury goods group LVMH, had established six domains on four continents, operating under the common name 'Chandon': Chandon Argentina (since 1959), Chandon California and Chandon Brazil (both of them since 1973), Chandon Australia (since 1986), Chandon China (since 2013) and Chandon India (since 2014). Other champagne houses have also expanded into new territories: Louis Roederer has an outpost in Anderson Valley in California, called Roederer Estate, which has been producing sparkling wines since 1982.

Photo 46. Underground cellars of the champagne producer Moët & Chandon in Épernay.
Photo: author, July 2024

Although now known for the exponential pace of growth in sustainable practices among its growers and wine producers, in the last quarter of the 20th century Champagne could hardly be considered a paragon of environmental responsibility. The use of sanitized urban waste, locally known as *boues de ville*, as fertilizer in the region's vineyards between the 1970s and its eventual prohibition at the end of the 20th century was one of the few events in the champagne industry's long history that can be seen as a blot on its reputation.

In the third decade of the 21st century, a growing number of sparkling wine producers are embracing environment-friendly production methods. This trend dovetails with a commitment on the part of authorities in major winegrowing regions to support producers in adopting sustainable practices. One example is Comité Champagne (CIVC), also called Comité Interprofessionnel du Vin de Champagne. Set up in 1941, this organization is charged with organizing, and controlling the production and distribution as well as promotion of the wines of Champagne, while also being responsible for conducting and supporting vitivinicultural research considered vital to the region. Between 2003 and 2022, the collective effort of Champagne's vine-growers and wine producers led to the reduction of the champagne industry's carbon footprint by 20%. This effort is visible in Maison Ruinart, which, in 2014, under its managing director Frédéric Dufour implemented a rational delivery policy, to the point of not using air freight. As a result, 85% of champagne shipments are made by sea and 15% of champagne deliveries are made over land. The concept of sustainable oenology with a focus on eco-designed packaging, one of the champagne industry's long-term commitments, is exemplified by Ruinart's 2019 launch of recyclable cases described as 'second skin'. Composed of 99% natural wood fibres and weighing 40g each, the cases designed for Ruinart's bottles replaced the firm's much heavier single-bottle (unitary) boxes in 2020.

Access to Champagne grapes is restricted by contracts between existing champagne producers and growers, production methods — strictly defined and monitored by Comité Champagne — require heavy investments, and export markets are controlled by the largest producers. These factors mean that new entrants to the champagne industry are very rare. One such anomaly is Champagne Barons de Rothschild. Founded in 2005, this new champagne house brings together three branches of the Rothschild family, now represented by: Baron Philippe Sereys de Rothschild, Baron Éric de Rothschild with his daughter Saskia de Rothschild, and Baroness Ariane de Rothschild.

Sustainability in viniculture and winemaking is now the watchword for a growing number of sparkling-wine producers in Spain, where no less than three types of traditional-method sparkling wines can be distinguished: cavas; premium sparkling wines of the Penedès region, known by the name Clàssic Penedès since 2014; and corpinnats. Corpinnat producers, who co-created the Associació de Viticultors i Elaboradors Corpinnat (Association of Wine Producers and Growers Corpinnat), also called AVEC, established in 2017, are committed to the following principles: organic viticulture, based on the use of mainly autochthonous grapes, with the historical varieties constituting a minimum of 90% of grapes used for production; long-term agreements with growers, ensuring fair prices for their grapes; and extended wine maturation periods, with the (vast) majority of corpinnat wines having matured in the bottle for more than 30 months. Makers of sparkling wines categorized as Clàssic Penedès apply organic grape-growing practices, with permitted methods for making sparkling wines including both the traditional method and the ancestral method. The cava industry, represented by the Cava Regulatory Council (Consejo Regulador de la Denominación de Origen Cava), made a commitment in 2022 to produce its cava wines in the Cava de Guarda Superior segment (including cava reserva, cava gran reserva and cava de paraje calificado) using solely organic methods, with the regulation to take full effect from 2025.

Spain's sparkling-wine-producing market is currently shared among three trades: the cava industry, the group of wine producers using the Clàssic Penedès designation, and producers of corpinnats. The largest of these is the cava industry, with 206 registered producers of cavas in 2022, compared with only 17 using the Clàssic Penedès label. The association of producers selling corpinnats is the smallest, with 11 members as of 2021: Gramona, Llopart, Nadal, Recaredo, Sabaté i Coca, Torelló, Can Feixes, Júlia Bernet, Mas Candí, Can Descregut and Pardas.

Most producers of sparkling wines in South Africa made according to the classical (bottle-fermentation) method (known as cap classiques), are members of Cap Classique Producers Association (CCPA). This association, established in 1992 to constantly improve the quality of the wines of its members, is one of many organizations with which cap classique manufacturers are currently associated.

Producers in the Western Cape's wine regions, such as the Coastal Region and the Breede River Valley, have participated in environmental and social sustainability schemes since the early 2000s. Committed to biodiversity-friendly viticulture and environmentally responsible winemaking practices, the companies Boschendal Estate (located in the Coastal Region's Franschhoek district), Graham Beck (located in the Robertson district of the Breede River Valley) and Spier Wine Farm (located in Stellenbosch, part of the Coastal Region) are among cap classique producers recognized by the World Wide Fund for Nature (commonly referred to as WWF) as the industry's conservation champions. The wine producer Ken Forrester Wines is a member of Old Vine Project (OVP), with its Sparklehorse Chenin Blanc Cap Classique 2018 permitted to carry the Certified Heritage Vineyards seal, issued by the OVP organization since 2018, certifying that the wine has been made from vines that are at least 35 years old. Other examples of environmentally and socially sustainable projects in the Stellenbosch district are provided by Spier Wine Farm. In addition to winning the Getaway Award for Leadership in Water Conservation at the Nedbank Green Wine Awards ceremony, held in October 2015, the wine producer launched a series of their own sustainable initiatives in 2018. Collectively known as Spier Growing for Good, these initiatives aim to foster environmental sustainability through community engagement and empowerment.

Anson 2015; Atkins [1994] 2015a: 531–2; Avellan 2022c; Donzé 2018; Dunstan 1983 [2006–2023]; Hryhorieva 2021; Jenster *et al.* 2008; Liem 2017: 38–9; McGregor 2016; Phillips 2016: 276–7, 2017: 171–2; Ryan 2020: 280–439; Sullivan 1998: 173–4; Visconti 2022

www.astidocg.it/en
www.australianwine.com
capclassique.co.za
www.cava.wine/en
www.champagne.fr/en
champagne-bdr.com/en
www.corpinnat.com/en
dopenedes.cat
franciacorta.wine/en
www.gancia.com/en
www.inniskillin.com
larmandier.fr/en
www.lvmh.com
maisons-champagne.com
www.magnotta.com
napavintners.com
oldvineproject.co.za
www.recaredo.com/en
roedererestate.com
www.ruinart.com
www.schramsberg.com
www.seppelt.com.au
vqaontario.ca
www.wineaustralia.com
winecountryontario.ca
www.winegrowerscanada.ca
www.wosa.co.za
boschendal.com
family.rothschildarchive.org
kenforresterwines.com

Photo 47. Vineyard workers hand-harvesting white-skinned grapes in Spier Wine Farm's vineyard in the Stellenbosch district (Western Cape's Coastal Region, South Africa). The grapes are transported to the pressing centre in 10 kg lug boxes.

Photo: courtesy of Spier Wine Farm (Lize-Marié Nieman and Johan Jordaan), January 2024

sparkling wine style

Distinctive features of a sparkling wine: the result of a winemaker's realized vision, ultimately related to the choice of grapes from which the wines are made. Sparkling wines are typically distinguished based on two main factors: year of harvest and place of origin. When dividing sparkling wines by harvest year, a distinction can be made between vintage (vintage-dated or single-vintage) and non-vintage wines. A growing number of producers also offer multi-vintage sparkling wines, a relatively new category, created by combining wines of several vintages, the resultant wines resembling vintage sparkling wines (rather than non-vintage sparkling wines) in style. While non-vintage sparkling wines are blends made for their year-on-year consistency, makers of multi-vintage sparkling wines want their wines to be recognized for their uniqueness.

Australia's sparkling-wine-producing companies are known for their multi-regional sparkling wines, i.e. wines for which grapes are sourced from a variety of winegrowing areas or regions. A notable example is provided by House of Arras, a wine-producing firm in Tasmania, its winemakers relying on fruit from vineyards located in Tasmania's six different areas: Tamar Valley, Pipers River, Coal River Valley, East Coast, Huon Valley and Derwent Valley. Made by the winemaker Ed Carr, House of Arras's vintage-dated sparkling wine E.J. Carr Late Disgorded 2007 was made from grapes grown in the island's four growing areas: Coal River Valley, Derwent Valley, Pipers River and Tamar Valley.

The criterion of origin allows distinctions between multi-regional sparkling wines, multi-vineyard sparkling wines, single-village and single-vineyard sparkling wines, and sparkling wines made from grapes coming from old vines, illustrative examples of which are *vieilles vignes* champagnes: Bollinger's Vieilles Vignes Françaises 2013 and Francis Boulard & Fille's Vieilles Vignes Blanc de Blancs 2018. The old-vine champagne produced by Bollinger belongs to the company's range Vieilles Vignes Françaises (VVF), of which the first cuvée was Bollinger's Vieilles Vignes Françaises 1969, launched in 1974. The VVF range wines have been produced from Pinot Noir grapes grown (using the old vine-growing method known as *provignage*) in vineyards that have survived the phylloxera plague, which ravaged Champagne's vineyards in the early 1890s.

For custodians of vines planted in the pre-phylloxera period, such as the makers of old-vine champagnes, phylloxera continues to be a serious problem. Pinot Noir grapes for Bollinger's VVF range were originally sourced from three vineyards:

Photo 48. Macabeo grapes from old vines are used in the production of Segura Viudas's flagship cava wine. The wine Segura Viudas Reserva Heredad has been made by Manel Quintana and Ton Raventós from 75% Macabeo and 25% Chardonnay.
Photo: Marta Ręgowska, April 2023

Croix Rouge (in Bouzy), Chaudes Terres and Clos Saint-Jacques (both at Aÿ). From the 2005 vintage onward, Aÿ's walled vineyards have been the sole source of grapes from old ungrafted vines destined for the company's prestigious range, after vinegrowers from Bouzy lost their battle of several years against the grape root aphid phylloxera, which first attacked the Bouzy vineyard in 1998.

Some producers differentiate between sparkling white wines based on the colour of grapes chosen for their production. Hence, the descriptor *blanc de blancs* appears on the labels of bottles containing white sparkling wine created exclusively from light-skinned grapes. By contrast, the descriptor *blanc de noirs* refers to white sparkling wines made using only black-skinned grapes. Primary examples of these wines are offered by the champagne industry. While *blanc de blanc* champagnes are usually made from Chardonnay grapes, some producers in the Champagne region also use other light-skinned grapes for their production, namely Arbanne, Petit Meslier, Pinot Gris and Pinot Blanc. *Blanc de noirs* sparkling wines created in Champagne are made from Pinot Noir and/or Meunier grapes.

Eugène-Aimé Salon is considered the pioneer of single-vintage single-variety *blanc de blancs* champagnes of a single cru. The origins of this rare wine style date back to 1905 when Salon began crafting these wines using Chardonnay grapes grown at Le Mesnil-sur-Oger — at first solely for personal consumption. Encouraged by his friends' enthusiasm for these wines, he founded the wine house Champagne Salon in 1920. The company has been releasing wines from only the greatest vintages, and only after having aged them for an average of 10 years. In 2023, the firm, now part of Laurent-Perrier Group, celebrated the release of Champagne Salon 2013, marking the 44th launch since the style's history began with the late-September 1905 harvest.

Lastly, two general styles of rosé sparkling wine have developed, mainly as a result of winemaking practices in the Champagne region, namely: *rosé d'assemblage* and *rosé de macération*. Wines from the first group are the most commonly produced rosé sparkling wines in Champagne, where these wines are created by blending white base wine with between 5% and 20% of still red wine, which makes a significant contribution to the colour, structure and aromatic expression of the resultant blend, usually described as relatively light and delicate. Wines from the second group are made by racking off tinted must from a load of (lightly) macerated black-skinned grapes. Depending on the time of drawing off the must and the fraction of must selected, two main types of *rosé de macération* (also known as *rosé de saignée*) can be distinguished. Wines made from free-run juice (or *jus de goutte*) drawn off early in the maceration process (typically within the first few hours) are usually described as bright and intense. By contrast, some producers make their base wines for *rosé de macération* by racking off two fractions of must: tinted free-run must (*jus de goutte*) and tinted must obtained from gentle pressing (called *jus de presse*). These fractions are vinified either together or separately, in which case the winemaker can make the decision to combine these batches in the desired proportions at a later stage, following the fermentation process. Sparkling rosé wines created in this way are often described as relatively more opaque, vinous and complex.

Examples of *rosé d'assemblage* champagnes are provided by the vast majority of producers of rosé champagnes, notable among them Deutz's Champagne 'Amour de Deutz' Brut Rosé 2009 and Pierre Gimonnet & Fils 'Rosé de Blancs' 1-er Cru Brut Rosé. A commendable rosé champagne made using the *saignée* method is Leclerc Briant's Rosé de Saignée, Extra Brut. Made from 100% Pinot Noir grapes collected during the 2014 harvest and macerated for two days, bottled in June 2015 and disgorged in November 2021, the wine has a solid structure and a luminous colour with ruby reflections. A similar example is offered by champagne producer Gaston Collard's Rosé de Macération, Brut, a wine made from 100% Pinot Noir grapes, macerated for two days, with the must pumped over two times in that period for optimal extraction.

Butler 2023; CIVC 2023b; Curtis 2020; Dupont 2023a; Faith and Robinson [1994] 2015a: 220; Gargett 2021; Köhler 2022; Lawrence 2022; Liem 2017: 220–1; Machard de Gramont 2022; Pessey 2002: 74; Rose 2021: 57–9; Skelton 2007: 44; Stevenson 2011: 168–75, 2015a; Tardi 2016: 185–9; Walkey 2022

houseofarras.com.au
www.champagne.fr/en
www.champagnegastoncollard.org/en
www.champagne-bollinger.com
https://www.champagne-deutz.com/en/intro/o
www.champagne-salon.fr
www.leclercbriant.fr/en
seguraviudas.com

sparkling winemaking

The production of effervescent wines containing natural (endogenous) carbon dioxide in which pressures, measured at 20°C, are above 2.5 bar (2.47 atm). The activities involved in the production process broadly encompass base winemaking, followed by preparation of sparkling wines. The former consists in rigorous selection of grapes and their subsequent vinification. The latter centres around the procedure of blending base wines and the choice of a sparkling method.

Enabled by the production of strong glass bottles capable of withstanding great over-pressure caused by the fermentation process, developed in England's glassworks by the mid-17th century, sparkling methods were initially (i.e. from the late 1720s) practised on a commercial scale mainly in northeastern France. From the early decades of the 19th century, these practices began to spread to other winemaking regions of Europe, in particular the Kingdom of Württemberg (from the mid-1820s), Piedmont (from the mid-1860s) and the Catalan region of Penedès (from the early 1870s). By the end of the 19th century, European sparkling wine producers had found their followers in Australia.

Nowadays, based on 2018 statistics, published by the OIV (Organisation Internationale de la Vigne et du Vin) in 2020, the world's top sparkling-wine-producing countries include Italy (around 5.3 million hectolitres), France (about 4.4 million hectolitres), Germany (around 2.8 million hectolitres), Spain (2.1 million hectolitres) and the United States (1.3 million hectolitres). Notably, up to two-thirds of Italy's total annual production volume of sparkling wines are made in the Prosecco zone using the closed-tank method, while France's traditional-method (bottle-fermented) sparkling wines, including champagnes and crémants, constitute more than 80% of the country's annual sparkling wine production.

In major wine-producing countries and regions, sparkling winemaking procedures are strictly regulated by designated authorities. An illustrative example is provided by Cava Regulatory Council (Consejo Regulador de la Denominación de Origen Cava), a body whose aim is to monitor the quality of each stage of the production process. Regulations set for the cava-producing region state include, *inter alia*, that base wines, obtained from authorized grapes as a result of the alcoholic fermentation of grape juice, must have an alcohol content in the range of 9.5–11.5% abv, and that all sparkling wines must be made following the traditional method.

📖 CIVC [2014] 2022: 8–25, 2023b; Faith and Robinson [1994] 2015a: 241; Gilby 2020; OIV 2020a: 6–8

💻 capclassique.co.za
www.cava.wine/en
www.champagne.fr/en
www.corpinnat.com/en
napavintners.com
www.prosecco.it/en

stabilization of base wines and partially fermented musts

A choice of winemaking operations conducted with the aim of preserving desirable features of base wines or partially fermented musts. These winemaking operations can include either: cold storage; or the use of fining agents; or filtration; or a combination of some or all of these activities. When the winemaker decides to apply all of these stabilizing treatments, the stabilization process usually begins with cold stabilization, preventing undesirable reactions that may impair the wine or partially fermented must.

📖 Croguennec 2016: 292; Enartis 2019: 10; Forbes [1967] 1989: 311; Halliday and Johnson [1992] 2013: 114–15; IOC 2018: 2; WSET 2020: 12

Photo 49. Protected from light, bottles of champagne rest on their sides in the cellars of Maison Ruinart in Reims.
Photo: author, July 2024

stacking *sur lattes*

A winemaking activity conducted during extended maturation of sparkling wine on lees; it involves laying bottles of new (bottle-fermented) sparkling wine horizontally (i.e. on their sides). Bottles are placed in rows — one on top of the other, with the rows separated by thin wooden slats. A space-efficient method of storage, using wooden laths also enables the winemaker to minimize damage should one of the bottles explode. In the course of the maturation process, bottles of sparkling wine are unstacked and restacked at least once a year, the aim being to remove from the pile both bottles that are broken or cracked and bottles whose corks are damaged.

📖 CIVC 2013: 29–32; Forbes [1967] 1989: 320–1; Parker 2008: 453; Stevenson 2003: 31; WSET 2020: 8–9

💻 maisons-champagne.com/en
www.ruinart.com/en-int

stacking *sur pointes*

Placing of bottles containing bottle-fermented sparkling wine in inverted positions (i.e. neck down) in rows — one on top of the other — so that the temporary stoppers of the bottles in top rows rest safely in the concave bases of the bottles placed directly below them. The bottles remain stacked with their necks down for various periods; some producers allow all their wines to continue their extended maturation from this point on with limited contact with the sediment of dead yeast cells, because they are trapped in the neck of the bottle; some use this maturation method for their best wines only. The length of the process is a decision for the winemaker, and can last from several months to several years. In extremely rare cases, the storage of bottle-fermented sparkling wines in inverted bottles, during which the lees continue to have a limited effect on the wine, can be extended beyond that period.

Faith [1989] 2016: 215–18; Fielden [1994] 1998: 291; Grainger and Tattersall 2005: 102; Grainger and Tattersall [2005] 2016: 141; Jeandet and Charters 2012: 20; Liem [2019] 2022; Pessey 2002: 76; WSET 2020: 9–10

maisons-champagne.com/en
www.champagne.fr/en

static settling of must

≈ natural clarification of coloured musts

≈ natural clarification of white must

sulphurization of coloured musts

The operation of adding small amounts of sulphur dioxide to pink-coloured and red musts because of its antioxidant and antimicrobial properties and with regard to the dissolvent effects of its use. The use of sulphur dioxide is usually considered necessary in the case of rosé must, which — as with white juice — will usually require some protection against exposure to oxygen. When producing rosé wines, sulphiting musts during the limited maceration of red berry skins is known to favour the dissolution of phenolic compounds, which has an impact on the extraction of natural colourants. By contrast, the addition of sulphur dioxide to red must tends to be considered less important, not least because red musts are known for their greater tolerance to oxygen. Their moderate exposure to oxygen is in fact considered necessary in some cases to enhance both the colour and aroma of the resultant wine. Nonetheless, recent oenological studies on musts obtained from Tempranillo, a grape variety used (increasingly) for the production of sparkling red wines, indicate that sulphur dioxide can be a good extractant of aromatic compounds when added in the process of maceration of red berry skins. The use of potassium metabisulphite at a dose of 50 mg/l has been found to cause a marked increase in the extraction of terpenes and terpenoid compounds, in particular linalool, α-terpineol, geraniol, p-cymene and citronellol. The use of sulphur dioxide during the maceration of Tempranillo grapes has also been found to have a positive impact on the extraction of esters, notably methyl salicylate.

Garde-Cerdán *et al.* 2021; Makarov and Lutkov 2021: 291; OIV 2020b: 3; Ribéreau-Gayon *et al.* 2006: 216, 449; Tarko *et al.* 2020: 3, 9

www.sutherlandestate.com.au

sulphurization of white must

The operation of adding small amounts of sulphur dioxide to grape juice intended for white base wine with the purpose of conserving the integrity of the juice. Given the antioxidant and antiseptic properties of sulphites, this operation serves to counteract the process of enzymatic browning while protecting the juice against microbial infections and thus safeguarding the sensory qualities of the resulting wine. The use of sulphur dioxide can make a significant contribution to minimizing the oxidative loss of aroma precursors. Moreover, the addition of sulphur dioxide to the juice at this stage, which usually ranges from 20–60 mg/l, facilitates the clearing of juice from sediment at a later stage of the vinification process. Sulphurizing is known to aid solubility and flocculation, thereby improving the natural precipitation of sediments in the juice, which facilitates their subsequent removal.

Alvarez 2018; Liem 2017: 302; Nagórka 2013: 108; Walters [2016] 2022: 197; Zoecklein 2002: 9

albetinoya.cat/en
champagne-telmont.com
giolitalia.com/en
maisons-champagne.com/en
villiera.com
www.champagne.fr/en
www.champagne-drappier.com/en
www.prosecco.wine/en

There is an ongoing trend among producers towards limiting or avoiding the use of sulphur dioxide in the vinification process. Some sparkling wine producers, champagne producers included, have abandoned its use in some of their wines in the belief that an unsulphurized wine shows a purer and more transparent expression of flavours. Producers showing commitment to limiting or avoiding the use of sulphur dioxide in the winemaking process can be found in places as far apart as Stellenbosch, a district of Coastal Region in Western Cape, RSA (e.g. Villiera Wines, producer of the cap classique wine Villiera Brut Natural Chardonnay 2019), the Penedès region in Catalonia (e.g. Albet i Noya, producer of the Classic Penedès ancestral-method wine Albet i Noya 'Efecte: Ancestral' Reserva Brut Natural 2020) and the province of Treviso, part of the Prosecco zone (e.g. Azienda Agricola Giol, producer of the semi-sparkling wine Giol Prosecco Col Fondo). Among the most lauded unsulphurized champagnes are: Drappier's Brut Nature Sans Soufre NV (100% Pinot Noir), Benoît Lahaye's 'Violaine' Brut Nature 2017 (50% Chardonnay, 50% Pinot Noir), Champagne Telmont's 'Sans Soufre' Extra Brut 2013 (58% Chardonnay, 42% Meunier) and Champagne Erick Schreiber's 'Astral' Brut 2014 (85.4% Pinot Noir, 14.6% Chardonnay).

T

taille

Juice extracted from clusters of grapes during the second stage of pressing. When making base wine in the Champagne region, the second fraction of pressed juice, referred to as *taille*, constitutes an additional 500 litres of juice extracted (from each 4,000 kg load of grapes) following the initial pressing of 2,050 litres of juice, known as cuvée. Notably, *taille* is richer in minerals (especially potassium) and lower in acids and sugars than cuvée, as the juice for *taille* is obtained from the flesh of the berries closer to the skins and seeds. *Taille* is allowed to be used for making champagne. While some champagne producers, who appreciate fruity aromas displayed by *taille*, decide to keep this juice, in most cases its very first part, known as *première taille*, with the intention of using it for softening the target blend, many of the leading champagne producers choose to sell their *taille* to other wine-producing companies.

📖 CIVC 2013: 39, 2023b; Croguennec 2016: 292; Forbes [1967] 1989: 297–8; Jackson [1994] 2008: 541; Liem 2017: 47; McKay *et al.* 2011: 328; Ribéreau-Gayon *et al.* 2006: 460; SL 2018/2019: 14; Zoecklein 2002: 6

💻 www.champagne.fr/en
maisons-champagne.com/en

tank method

An industrial method of sparkling wine production involving the use of hermetically closed pressure-resistant tanks as vessels for the storage of the wine during the process of secondary fermentation. Two (general) variations of this method have developed: the Charmat method (also known as the Martinotti method) and the continuous method (also known as the Russian method). Both variations are regarded as suitable for the production of large volumes of mid-range sparkling wine. The Charmat method is now prevalent, its advantage over the Russian method being that it uses cheaper equipment. The continuous method appears to be an option for producers capable of achieving economies of scale through a large and loyal customer base.

📖 Fielden [1994] 1998: 296–7; MacNeil [2002] 2015: 815–16, 1,109–10; OIV 2022: II.4.1–12, II.4.3–1, II.4.4–1; Rose 2021: 26–7; Seidanova 2022; Sussman 2021: 11, 13; WSET 2020: 11–12

tank-fermented sparkling red wines

≈ Charmat-method sparkling red wines

tank-fermented sparkling rosé wines

≈ Charmat-method sparkling rosé wines

tank-fermented sparkling white wines

A category of double-fermentation sparkling white wines encompassing wines that have been made effervescent by means of closing individual tanks (or a system of interconnected pressure-resistant tanks) containing base wine undergoing second fermentation. These wines can be divided into Charmat-method sparkling white wines and sparkling white wines produced using the continuous method.

📖 Chroman 1986; Cowderoy 2021: 36–41; JMF 2017; Mayson [1994] 2015: 471; Pomarici *et al.* 2019: 421–3; Rose 2021: 26–7; Speller [1994] 2015c: 647–8; TFA 2022; Walkey 2019

tartrate stabilization by cooling

A practice of holding blended base wines, known as cuvées, at temperatures close to or a little lower than the water-freezing point for a period of two to three weeks, the aim being to induce precipitation of tartrates so that they could be separated from the wine before bottling. The formation of tartrate (calcium tartrate and/or potassium bitartrate) crystals in the bottle is usually considered undesirable in sparkling wines, as the crystals can serve as (additional) nucleation sites for carbon dioxide bubbles, which might cause the final wines to gush, and quickly lose their effervescence.

Purified potassium bitartrate, commonly known as cream of tartar, is often added to the cuvée before the production of effervescence in the wine. This facilitates the formation of potassium bitartrate crystals while accelerating their sedimentation, a process followed by their removal by physical techniques. Cream of tartar is added to the cold wine, which ought to be stirred regularly to ensure that the salt of tartaric acid added is distributed evenly.

📖 Cisilotto *et al.* 2023; Cordingley 2020; Croguennec 2016: 292; Jackson [1994] 2008: 427,

Photo 50. A half-bottle of base wine with tirage liqueur added to activate the secondary fermentation: house of Champagne Charles de Cazanove, located at Place de la République in Reims.

Photo: author, July 2024

542; Kojder 2022; Nagórka 2013: 110; OIV 2022: II.3.3–6; SL 2018/2019: 33; Zoecklein 2002: 10

tirage

An operation crucial to the production of sparkling wines using secondary fermentation, which involves placing the cuvée in a sealed container — usually a bottle or tank — in order to initiate the second fermentation with the aim of obtaining effervescence. When bottles are used for the purpose, the process is sometimes called *tirage en bouteilles*. When tanks are employed, the process can be referred to as *tirage* in closed tanks (or *tirage en cuve close*). In either case, this operation is typically accompanied (or preceded) by the addition of *tirage* liqueur (*liqueur de tirage*), a solution that helps activate and carry out the second fermentation process. In some wine regions, bottling of wine to create effervescence is strictly regulated; for example, in Champagne new wines may not be bottled until the first day of January following the harvest.

Cisilotto *et al.* 2023; CIVC 2023b; Guy 2003: 69; McCann 2023; OIV 2022: II.4.1–10, II.4.1–3, II.4.2–1; Rand 2022: 19; Rose 2021: 35–6

maisons-champagne.com/en
champagnedecazanove.com

tirage en bouteilles

≈ tirage in bottles

tirage in bottles

A winemaking operation in the production of bottle-method sparkling wines. This winemaking activity involves bottling the prepared cuvée after a small sample of the blend (which in many cases has been resting in the bottling vat), has been examined to see if the wine meets all the specified criteria. Usually, when bottles are filled with cuvée, the latter already contains *tirage* liqueur, a solution added to the wine to help activate and conduct the secondary fermentation process, which can also slightly modify some of the wine's (sensory) characteristics.

Forbes [1967] 1989: 312–13; Guy 2003: 69; McCann 2023; OIV 2022: II.4.1–10

maisons-champagne.com/en
www.champagne.fr/en

tirage in closed tanks

A standard winemaking operation in the production of tank-method sparkling wines. The operation involves putting the prepared cuvée, usually blended at this stage with *tirage* liqueur, into pressure-resistant tanks after a small amount of the blend (which has normally rested in the bottling vat), has been examined to see if the wine meets the specified criteria. Some (sensory) features of the wine that is placed in tanks for the second fermentation can still be slightly modified by the composition of the *tirage* liqueur that it (now) contains.

Buxaderas and López-Tamames 2012: 14; Cisilotto *et al.* 2023; OIV 2022: II.4.3–3, II.4.4–3

traditional method

The most fastidious and time-consuming method of transforming cuvée into sparkling wine, classified as one of the two known methods in which the base wine ferments in bottles. The origins of this sparkling method trace to the practices of English coopers in the mid-17th century and winemakers in the Champagne region in the first decades of the 18th century. It is now used widely by quality-conscious producers. It involves the fermentation of base wine induced by the addition of *tirage* liqueur, typically composed of wine, a sweetening agent, (selected) yeast cultures, nutrients and adjuvants (sometimes referred to as 'riddling aids', as their presence is known to assist in the removal of lees following the maturation process). The key steps differentiating this bottle method from the transfer method, which also uses bottles for both the second fermentation and maturation, include the length of the maturation process and the method for eliminating the yeasty sediment from the bottles of sparkling wine before the wine is dosed with *liqueur d'expédition*. The maturation process (which involves contact with the lees) usually takes (much) longer in the traditional method than in the transfer method.
In the classical (traditional) method (as opposed to the transfer method) bottles containing sparkling wine are riddled and the sediment is removed from the neck of the bottle by disgorgement, a technique that uses pressure accumulated inside the bottle to eject the lees.

CIVC [2014] 2022: 14–15; Grainger and Tattersall [2005] 2016: 137; Johnson [1989] 2021: 109–16; Kramer 2023; Liem 2017: 44–69; Rand 2022: 19–20; Wallner 2018: 36

Photo 51. In the traditional method of producing sparkling wine, the operation of *tirage* initiates the secondary in-bottle fermentation followed by prolonged maturation on lees. At the end of this maturation period, bottles with Boizel's champagne are placed in A-shaped wine racks called *pupitres* in the dark cellars of Maison Boizel in Épernay (at Avenue de Champagne), which begins the lees removal process.
Photo: author, July 2024

traditional-method fermentation

A winemaking process initiating the traditional method of preparing effervescent wines. This involves induced conversion of sugar (contained in *tirage* liqueur) into alcohol and carbon dioxide, typically in the presence of clarifying agents. Taking place in closed bottles, this process leads to an increase in the wine's alcohol content, usually by 1.2–1.5% abv, and the production and (gradual) dissolution of carbon dioxide in the base wine. The length of this process depends on a number of factors, including fermentation temperatures, with temperatures around or lower than 10°C slowing down the second fermentation, and temperatures around or higher than 12°C accelerating the process. At 9–12°C, which is a typical temperature range for running the second fermentation in the region of Champagne, the sparkling fermentation is usually completed within 50–60 days.

Halliday and Johnson [1992] 2013: 116; IOC 2018: 4; Jackson [1994] 2014: 710–12; Liger-Belair and Cilindre 2021: 24–5; Liger-Belair *et al.* 2023: 22,845; Pessey 2002: 39; Schiessl 2017a

www.cremants.com/en
franciacorta.wine/en
maisons-champagne.com/en
vins-cremants.lu/en
www.boizel.com/en
capclassique.co.za
www.cava.wine/en
www.champagne.fr/en
www.corpinnat.com/en

traditional-method sparkling red wines

A category of sparkling red wine produced using the double-fermentation method with the second fermentation taking place in the bottle. Wines in this category include (among notable others) sparkling red wines made by quality-conscious producers in Australia and England. Among the traditional-method sparkling red wines that in the past several years have received acclaim from critics are Shiraz-based wines Rockford Sparkling Black Shiraz NV and Teusner MC Sparkling Shiraz Late Disgorgement–2005, made (respectively) by Rockford Wines and Teusner, producers from the Barossa Valley (South Australia). Among England's notable classical-method fully sparkling red wines are the vintage-dated wine Bolney Estate Cuvée Noir 2014, made from Dornfelder grapes, and the non-vintage wine Ridgeview Sparkling Red Reserve–Limited Edition, crafted from Pinot Noir grapes.

Hill 2021; Jeffery and Wilkinson 2014: 81; Stelzer 2018: 61; Warren-Adamson 2020; WSET 2020: 111

bolneywineestate.com
teusner.com.au
www.ridgeview.co.uk

traditional-method sparkling rosé wines

A category of sparkling rosé wines made using the double-fermentation method with the second fermentation taking place in the bottle. Depending on the method of imparting colour to the base wine, wines in this category can be divided into blended rosé wines (known in French as *rosé d'assemblage*) and macerated rosé wines (known in French as *rosé de macération* or *rosé de saignée*). In the Champagne region, when creating champagne classified as *rosé d'assemblage*, winemakers tend to add from around 5% to around 20% of red base wine to white base wine at the blending stage. A notable example is provided by Pierre Gimonnet et Fils 'Rosé de Blancs', a rosé champagne characterized by a light pink colour achieved by its composition (93% base wines made from Chardonnay grapes from Côte des Blancs and 7% base wines created from Pinot Noir grapes sourced from Bouzy [Montagne de Reims]). Another interesting blended rosé is Champagne Ruinart Brut Rosé, in which the aromatic freshness of the Chardonnay grapes (45%), sourced from Côte des Blancs and Montagne de Reims, is balanced by the

Photo 52. Interesting attempts at producing sparkling red wine can now be found in Central Europe. Krzysztof Fedorowicz, a winemaker and owner of Winnica Miłosz — a small-scale producer near Zielona Góra (Western Poland) — has used estate-grown Pinot Noir grapes to craft the vintage-dated traditional-method sparkling red wine Grempler Sekt–Pinot Noir 2021.
Photo: author, August 2024

fruity character of the Pinot Noir grapes (55%), grown in Montagne de Reims and Vallée de la Marne, of which 12% were vinified to make red base wine needed for the final blend.

The history of the production and sales of traditional-method sparkling rosé wines can be traced back to the mid-18th century. The champagne houses Ruinart and Veuve Clicquot may have been among the pioneers in making and commercializing these wines. As reported by the historian Isabelle Pierre, archival documents of the house of Ruinart indicate that in March 1764 the house shipped 60 bottles of rosé champagne to the Roman German Empire. The type of wine shipped was known by the name *oeil de perdrix*, still used today to refer to sparkling wines characterized by pale salmon and copper hues, obtained by tinting the juice of red grapes by very brief skin contact. That wine seems to have been the forerunner of today's macerated rosés. In 1818, Barbe Nicole Clicquot, the head of the house now known as Veuve Clicquot, decided to blend red wine made from grapes from Bouzy with white base wines, which led to the development of a new method for crafting rosé champagnes, known today as blended rosés or *rosés d'assemblage*. The method of crafting rosé champagnes by blending white base wines with still red wine prior to the second fermentation, resulting in sparkling wines with increased complexity, is now prevalent among champagne producers.

Production of traditional-method sparkling wine classified as *rosé de macération* involves closely monitored contact between the skins of destemmed (or partially destemmed) dark-skinned grapes, such as Meunier or Pinot Noir, and their juice, thus enabling limited release of colourants, tannins, flavour substances and aromatic compounds into the juice. This controlled process usually takes between one and three days. The coloured must is then separated from the skins and racked into fermentation vessels. Importantly, coloured musts for wines of this style can be obtained using either free-run juice from dark-skinned grapes or a combination of free-run juice and must resulting from light pressing. Illustrative examples of rosé champagne made using the *saignée* method include Larmandier-Bernier 'Rosé de Saignée' Premier Cru, Extra Brut, NV, and Champagne H. Blin 'Rosé de Saignée'–Sélection Parcellaire–2015. The first of these wines has been made from two grape varieties used in different quantities: 90% Pinot Noir grapes and 10% Pinot Gris grapes, with the grapes macerating for the period of 2–3 days before the onset of natural fermentation and malolactic conversion. Created exclusively from Meunier grapes, H. Blin's wine boasts a deep pink colour combined with rich and concentrated flavours, a result of brief, albeit intense, maceration of skins.

Since the early 2000s, the quality of sparkling rosé wines produced by the traditional method has improved significantly, a result of a change in attitude to the concept of creating these wines among producers and changing expectations about the quality of rosé sparkling wines among consumers. The mid-2010s saw the birth of Fleur de Miraval, a champagne producer devoted exclusively to rosé champagnes. The growth in quality of rosé sparkling wines, notably champagnes, explains the marked increase in demand. Since the beginning of the third decade of the 21st century, in each consecutive year rosé champagnes have accounted for no less than 10% of all champagne shipments outside France, with their share in the total champagne export volume in 2022 reaching 10.6%. Rosé champagnes have a faithful following in the United States, the largest export market for champagne: since 2020, annual sales of rosé champagne in the US have comprised over 17% of the American market for champagne by volume, the second-largest share of the market (behind non-vintage *brut* champagnes).

Avellan 2023a; Benucci 2020; CIVC 2019, 2020, 2021, 2022, 2023a: 10, 20–1, 2023b; Edwards 2009: 115; Grainger 2021: 243–5; Halliday and Johnson [1992] 2013: 116; Hudin 2016; Hunt 2016; Hyland 2019; Jackson [1994] 2014: 710–12; Köhler 2022; Liem 2017: 302; Medina-Plaza *et al.* 2024; Napjus 2014; Robinson 2011, 2019b; Schiessl 2023; Stávek *et al.* 2012: 1,133–45; Tardi 2016: 116–17; TFB 2017a: 11; WA 2022: 7–10, 12–15, 17–22

www.cremants.com/en
franciacorta.wine/en
larmandier.fr/en
maisons-champagne.com/en
vins-cremants.lu/en
www.austrianwine.com
capclassique.co.za
www.cava.wine/en
www.champagne.fr/en
www.champagne-blin.com
www.champagne-gimonnet.com
www.corpinnat.com/en
www.ruinart.com/en-int
www.veuveclicquot.com/en-int
www.winesofgermany.com
www.wosa.co.za

While the lion's share of sparkling wines, especially rosés, are packaged in clear-glass bottles for marketing purposes, a small but increasing number of quality-conscious producers are selling their most prestigious wines in opaque bottles to provide protection against ultraviolet (UV) radiation from both sunlight and artificial sources. Oenological studies on the effects of storage conditions on sparkling rosé wine demonstrate that the problem of excessive exposure of the wine to light (called lightstrike), which compromises the wine's quality by causing reactions from highly photosensitive compounds, in particular riboflavin, affects mainly wines packaged in clear-glass bottles, as these are capable of blocking only up to around 10% of damaging rays. Lightstrike results in the transformation of amino acids present in the wine into sulphur compounds (notably dimethyl sulphide and dimethyl disulphide) responsible for unpleasant flavours detected at tastings. These wine defects are considered irreversible. A growing number of wine professionals are recommending that clear-glass sparkling wine bottles be replaced with brown glass, which provides the most effective protection against harmful light. It is also recommended that clear-glass bottles of sparkling wine already on the market, whether on a shelf in a wine shop or on a wine rack in a private home, ought to be sheltered from light and kept in cold storage, preferably at temperatures of 5–12°C.

Table 2 Annual exports of rosé champagne against all champagne shipments outside France by volume and by value for the period 2018–2022, expressed as percentages

Source: www.champagne.fr

Export years	Share of rosés in champagne exports by volume	Share of rosés in champagne exports by value
2018	9.9%	11.8%
2019	9.7%	11.5%
2020	10%	11.6%
2021	10.9%	12.6%
2022	10.6%	11.9%

traditional-method sparkling white wines

A category of sparkling white wines made using the double-fermentation method, with the second fermentation taking place in the bottle and the subsequent maturation period usually allowed to last longer than the period of maturation on lees in the production of transfer-method wines. Compared to sparkling white wines crafted by the transfer method, these wines are usually characterized by more pronounced autolytic features. Originating from Europe, traditional-method sparkling white wines are now produced on all continents where grapes are grown. Differences between sparkling wines in this category derive from a variety of factors. These include diverse terroirs, different rules adopted for the production of traditional-method sparkling wines in regions (resulting in differences between regions in terms of grape varieties authorized for production and the minimum period of maturation on lees), and also autonomous decisions by the winemaker based on their individual experience. Among the most recognized groups of quality sparkling wines made by the traditional method, each defended by the relevant authoritative body against imitation by strengthening the unique association between the wine and its area of origin, are: champagnes (France), crémants (France and Luxembourg), cavas and corpinnats (Spain), cap classiques (South Africa), franciacortas (Italy) and winzersekts (Germany).

A significant increase has recently been observed in the demand for crémant, a French and Luxembourgish quality sparkling wine made by the traditional method. In France, founded in 1982, the National Federation of Crémant Growers and Producers (Fédération Nationale des Producteurs et Élaborateurs de Crémant) brings together eight appellations: Alsace, Bordeaux, Burgundy, Die, Jura, Limoux, Loire and Savoy. The Alsace region remains the most prolific. Producers of *crémant d'Alsace* also have the greatest share of the market, with the annual sales of their wine(s) consistently accounting for more than 30% of the entire volume of crémant (from all regions) sold each year. The total amount of France's crémant commercialized in 2022 reached 102.3 million bottles, surpassing the 100-million-bottle mark for the first time. This constitutes less than a third of the total volume of champagne sold that year, which amounted to just under 326 million bottles.

Significant differences are observed not only between but also within these wine groups. Suffice it only to note the differences in grape varieties used to produce crémants. Producers in each of the eight crémant-producing regions (Alsace, Bordeaux, Burgundy, Die, Jura, Limoux, Loire and Savoy) have their own selection of authorized varieties. While white *crémant d'Alsace* can be made by a harmonious combination of grapes such as Auxerrois, Chardonnay, Pinot Blanc, Pinot Noir and Riesling, grapes authorized for the production of white *crémant de Bordeaux* include (among others) Sémillon and Sauvignon Blanc, with Chenin Blanc being the most common dominant component in blends intended for *crémant de Loire*. Local grape varieties Altesse and Jacquère are obligatory in the blends used to make *crémant de Savoie*, with other authorized varieties in the Savoie region including Chardonnay, Chasselas, Gamay and Pinot Noir. Yet another set of grapes is used by makers of *crémant de Luxembourg*. In addition to the classic sparkling wine grapes Chardonnay and Pinot Noir, the father-and-daughter team Laurent and Corinne Kox of the family-owned wine-producing company Domaine Laurent & Rita Kox from Luxembourg's wine region Moselle make their white crémants using Auxerrois, Pinot Blanc, Pinot Gris and Riesling.

Bell 2015a; Bertrand 2022; Currin 2022; Feilden 2024; FNPEC 2018/2019: 3–11; Halliday and Johnson [1992] 2013: 116; Howell 2013; Hudin 2016; IOC 2018: 4; Jackson [1994] 2014: 710–12; Johnson [1989] 2021: 106–16; Liger-Belair and Cilindre 2021: 24–5; Pessey 2002: 7–19, 46–7; Rand [2018] 2022: 226–9, 2022: 18–20; Schiessl 2023; Stevenson [2015] 2022: 86–7; Sussman 2021: 20–47, 61–7, 92–7; WA 2022: 7–10, 12–15, 18–22; Waterworth 2023; Welke *et al.* 2022: 108–12; WSET 2020: 15–26, 32–53

cremants.com
franciacorta.wine/en
maisons-champagne.com/en
vins-cremants.lu/en
www.austrianwine.com
www.bordeaux.com/gb
www.cava.wine/en
www.champagne.fr/en

Photo 53. A bird's-eye view of Ried Steinsetz, a source of Grüner Veltliner for Schloss Gobelsburg's base wine intended for the company's wine Schloss Gobelsburg Brut Reserve NV. Run by Michael Moosbrugger, Schloss Gobelsburg is an acclaimed producer of quality traditional-method sparkling white and rosé wines, located in the winemaking region of Kamptal, Lower Austria.
Photo: courtesy of Schloss Gobelsburg (Michael Moosbrugger)

capclassique.co.za
www.corpinnat.com/en
www.domainekox.lu/en
www.gobelsburg.at/en
www.winesofgermany.com
www.winesofsa.co.uk
www.wosa.co.za

transfer method

A simplified version of the classical method of crafting sparkling wines, most probably devised in Germany in the third decade of the 20th century. It is now being applied less and less as it lacks the prestige and pricing advantage of the traditional method; yet it remains popular among large producers in Germany, the United States, Australia and New Zealand, as it enables them to avoid costs associated with manual riddling and disgorgement, the two labour-intensive and time-consuming steps used in the traditional method of creating sparkling wines. As in the case of the classical method, the transfer method involves managing the secondary fermentation in the bottle, followed by a period of maturation. The main factor differentiating these operations from those performed by winemakers in Champagne is the stage that follows maturation of sparkling wine on lees: namely, the transfer of the contents of bottles to bulk pressure tanks, where the wine is clarified, usually by filtration. Bottles of sparkling wine made using the transfer method are often labelled 'fermented in the bottle' or 'bottle-fermented', which — by implication — distinguishes them from bottles of quality sparkling wine made using the traditional method; only the latter can carry the label 'fermented in this bottle', provided it has not been subjected to the process of *transversage*.

📖 Bakker and Clarke [2004] 2012: 33; Gardner 2015a; Howe 2003: 149; Jackson [1994] 2014: 715; McKay *et al.* 2011: 342; Schiessl 2023; WA 2022: 10–14; Wasserman and Wasserman 1984: 94; WSET 2020: 10

transfer of clarified bottle-fermented sparkling wine to pressure tanks with dosage liqueur

A winemaking operation in the production of transfer-method sparkling wines, which takes place after bottle-fermented wine, following its time on lees, has been cleared of its sediment. The operation involves adding to a pressure tank to which bottle-fermented sparkling wine is about to be transferred under pressure — or in which it already resides — a solution (often referred to as dosage liqueur) the main component of which is wine, or grape must, or a mixture of wine and must, where the must can be either unfermented, or partially fermented, or concentrated, or concentrated and rectified. The sugar content of the liqueur varies from producer to producer, determining the wine's final sweetness. Once the sparkling wine has been blended with the liqueur, and then subjected to sterile filtration, it is (usually) considered ready for rebottling.

📖 Bakker and Clarke [2004] 2012: 33; Fielden [1994] 1998: 295–6; Hornsey 2007: 4.3.3; Jackson [1994] 2014: 715; Panesar *et al.* 2017: 504; Skelton 2021: 152; UGA 2019; WA 2022

transfer of disgorged bottle-fermented sparkling wine to non-standard-sized bottles

A winemaking practice used occasionally in the traditional method of creating effervescent wines which involves removing disgorged bottle-fermented sparkling wine from a pressure tank in which the dosage liqueur and the wine have been blended in order to place the resultant wine in bottles of sizes other than 75 cl. This practice is the second (and final) part of the complex operation known as *transversage*. Among the small-format non-standard bottles that sparkling wine producers often select are quarter-bottles (18.75 cl or 20 cl) and half-bottles (37.5 cl). The large-format bottles typically include magnum (1.5-litre bottle equivalent to two standard bottles), jéroboam (3-litre bottle equivalent to four standard bottles) and rehoboam (4.5-litre bottle equivalent to six standard bottles). Bottles with even greater holding capacities, such as methuselah (6-litre bottle equivalent to eight standard bottles) and salmanazar (9-litre bottle equivalent to 12 standard bottles) are made to order (see Table 3). One reason producers might decide to age their wines after disgorgement in magnums rather than standard bottles is a greater potential for reductive ageing owing to the wines' relatively lower exposure to oxygen present in the headspace as a result of oxygen ingress both at disgorgement and during the period of post-disgorgement ageing.

Avery [2010] 2012: 47; Burton and Flewellen 2013; Forbes [1967] 1989: 338–9; Liem 2017: 297; Stevenson [2015] 2022: 85–7; TFB 2016

cremants.com
franciacorta.wine/en
maisons-champagne.com/en
www.champagne.fr/en

Table 3 Sparkling wine bottles by type, from least to most capacious

Sparkling wine bottles by type	Holding capacity	
	litres	standard bottles
quarter-bottle	0.1875	0.25
half-bottle	0.375	0.5
standard bottle	0.750	1
magnum	1.5	2
jéroboam	3	4
rehoboam	4.5	6
methuselah	6	8
salmanazar	9	12
balthazar	12	16
nebuchadnezzar	15	20
solomon	18	24
souverain	26.25	35
primat	27	36
melchizedec	30	40

transfer of disgorged bottle-fermented sparkling wine to pressure tanks with dosage liqueur

A winemaking practice used occasionally in the traditional method of creating effervescent wines that involves decanting sparkling wine from bottles into a bulk pressure tank following disgorgement in order to evenly distribute dosage liqueur in the wine. The transfer of the wine to pressure tanks and its blending with dosage liqueur is seen as the first part of the complex operation known as *transversage*. The dosage liqueur added to the tank is a solution the main component of which is either wine or grape must, or a mixture of wine and must, where the must can be either unfermented, or partially fermented, or concentrated, or concentrated and rectified. The sugar content of the liqueur varies from producer to producer, determining the wine's final sweetness.

Campbell 2021; Howe 2003: 149; Juhlin 2022; LaVilla 2010: 494; TFB 2016; UGA 2019

cremants.com/en
franciacorta.wine/en
maisons-champagne.com/en
www.champagne.fr/en

transfer-method fermentation

A winemaking process initiating the transfer method of preparing effervescent wines. This involves induced conversion of sugar (contained in *tirage* liqueur) into alcohol and carbon dioxide, typically without the presence of clarifying agents. As in the production of traditional-method wines, this process takes place in closed bottles (often magnum bottles, each holding 1.5 litres), leading to both an increase in the wine's alcohol content, usually by 1.2–1.5% abv, and the production and (gradual) dissolution of carbon dioxide in the base wine. The length of this process depends largely on fermentation temperatures, with those around or lower than 10°C slowing down the second fermentation, and temperatures around or higher than 12°C accelerating the process.

Bakker and Clarke [2004] 2012: 33; Jackson [1994] 2008: 9, 546–7; Nagórka 2013: 113; Ribéreau-Gayon *et al.* 2006: 468; Skelton 2021: 152

transfer-method sparkling white wines

A category of sparkling white wines made using the double fermentation method that involves the use of bottles for the second fermentation and subsequent maturation on lees followed by the transfer of the sparkling wine to a pressure-resistant tank for the removal of sediment before putting this wine into fresh bottles. Compared with sparkling white wines crafted using the traditional method, these wines are usually characterized by less pronounced autolytic features, as in most cases their production methods involve a shorter period of maturation on lees than is the case in the production of sparkling wines made using the traditional method. Examples of quality transfer-method wines are provided by the Hungarian producer Törley (Pezsgőpincészet). The firm's Tokaji range includes two bottle-fermented wines made using the transfer method: Törley Tokaji Brut and Törley Tokaji Doux.

Photo 54. Maison Ruinart offers champagne in bottles of various sizes. Tasters at Maison Ruinart confirm that the size of the bottle has a huge impact on storage lifetimes. The shortest storage is recommended for sparkling wines in bottles with a holding capacity below 75 cl, whereas the longest storage is recommended for wines purchased in bottles with a holding capacity exceeding 75 cl.
Photo: author, July 2024

Widely adopted in Australia, the transfer method is applied for a large proportion of sparkling wines made in this country.

Ellis 2012: 167; Gilby 2020; Howell 2013; LaVilla 2010: 264; Schiessl 2023; Small and Couturier 2011: 288; Sullivan 1998: 341; WA 2022: 10–14; Walkey 2023

torleypezsgo.hu/en

transversage

A winemaking process whereby, immediately after disgorgement, bottle-fermented sparkling wine created using the traditional method is transferred from bottles into a pressure tank so that it can be blended with dosage liqueur and rebottled under pressure using non-standard-sized bottles, mainly quarter-bottles with a holding capacity of 18.75 cl or 20 cl, which are sold by airlines, and bottles larger than a jéroboam (3 litres), which are used for celebrations; the largest bottle offered, albeit very rarely, by some champagne producers is called a melchizedec, having the holding capacity of 30 litres (equivalent to 40 standard bottles). Importantly, the bottle-fermented wine that has been disgorged and transferred to pressure tanks for blending with the dosage liqueur and subsequently transferred under pressure to bottles of non-standard sizes can be described as 'bottle-fermented' or 'fermented in the bottle' but not 'fermented in this bottle', which is a descriptor reserved for traditional-method sparkling wines sold in the bottles in which they have been fermented.

Campbell 2021; CIVC 2013: 38; Howe 2003: 149; Liem [2019] 2022; TFB 2016; UGA 2019

maisons-champagne.com/en

treatment of sediment in ancestral-method sparkling wines

A winemaking process resulting from the selection of the appropriate method of managing the yeast deposit formed in bottles containing effervescent wine obtained from a single fermentation. The winemaker can choose to decant the wine off the deposit before rebottling it under pressure, or to keep the deposit of sediment in the wine as part of the wine's style, leaving the decision to the consumer about whether to decant or not.

📖 Rose 2021: 27–8; Sussman 2021: 54–6; Teclemariam 2023; Walkey 2021; WSET 2020: 11, 72

V

vertical blending

A winemaking procedure of combining base wines across vintages, which means creating a blend of wines from grapes harvested in different years. Thus, the final blend, sometimes also referred to as a cuvée, can include base wines of (many) various vintages: not only wines of previous vintages, also known as reserve wines, but also wines from grapes of the most recent vintage. While small wine-producing companies tend to keep reserve wines of the last several vintages, the largest wine houses, such as those in Champagne, maintain vast collections of reserve wines. Cellar masters blend base wines of different vintages to produce standard non-vintage sparkling wines (NV), appreciated for the constancy of their style and quality, and multi-vintage sparkling wines (MV), often called prestige cuvées for their exceptional quality. The latter include, champagnes from Laurent-Perrier's range Grand Siècle, of which recent examples are: Laurent-Perrier Grand Siècle N° 23 (a blend composed of wines from three vintage years 2006, 2004 and 2002); Laurent-Perrier Grand Siècle N° 24 (a quality blend of wines from harvest years 2007, 2006 and 2004); and Laurent-Perrier Grand Siècle N° 25 (a blend of wines from vintage years 2008, 2007 and 2006).

📖 Burton [2014] 2016; Burton and Flewellen 2013; Field 2022b; Lawrence 2022; TFB 2016

💻 www.laurent-perrier.com/en

The range of unique (multi-vintage) champagnes known by the name of Grand Siècle was conceived in the early 1950s by the then Laurent-Perrier head Bernard de Nonancourt, after he had assumed control of the family business in October 1948. The idea of multi-vintage champagnes as defined by Maison Laurent-Perrier involves blending reserve wines made from grapes from three exceptional vintages seen as complementary. The first prestige cuvée champagne of the house, initiating Laurent-Perrier's Grand Siècle range, was launched in 1959.

vinification of grapes for red and rosé base wines and partially fermented musts

One of the key stages in the production of red and rosé base wines and musts intended for effervescent wines. As with the vinification of grapes for white base wines and musts, this stage centres around the management of the fermentation process, but differs significantly from that vinification method in that it begins with the maceration of red berry skins (and not with the pressing of grapes).

📖 Barros *et al.* 2022: 3,246–7; CIVC [2014] 2022: 9; Demmond 2023; Liem [2019] 2022; Liger-Belair and Cilindre 2021: 23; Smith 2020; Stelzer 2018: 67

💻 www.champagne.fr/en
www.ridgeview.co.uk
www.seppelt.com.au

vinification of grapes for white base wines and partially fermented musts

One of the key stages in the production of white base wines and musts intended for effervescent wines. Following the pre-treatment of grapes and preceding the post-fermentation maturation, this stage centres around the fermentation process. However, this stage differs from the vinification of grapes for red and rosé base wines and musts in that, in the vinification managed with the aim of obtaining white base wines or partially fermented musts, the sequence of activities and processes aimed at converting grapes into base wine or must usually begins with the pressing of grapes.

📖 CIVC [2014] 2022: 9; Forbes [1967] 1989: 295–306; Halliday and Johnson [1992] 2013: 118–19; Liem 2017: 44–51; Whitney 2022; WSET 2020: 4–5, 23, 51, 61, 65, 75–6

vinification of sparkling wine grapes

One of the key stages in the production of base wines and musts destined for effervescent wines. This stage, following pre-treatment of grapes and preceding post-fermentation maturation, involves a sequence of operations and processes regarded as necessary in obtaining base wines or partially fermented musts with desired properties, which include and concern the process of primary fermentation. These operations and processes, as well as their sequence, vary depending on the base wine or partially fermented must to be obtained. The main dividing line is usually between the operations and processes aimed at obtaining white base wines or musts and those required to obtain red and rosé base wines or musts.

📖 Barros *et al.* 2022: 3,247; Broggi 2023; CIVC [2014] 2022: 9; Halliday and Johnson [1992] 2013: 114, 118–19; Liem 2017: 47–51; Schiessl 2023; Sussman 2021: 11–15; Whitney 2022; WSET 2020: 5–6; Zoecklein 2002: 5–10

💻 franciacorta.wine/en
maisons-champagne.com/en
www.champagne.fr/en

W

whole berry pressing

≈ whole cluster pressing

whole bunch pressing

≈ whole cluster pressing

whole cluster pressing

Extraction of white juice from whole bunches of grapes, whether light-skinned or black-skinned, without previously destemming or crushing them. This winemaking activity, also known as fractional pressing, is now the norm in many regions in which quality sparkling wines are produced, including (but not limited to) Champagne, Franciacorta and the Western Cape regions in South Africa. It involves loading freshly harvested grapes into the press and then allowing the juice to flow freely from the load of grapes. Before physical pressure is applied to the grape load in the press to obtain quality juice for base wine, the free-run juice is typically checked for impurities or signs of early oxidation, as in such cases the juice would fail to meet the standards set for quality wines and would need to be discarded.

This winemaking activity results in obtaining grape juice press fractions

Photo 55. Vineyards of Champagne Augustin at Avenay Val d'Or (Vallée de la Marne). Whole clusters of grapes from these vines were pressed in 2016 for the production of the champagne producer's prestige cuvée Champagne Augustin 'Élément AIR' Brut, which was disgorged and released in 2022.
Photo: author, July 2024

differing in quality. Fractional pressing for sparkling wines made in Champagne is regulated by volumetric requirements for the juice obtained. In addition, winemakers discriminate juice press fractions by sensorial assessments of phenolic content, a method applicable especially in small-scale juice production. Componential analyses of grape juice fractions destined for the production of sparkling white wines using ultraviolet-visible spectral phenolic fingerprinting (also called SPF) seem a promising method in the case of high-volume juice production. Recent oenological research into grape juices obtained from pressing the grapes of Pinot Noir and Chardonnay suggests that the method enables objective real-time discrimination of press fractions based on precise determination of their content.

Boulton *et al.* [1996] 1999: 68; Cloete 2017; Considine and Considine 1982: 2,150; Halliday and Johnson [1992] 2013: 114–19; Joshi *et al.* 2017: 378; Kerslake *et al.* 2018; Krebiehl 2023; Ring 2023; Speller [1994] 2015a: 357; Zoecklein 2002: 5–9

franciacorta.wine/en
maisons-champagne.com/en
www.sawis.co.za

wirehooding

The winemaking operation of securing the cork in the bottle neck using a wire cage rather than string, the invention of which (at the end of the first half of the 19th century) is credited to Adolphe Jacquesson. The first wire cages similar to those

Photo 56. A wire cage tested for efficiency at Adoria Vineyards (in southwestern Poland), run by the firm's owner and manager Mike Whitney. Wire cages such as this are used to secure corks placed in the necks of bottles containing traditional-method sparkling wine.

Photo: author, August 2023

used today, also known as *muselets*, began to be manufactured for the purpose of securing the cork in the last quarter of the 19th century. Nowadays, the *muselet* typically consists of three parts: the lower ring made of soft wire, twisted into a loop; the body, made up of four twisted strands of wire, sometimes also referred to as legs; and the circular metal cap (called *plaque de muselet*) attached to the top of the cage. The loop can be easily untwisted by hand to access the cork and open the bottle.

Placing a metal cap between the top of the cork and the upper girdle of the wire cage serves the purpose of balancing the physical forces acting on the cork while protecting it from damage that could be caused by direct contact between the cork and the wire.

CIVC 2023b; Halley 2005: 419; Liem 2017: 65, 299; MacNeil 2019; Phillips 2019; SL 2018/2019; TFB 2016; Tirabassi 2020: 56; Walkey 2014

adoriavineyards.com
www.champagne.fr/en
maisons-champagne.com

WINE AND FOOD

SPARKLING WINE PROFILE

Colours

Sparkling wines, made not only in Europe but on all continents where the climate allows the cultivation of vines, are a result of the choice of grapes combined with various production methods, which gives these wines an array of colours. Taking into account the colours exhibited by effervescent wines, a distinction is (typically) drawn between sparkling white wines, sparkling rosés and sparkling red wines.

Sparkling white wines, which dominate in the sparkling wine category, vary widely in hue, depending — among other factors — on grape variety, the grapes' provenance and acidity, method and manner of pressing, skin contact, the choice of fermentation vessels, length of a wine's contact with yeast lees, and the age of the wine. Hence, many young quality sparkling white wines crafted from Riesling grapes grown in the wine region of Pfalz (southwestern Germany), such as vintage-dated sparkling wine Reichsrat von Buhl Riesling Brut–2019, can be expected to be pale lemon or light yellow with a greenish tinge, while the colour of pale gold can be identified in many Chardonnay-based traditional-method wines, which is the case with the non-vintage wine Terres Secrètes Crémant de Bourgogne Blanc de Blancs Brut, with the Chardonnay grapes sourced from vineyards in the area of Mâconnais (the southern edge of Burgundy). Highly acidic sparkling white wines (with very low pH) tend to be characterized by pale hues. This is the case with Iron Horse Classic Vintage Brut–2019 (Magnum), a quality vintage-dated sparkling wine made in Green Valley, an area known as the foggiest in Sonoma County's Russian River Valley. Light straw in hue, the wine boasts an acidity of 7 g/l, with a pH of 2.99. A brighter (and more vibrant) appearance is presented by Klein Constantia Cap Classique 'Blanc de Blanc'–2020, a wine crafted in Constantia (Western Cape) from Chardonnay grapes. That wine has an acidity of around 6.1 g/l, with a pH of 3.32.

Differences in the colour of sparkling white wines can also be a result of the manner of juice extraction. Pressing whole clusters, a method used in Champagne involving the presence of stems that shield berries from excessive pressure, reduces the contact time of the white juice with (natural) colourants present in the skins, with the juice of an early pressing, considered to be the most valuable for champagnes, being less tainted with colour than the juice of a later pressing. Notably, giving white musts prolonged skin contact, which can overlap with (or even extend beyond) the fermentation process, results in significant changes of colour, producing wines with hues ranging from light orange to dark shades of amber. Pétillant naturel wine Jadwiga Orange Pét-Nat 2019, produced by Winnica Jadwiga, a small wine manufacturer from Wzgórza Trzebnickie (Lower Silesia), was crafted by leaving the must from the Johanniter grapes on the skins for 10 days, resulting in an orange colour of medium intensity. By contrast, the colour of Podere Pradarolo 'Vej' Metodo Classico, Brut–2012, a fully sparkling wine produced in the Italian region Emilia-Romagna from the grapes of Malvasia di Candia Aromatica — described as (deep) amber — is the result of extended skin contact, lasting around 60 days.

The hue of a sparkling white wine is also influenced by the presence of yeast cells in the liquid, both during the initial fermentation process and in the subsequent stages of the wine's development. These cells are capable of absorbing some of the elements responsible for the pink and brown shades in the liquid, which — in the traditional method of sparkling winemaking — are eventually removed with the lees. This enables winemakers in Champagne and other regions to use pale pink base wines made from dark-skinned grapes (such as Pinot Noir and Meunier) in their blends with the intention of producing a traditional-method sparkling white wine, because — following disgorgement — the resultant wine will (in most cases) turn out white.

Photo 57. The sparkling wine producer Ridgeview from East Sussex (southern England) colour-codes their wines with differently coloured metal caps that hold corks in place. Pale pink is used for Ridgeview Fitzrovia Rosé NV, dark red caps for Ridgeview Sparkling Red Reserve–Limited Edition.
Photo: author, March 2024

Finally, wood-fermented base wines added as part of a cuvée are reported to intensify the colour of the final sparkling white wine. Moreover, with time, phenolic compounds in white wines responsible for their colour may brown under the influence of very small amounts of oxygen as the wines mature, thus deepening a final sparkling white wine's colour, an effect that can be further enhanced by a variety of Maillard reaction-associated products. An example is the vintage-dated traditional-method sparkling white wine House of Arras Grand Vintage 2009 made by Ed Carr, with 10% of the must for this wine fermented in French oak barriques. Crafted from 66% Chardonnay and 34%

Pinot Noir, the blend rested in bottles for 12 years before it was disgorged, giving this wine the colour of yellow gold. Bottle-stored for several decades, old and very old champagnes tend to develop amber hues, the intensity of which usually increases with age. Based on reports by esteemed wine critic Essi Avellan, the vast majority of the over-30-year-old vintage-dated Krug champagnes from Marina Olssen's collection tasted in December 2021 at Vinkällaren Grappe in Stockholm, including (among others) Krug Champagne Vintage 1981, Krug Champagne Vintage 1985 and Krug Champagne Vintage 1989, exhibited intense orange-toned hues.

Sparkling rosés, which owe their pink colour either to a (very) brief contact between grape juice and the skins of dark-coloured grapes, or to the blending of red and white base wines before the fermentation that transforms the cuvée into an effervescent wine, can vary enormously in hue and intensity. These sparkling wines are produced in various shades of pink, from barely perceptible to salmon pink to raspberry pink to pale red, with the colour and intensity depending largely on the choice of grape varieties. To attain a desired shade, juices for sparkling rosés extracted from grapes with relatively low pigment contents, such as Meunier, may need a longer skin contact time than those from thick-skinned grapes, such as Monastrell or Syrah.

The same grape variety and different skin contact times will also produce rosé wines characterized by various shades of (orangey) pink. An unusually subtle light amber tint typical of Jean Vesselle Oeil de Perdrix–Brut, a champagne crafted using 100% Pinot Noir grapes, subjected to slow pressing, contrasts with the bright cherry red hue observed in the cava wine Juvé & Camps Pinot Noir Brut Rosé–NV, which is the result of up to 12 hours of contact between the red grape must and the skins of Pinot Noir grapes. Moreover, as with sparkling white wine, the colour of sparkling rosé depends on the wine's pH levels, with low pH reported as having a bleaching effect.

As for blended rosé wines, the degree to which the final sparkling rosé is coloured depends on the decision on the proportion of red wine in the final cuvée, which is the preserve of the cellar master. Still, the decision is often influenced by a wine's style developed over many years, which may be unique to a particular house or region. Numerous houses in Champagne tend to release non-vintage champagnes characterized by the colour of pale pink. A notable example is the rosé champagne Charles Heidsieck Rosé (Brut) Réserve, often described as presenting a powdery pink robe with luminous tones reminiscent of a wild rose.

Like sparkling white wines, sparkling rosés darken with age. Hence, a marked contrast can be observed between relatively young crémants and moderately old bottlings of vintage champagne. While the wine Langlois Crémant de Loire Brut Rosé, made using the grapes of Cabernet Franc and Pinot Noir, which rested in contact with lees for around 18 months, has the colour of pale (salmon) pink with discreet copper highlights, the wine Champagne Charles Heidsieck (Brut) Rosé Millésimé 2012, a blend of 62% Pinot Noir (including 9% red wine) and 38% Chardonnay, disgorged in January 2021 after eight years of contact with lees, boasts a coppery golden robe with glints of tangerine. An even deeper robe has been developed by the 10-year-old champagne Laurent Perrier 'Alexandra' Grande Cuvée Rosé Millésimé

2012. Made from 80% Pinot Noir and 20% Chardonnay, the wine displays a crimson colour, similar to the colour of the flesh of blood oranges.

Sparkling red wines range in colour from (light) ruby in the case of wines offered by European producers, notably those in Italy and England, to various shades of deep dark red or dark garnet in the case of many sparkling red wines crafted in Australia. Colours and hues in sparkling red wines are influenced (to a greater or lesser degree) by many factors, including (among others) the choice of grape variety (or varieties), environmental conditions in the vineyard in the year of the harvest, the wine's acidity, maceration times and temperatures, fermentation method, and maturation and ageing periods.

Sparkling red wines with moderate to high acidity are likely to develop a bright ruby colour, and less acidic red wines tend to take on purple hues. For example, Ridgeview Sparkling Red Reserve–Limited Edition, crafted in East Sussex (southern England) from Pinot Noir, a grape variety considered relatively low on pigments and tannins, especially when compared with Malbec or Syrah, has an acidity of 7.1 g/l (at a pH of 3.19) and a ruby red colour. By contrast, the vintage-dated sparkling red wine Williamson Wines 'Bubbles' Sparkling Malbec 2008, made in Sonoma County (northwestern California) from the Malbec grapes, known for their low to moderate acidity, has a pH of 3.4. The medium pH level along with the thick (highly pigmented) skins of the ripe Malbec grapes may have had a considerable influence on this wine's deep purple red colour, which the producer refers to as a dark garnet.

Photo 58. A glass of G.H. Mumm Champagne 'Cordon Rouge' Brut served during a champagne tasting on the premises of Maison G.H. Mumm in Reims. The robe of this wine, dominated by a golden-yellow hue, is enlivened by gentle streams of fine bubbles.
Photo: author, July 2024

Importantly, the intensity of a sparkling red wine's colour is largely dependent on the choice of practices used in the winery. Oenological studies on the effect of winemaking practices on the colour of sparkling red wines produced from Syrah grapes consistently show that the pre-fermentative cold maceration will lead to a more intense red colour in the target wines and greater colour stability. Interestingly, many winemakers who aim to produce sparkling red wines from Pinot Noir grapes choose to remove the stems prior to the onset of fermentation, one of the reasons being that the stems absorb colour. Incorporation of wood-matured base wines into the final blend adds to the stability of the final wine's colour. Moreover, as sparkling red wines mature, they tend to take on brick-red and amber tones, lightening with time.

Angelosante 2020; Avellan 2022a: 104–6; Barros *et al.* 2022: 3,254; Charnock *et al.* 2022; Field 2021; Forbes [1967] 1989: 297–8; Hale 2023; Liem 2017: 174; Pessey 2002: 43; Robinson and Cheynier [1994] 2015: 259–60; Williams [1994] 2015: 622

charlesheidsieck.com/en
en.poderepradarolo.com
terres-secretes.com/en

von-buhl.de/en
williamsonwines.com
amistavineyards.com
www.bayofshoalswines.com.au
www.bourgogne-wines.com
www.champagne.fr/en
www.champagne-laherte.com
www.champagnejeanvesselle.com/en
www.greenvalley-russianriver.com
houseofarras.com.au
www.ironhorsevineyards.com
www.kleinconstantia.com
www.langmeilwinery.com.au
www.laurent-perrier.com/en
www.mumm.com/en-ww
www.recaredo.com/en
www.ridgeview.co.uk

Mousse

The frothy head forming at the surface of sparkling wine, known as mousse or effervescence, is the most distinctive feature of the sparkling wine category, setting it apart from other wines. Bubbles in a glass initiate as a result of the energy released during the pouring of sparkling wine, causing a weakening of the bonds between the liquid and dissolved carbon dioxide. When the wine is in the glass, continuous chains of bubbles are produced as a consequence of dissolved carbon dioxide diffusing into various bubble nucleation sites. Bubble nucleation centres in a sparkling wine glass often include dust particles from the air or cellulose fibres from the cloth that has been used to dry the glasses. Some sparkling wine glasses, many of which are used by a number of houses in Champagne during tastings, have artificial nucleation sites engraved with lasers. The bubbles rise to the wine's surface from a number of sites, and — as they burst — create a mist of volatile aromatic compounds in the space above the glass, enhancing the sensory impact of the wine during tastings.

When a bottle of sparkling wine is opened, the wine begins to de-gas as a result of Henry's Law, a natural process in which an equilibrium is regained between the carbon dioxide (CO_2) dissolved in the wine and the rapidly decreasing amount of gaseous CO_2 molecules in the headspace — initially of the bottle and then of the glass. Understood as one of the two known mechanisms responsible for gradual loss of carbon dioxide during tastings, effervescence is the release of CO_2 in the form of bubbles (the other mechanism for the release of CO_2 molecules — invisible to the eye — is the diffusion of these molecules directly through the wine's surface). The effervescence process is the effect of an interplay between the sparkling wine properties, carbon dioxide molecules dissolved in the wine, and small air pockets entrapped (in the process of pouring the wine to glasses) in micro-scratches found on the glass wall and microparticles that have adhered to the wall of the glass under electrostatic forces. Gas molecules present in the liquid cluster together, overcoming the nucleation energy barrier and rising to the surface. The results of research conducted by wine scientist Gérard Liger-Belair of Université de Reims Champagne-Ardenne — published in the *Journal of Physical Chemistry* in 2014 — show that around one million bubbles are likely to form in 100 ml of champagne served in a flute under standard tasting conditions.

With the persistence of effervescence in sparkling wine and the small size of bubbles viewed as signs of the wine's quality, a number of methods have been devised to create the finest bubbles possible and to extend the effervescence time. The main factors that have been found to positively influence these parameters in the case of traditional-method wines include: moderate fining in the making of base wines which allows the winemaker to retain in the wine proteins capable of reinforcing the formation of bubbles; low temperatures (around 13°C) during the *prise de mousse* process; prolonged maturation of sparkling wine to guarantee the wine's extended contact with lees; and careful preparation for the tasting, which includes (among other precautions) cooling the wine before serving and pouring it into glasses slowly and gently to avoid unnecessary turbulence of the wine, as this might entrap bubbles of air, causing dissolved CO_2 to escape more rapidly.

The potential for effervescence (understood as the degree of bubble activity) and the development of foam (collar) depends on the pressure (measured in bars or atmospheres) exerted on the wine inside the bottle. Pressure requirements for the category of sparkling wines in the broad sense, related to their carbon dioxide contents, are set at above 2.5 bar (2.47 atm) when measured at a temperature of 20°C. Lightly sparkling wines, often referred to as pét-nats or pétillant naturel wines, typically have pressures in the 2.5–3.5 bar (2.47–3.45 atm) range. In practice, some producers offer ancestral-method wines with slightly lower or slightly higher pressures. Prominent examples include Di Giovanna 'Rufulia' Pét-Nat Catarratto (IGP Terre Siciliane) Metodo Ancestrale–2020, a wine (crafted by Gunther Di Giovanna from the Sicilian grape variety Catarratto) with a pressure of 2.2 bars, and Werkstatt Pétillant Naturel Riesling 2023, a semi-sparkling white wine (crafted by Bridget Mac, using grapes sourced from Mount Gambier, South Australia) with around 4 bars of pressure. Pressures in bottles of sparkling wines proper range from 3.6 bars (3.55 atm) to 6 bars (5.92 atm) or more. Pressures that at the upper end of this scale (or higher) are typically found in bottles of many sparkling wines made using the traditional method, including cavas and corpinnats (Spain), cap classiques (South Africa) and champagnes (France). By contrast, fully sparkling wines produced by the tank method are usually subjected to pressures no higher than 5.5 bars.

Adams 2018; Cheurlin 2021; Cilindre *et al.* 2010: 164–9; Cilindre *et al.* 2021; Gawel [1994] 2015: 509; Hewson 2021; Jackson [2002] 2017: 35–6; Kemp *et al.* 2015: 30–1; Liger-Belair 2014: 3,161–2, [2004] 2022: 219–20; Liger-Belair *et al.* 2008: 2,491, 2,494, 2,497–8; Liger-Belair and Cilindre 2021: 22–40; Polidori *et al.* 2009: 294–5; Robinson and Jordan [1994] 2015: 345; Roth 2010; Teclemariam 2023; WSET 2020: 7, 32

www.champagne.fr/en
www.di-giovanna.com
www.werkstatt.wine

Flavours

Sparkling wine flavours are influenced by a number of factors, both natural and human. A sparkling wine's flavour, understood broadly as the overall sensory impression of that wine's aroma (or bouquet) along with its taste components, depends (among other factors) on the climatic conditions in which the vine has been cultivated, grape varieties, the choice of vessels for the initial fermentation, *assemblage*, the method of transforming base wines (or musts) into effervescent wines, the length of time that the wine has rested in contact with lees, and — finally — the conditions of storage and tasting.

Climatic conditions in which the grapes are grown have a fundamental effect on the flavour character of the resultant wines, with the task of locating the site for wineries and vineyards being entrusted to the most experienced professionals, as in the case of Tony Jordan, tasked with finding a suitable location for Chandon Australia in the mid-1980s, and — later on — appointed lead consultant for locating the sites for both Chandon China (founded in 2013) and Chandon India (2014). Sparkling wines produced in wine regions described as relatively cool, such as those of Tasmania and Yarra Valley, especially when young, can be expected to show fresh floral aromas, citric notes, and flavours similar to those found in tart fruits, characterized by high intensity and balance, features resulting from long and slow ripening of the grapes. A notable example is the traditional-method sparkling wine Kreglinger Brut de Blancs–2017. Made in Tasmania from Chardonnay grapes, following five years in the bottle in contact with lees, the wine presents balanced flavours of lemon, white stone fruits, brioche and white flowers, with fresh hints of hazelnut and white pear.

Sparkling wines made in warmer regions, including those that are young and those that show maturity but have been well preserved, tend to demonstrate flavour profiles dominated by ripe fruit notes, sometimes accompanied by tropical fruit flavours. An example is the single-vineyard traditional-method wine Cruzat Finca Jose Blanco 'Orgánico' Nature. Made from Chardonnay grapes sourced from high-altitude vineyards located in Luján de Cuyo (the southern part of Argentina's wine-producing province of Mendoza), this sparkling wine showcases ripe notes reminiscent of peach, honeysuckle and toasted bread. Grapes grown at high altitudes are reported to result in wines with increased freshness and flavour intensity, as with the traditional-method Cruzat Cuvée Extra Brut, made from the grapes of Pinot Noir and Chardonnay grown in the vineyards of the Tupungato area, located at an altitude of around 1,400 metres. Showing tropical fruit flavours, similar to those of pineapple and passion fruit, this

Photo 59. Flavour profiles of sparkling wines made from classic grapes of Chardonnay, Pinot Noir and Meunier largely depend on the time that these wines have rested in bottles on lees. Wines from the Pinot Noir grapes, as these — grown in the village of Zachowice (southwestern Poland) — can be expected to show raspberry and strawberry flavours following around two years of maturation in contact with lees. Matured for around four to six years, these wines will likely demonstrate notes of truffles. After eight or more years of maturation on lees, sparkling wines from the Pinot Noir grapes tend to display spicy notes of ground pepper.
Photo: author, September 2022

wine has been described as structured and crisp.

Varietal aromas found in Asti-method wines result from highly aromatic grapes used in the production process. The grapey and floral flavours typical of Asti wines made in Piedmont come (predominantly) from linalool, geraniol and nerol, terpenes contributed by Moscato Bianco, a variety used to produce these wines. The significant presence of linalool can explain the fragrance of orange blossom, identified in Tojo 'Melis' Moscato d'Asti DOCG. Floral notes with peach overtones found in the lightly sparkling wines Coppo 'Moncalvina' Moscato d'Asti DOCG (Canelli) and Vietti Moscato d'Asti DOCG–2023 can be attributed to geraniol and nerol, flavour compounds associated with intense sensory features evoking rose petals and ripe peach.

Typical fermentative flavour compounds, which are commonly found in young sparkling wines, include acetate esters and ethyl esters, with their synergistic effect shaping the overall fruity and floral character of these wines (even if — individually — some of these esters are below their natural odour thresholds). Notable among many examples of these flavourants are ethyl hexanoate (often reported as reminiscent of green apple peel and strawberry), ethyl octanoate (carrying aromas similar to those of pineapple and pear), 2-phenethyl acetate (responsible for pleasant floral characteristics with notes of honey), and isoamyl acetate (the presence of which, known to diminish with time, can explain banana notes detected — by the nose and by the palate — not only in some young Charmat-method wines, including proseccos, but also in some young cavas).

Importantly, a number of producers of sparkling wines, including those in Champagne, see the use of wooden vessels for the primary fermentation as an opportunity to add to the flavour complexity of the final blend. Oak influence, which often manifests itself through the fermentative flavours of cream and butter, is reported as evident in many bottlings crafted at AR Lenoble by Antoine Malassagne. Subtle toasty (and spicy) flavours can usually be identified in a number of champagnes made using base wines that rested in oak. Makers of Taittinger 'Comtes de Champagne' Grands Crus Blanc de Blancs–2011 allowed 5% of base wine intended for this champagne to mature for four months in relatively fresh oak barrels to give this wine its characteristic slightly toasty flavour.

Some champagne makers opt for wooden vessels for the maturation of their base wines, as these provide an environment with limited exposure to oxygen, which contributes to further development of flavours. One can appreciate the results of such an approach by tasting champagnes made by Laurent Champs of Vilmart & Cie from Rilly-la-Montagne (several miles south of Reims). A standard period of around 10 months in oak for all of Laurent Champs' wines now available (Champagne Vilmart & Cie 'Grand Cellier,' Champagne Vilmart & Cie 'Coeur de Cuvée'–2016, Champagne Vilmart & Cie 'Grand Réserve' and Champagne Vilmart & Cie 'Grand Cellier d'Or'–2019, among others) has afforded each of his champagnes a rare balance between richness, intensity and freshness.

Maturation on lees, during which the wine is protected from light, combines the processes of yeast autolysis and slow oxidation, imparting to sparkling wines an array of tertiary flavours. These flavours can differ markedly depending on

Photo 60. Small wooden barrels at Winnica Turnau (in the Polish part of Pomerania, northwestern Poland) containing base wines that are held in reserve until they are needed to add nuance to a particular blend. So far, the company's winemakers have been focused mainly on unoaked styles.
Photo: author, June 2022

the length of maturation and are widely regarded as indicators of quality, especially for traditional-method wines, often increasing the potential for matching these wines with foods. The age-based differences between sparkling wines made by the traditional method has been revealed through molecular studies on aromatic affinities between wine and food, conducted by François Chartier since 2002. The results of these studies demonstrate the aromatic profiles of cava wine categories.

The youngest wines in the cava range, called Cava de Guarda, required to rest at least nine months in the bottle in contact with lees, demonstrate a range of dominant flavours, the main ones including acacia (resulting from the presence of methyl p-tolyl molecules), apple (caused primarily by methyl 2-methylpropanoate) and apple bloom (caused by the presence of amyl acetate). Among various aromatic molecules also typical of young cavas are: anethole (carrying the aroma of anise), linoleic acid (expressing herbaceous notes), ethyl hexanoate (showing fruity notes), linalool (responsible for the aromas of lavender and orange blossom), phenethyl alcohol (reminiscent of roses), palmitoleic acid (contributing notes of walnuts), ascorbic acid (adding the aroma of green apple), myristoleic acid (evoking butter) and methyl dihydro-jasmonate (exhibiting floral notes, most notably jasmine).

Stored in cellars with little light for a minimum of 18 months, wines categorized as Cava Reserva are characterized by a greater depth and complexity than those from the Cava de Guarda segment, with the dominant role in shaping the flavour profile of these wines played by γ-decalactone (a contributor of peach and coconut aromas), methyl propanal (exhibiting floral and spiced aromas), decanoic acid (responsible for toasted notes) and acetanisole (a compound with an anise-like aroma, sometimes described as nutty and evoking vanilla). Flavour compounds that can be viewed as complementing the profile of reserve cavas include (but are by no means limited to): isovaleraldehyde (showing malt flavours), benzyl mercaptan (exhibiting an aroma reminiscent of coffee), ethyl myristate (a compound characterized by a pleasantly sweet fragrance similar to that found in violets) and palmitic acid (desirable in sparkling wines for its creamy flavour).

After having rested in the bottle on lees for more than 30 months, as is the case with a great number of non-vintage champagnes (typically allowed to mature for between two and three years) and all Spanish traditional-method wines labelled 'Cava Gran Reserva', sparkling wines tend to retain some aromatics found in younger wines while also exhibiting a considerable amount of

Maillard reaction-related compounds. Flavourants typical of wines from this age group are dimethyl pyrazines (presenting toasted aromas), 2-acetylthiazole (displaying aromas of roasted hazelnuts), furfuryl mercaptan and benzyl mercaptan (showing aromas of roasted coffee), β-damascenone (demonstrating notes of baked apple), amyl propanoate (exhibiting an aroma similar to that of apricots), tetracosanoic acid (a molecule also found in peanut oil), palmitoleic, linoleic and oleic acids (compounds also detected in walnuts), and sotolon (3-hydroxy-4,5 -dimethyl-2(5H)-furanone) (an odorant lactone imparting powerful fragrance notes, reminiscent of curry, fenugreek and maple syrup). Among other flavour compounds listed in the molecular profile prepared by François Chartier for cavas that have rested in bottles for at least 30 months before their disgorgement and release are: ethyl furan 2-carboxylate (an aromatic contributing balsamic notes), 2-octanone (displaying notes evoking cheese), 2-acetylfuran (responsible for flavours reminiscent of cocoa and caramel), 5-methyl-furfural (exhibiting notes of caramel and coffee), octanoic acid (showing notes associated with coconut milk), myristoleic acid (showing buttery notes) and ethyl decanoate (an aromatic responsible for a fragrance of brandy and cooked apples).

Sparkling wines subjected to prolonged maturation in contact with lees tend to retain some of the features found in wines maturing for slightly shorter periods. These flavour features include (among others) the aromas of cocoa (ascribed to the presence of 2-acetylfuran) and caramel (detectable in long-aged sparkling wines containing 5-methyl-furfural). Some old sparkling wines exhibit vibrancy and freshness despite a significant loss of esters and terpenes during the second fermentation and maturation on lees. This freshness results from the lees' capacity to protect wines against oxidation, an ability that diminishes over time. Traditional-method sparkling wines that have rested on lees for the longest periods can be expected to show the greatest complexity of flavours. Wines that exhibit these characteristics include (among others) some quality non-vintage champagnes, all vintage champagnes (required to rest in contact with lees for at least three years) and cavas labelled 'Cava de Paraje Calificado' (a category of cavas that have matured on yeast lees for a minimum of 36 months). Dominant flavours in these wines often include those of very ripe fruit, cooked apples, toasted nuts, dried figs and apricots, as well as hints of aromatic herbs and truffles. The molecular base for this flavour profile typically consists (among other flavourants) of marked concentrations of 2-furanmethanethiol (showing notes of roasted coffee), ethyl-3-mercaptopropionate (an aromatic compound responsible for the aroma of cooked apples and animal flavours) and benzenemethanethiol (displaying notes of burnt wood or gun flint). Other flavourants that are often present in old quality sparkling wines include: benzyl mercaptan and furfuryl mercaptan (carrying notes of coffee), ethyl propionate and isobutyl acetate (contributing to flavour notes of figs), δ-decalactone (known for an intense peach flavour), heliotropin (likely to add creaminess and notes of black pepper), ethyl formate (contributing the aroma of rum and raspberries) and dimethyl sulphide (exhibiting notes of black truffle).

A Maillard reaction is a broad category of biochemical events initiated through interactions between reducing sugars and amino acids or proteins (partly formed during the sparkling wine's extended maturation on lees). Particularly welcome during the elaboration of traditional-method wines, such as cavas, corpinnats and champagnes (among others), these interactions can take place both during sparkling wine's prolonged maturation in contact with lees and during a long period of post-disgorgement ageing. Among a large number of compounds associated with these interactions, which in sparkling wines occur at low to moderate temperatures over long periods (usually at least several years), are thiols, thiazoles, furans and pyrazines. These are responsible for the rich bouquets of (very) old traditional-method sparkling wines, including flavours of bread, (toasted) nuts, roasted coffee and caramel.

The flavour complexity of old sparkling wines can be fully appreciated when tasting Llopart Original 1887 'Les Flandes' Brut Nature–2011, a vintage-dated single-estate corpinnat. Described by its authors as showing a combination of vibrancy, subtle freshness and depth, this wine — which has rested on lees for more than nine years — boasts rich flavours of ripe white fruit and honey, with notes of brioche, butter and nuts. Another example is the Pinot Noir-focused wine Schramsberg Reserve Late Disgorged–2004, made from California grapes (84% Pinot Noir and 16% Chardonnay) sourced from vineyards of Sonoma County, Mendocino, Napa Valley and Marin. Bottled in May 2005 and disgorged in October 2022, this multi-regional traditional-method wine has gained a unique complexity, boasting a rich bouquet of flavours dominated by Bosc pears and hazelnuts, while also — as described by Schramsberg's winemakers Sean Thompson, Jessica Koga and Hugh Davies — exhibiting tones of honey, candied walnut, brown sugar and vanilla against the background of sumptuous notes evoking lemon peel, baked orange, dried apricots, apple crumble, ginger biscuits, almond paste and graham crackers. Many of these flavour components can be linked to compounds associated with the Maillard reaction, formed in sparkling wine as a result of its prolonged contact with lees.

Cork and crown closures used during the production process have been found to differ in their impact on the resultant sparkling wine's sensory characteristics. Results of the most recent oenological analyses of sensory features of traditional-method wines (such as cap classiques, champagnes and cavas) produced by the second fermentation and maturation under corks and traditional-method wines produced by the second fermentation and maturation under crown caps demonstrate that the use of corks instead of crown cap closures during the second fermentation and post-fermentation contact with lees can bring about a sensorial change in the final sparkling wine, which tends to increase with the maturation period. Data generated so far indicate that — compared with traditional-method wines that have matured under crown caps — cork-matured traditional-method wines tend to show a relatively greater concentration of esters (responsible for fruity and floral flavours), with autolytic features present but less pronounced. The processes of second fermentation and maturation under cork are known to not only contribute to the final wine's greater complexity of flavours but also prolong its aftertaste, a feature associated with the smaller size of bubbles, greater foam stability, and increased bubble retention times at tastings, with dissolved carbon dioxide released more slowly.

Two other factors that may have an influence on the flavour profile of a sparkling wine after its launch, contributing much to the tasting experience, include storage and serving methods. Sparkling wines that have been released in clear-glass bottles need careful protection against light to prevent the transformation of the wine's amino acids into malodorous sulphides, such as dimethyl sulphide and dimethyl disulphide, the presence of which spoils the tasting experience. At low concentrations, they dull fruity flavours; at high levels, these compounds are responsible for disagreeable odours, which are often likened to those of cooked cauliflower, rotten cabbage, dam cardboard and sauerkraut. Brown bottles offer the most effective protection against light-strike, thus promoting the preservation and development of the wine's desirable flavour characteristics. Finally, pouring sparkling wines in a way that promotes slow release of dissolved carbon dioxide into the headspace of the glass has been reported to prolong the sensual experience of the wine, as bubbles ascending in the

Photo 61. Ever since 2010 at Maison Ruinart, entire batches of the house's vintage-dated *blanc de blancs* champagnes have been matured in bottles under natural cork rather than crown closures, a technique proven to result in enhanced flavour complexity and depth.

Photo: author, July 2024

liquid and bursting at the wine's surface play an important part in flavour release, helping diffuse aromatic compounds above the wine's surface over a long period of time, and thus propelling them into the taster's nose and mouth.

An increased complexity of flavours, part of which could be attributed to the second fermentation and prolonged lees contact under cork, can be fully appreciated in wines that have remained in contact with lees and cork for sufficiently long periods. One example is Champagne Telmont Blanc de Blancs 'Vinothèque' OR 2005. Crafted by the cellar master Bertrand Lhôpital from Chardonnay grapes, this champagne was disgorged in 2021. Its release marks the fruit of a long-standing commitment by the Telmont owners to launch their champagnes in opaque bottles to protect them from overexposure to light once they have left the producer's dark cellars.

Avellan 2023b; Campo *et al.* 2008; Charnock *et al.* 2022; Costantini *et al.* 2017; Cravero 2023; Dublin 2016; Escudero *et al.* 2000: 337–9; Gabbardo *et al.* 2016; Hall 2021: 151; Heelan 2015; Jackson [2002] 2009: 373; Jolly *et al.* 2021: 136–52; Kemp *et al.* 2015: 28–9; Kermode 2023; Kirkland 2016; Lhôpital 2022; Liem 2017: 65–8, 280–2; Pàmies 2021; Polidori *et al.* 2009; Pons-Mercadé *et al.* 2021; Rand [2018] 2022: 227–33; Riu-Aumatelle *et al.* 2006: 237, 241; Riu-Aumatelle *et al.* 2013: 51–9; Schmitt 2020; Smart *et al.* [1994] 2015: 346–7; Soto 2020; Stevenson 2019: 44–5; Ubeda *et al.* 2019: 554–9; Welke *et al.* 2022: 112–21, 124–38; WSET 2020: 64

bodegacruzat.com/en
champagne-telmont.com
francoischartier.ca/en
tojovini.it/en
www.astidocg.it/en
www.cadirajo.it/en
www.cava.wine/en
www.champagnevilmart.fr/en
www.chandon.com
www.coppo.com/en
www.kreglingerwineestates.com
www.llopart.com/en
www.ruinart.com/en-int
www.schramsberg.com
www.taittinger.com/en
www.vietti.com/en

Body

The body of sparkling wine, also referred to as its weight, can be numbered among its most fundamental structural features. In terms of mouthfeel, understood as the sensation of fullness experienced by tasters on their palates resulting from a wine's density and viscosity, sparkling wines vary widely, from light-bodied to medium-bodied to full-bodied. The main contributors to a sparkling wine's body include: alcoholic strength; varieties of grapes used; climate in which the grapes have grown; the wine's final sugar content; the use of skin contact and wood in the production process; and the length of time the wine has been allowed to rest in contact with lees, which may have an impact on the perceived intensity of the wine's sensory attributes. The most viscous wines are those with high alcohol concentrations and high levels of residual sugar. A wine from grapes grown in a cool climate is usually characterized by a leaner body than a warm-climate wine of the same variety.

Alcohol content and maturation times in contact with lees are viewed as key determinants of the weight of a sparkling wine; as such, the category of light-bodied sparkling wines are generally viewed as those with relatively low alcohol content, typically between around 5.5% abv and around 11.5% abv, with some rare bottlings classified as light-bodied reaching almost

Photo 62. The majority of champagnes offered by Boizel are medium-bodied sparkling wines, matured in bottles on lees for three to five years. This vintage car (a Renault Juvaquatre), seen here in the courtyard of Maison Boizel in Épernay, was formerly used regularly to deliver these wines to customers. Today, Boizel employees use this car only occasionally, mainly for deliveries to local restaurants in Épernay.

Photo: author, July 2024

12.5% abv. Importantly, these are mostly wines with little lees contact. Hence, Asti-method and Charmat-method wines tend to be lighter than wines crafted using the traditional method, not least because extended maturation on lees is considered obligatory in the production of the latter. Allowing sparkling wines with low alcohol content to mature on lees for a period from several months to a few years usually results in rounder and richer mouthfeel, qualifying such wines for the upper end of the light-weight category. A notable example is Marchiori 'Integrale' Sui Lieviti Valdobbiadene Prosecco Superiore–Brut Nature 2020, a moderately structured blend crafted from the grapes of Perera, Bianchetta, Verdiso, Glera Lunga and Glera Tonda, made effervescent by refermentation in the bottle and released with a natural sediment and an alcohol content of 11% abv. The choice of grape varieties also influences a sparkling wine's body, albeit to a limited extent. Traditional-method sparkling white wines made from white grapes, called *blanc de blancs*, tend to be lighter than those crafted using exclusively red grape varieties.

Alcohol concentrations in medium-bodied sparkling wines can be expected to be (at least a little) higher than in most light-weight sparkling wines, with a typical range between around 11.5% abv and around 13.5% abv. However, some of the sparkling wines classified as medium-weight, notably sparkling ice wines, have much lower alcohol concentrations, in some cases even around 10% abv, a feature compensated by their robust sweetness. Alcohol content at the lower end of (or even slightly below) the scale for sparkling wines classified as medium-weight are also in many cases offset by the significant complexity of these wines, gained during various stages of their production. These stages include skin contact before the primary fermentation (not least in the case of macerated rosés), the use of wooden vessels for primary fermentation, maturation of base wine in wood, and prolonged lees contact following the second fermentation, facilitating the release of polysaccharides (mainly mannoproteins) and amino acids into the wine, a highly desirable process enhancing the wine's mouthfeel. An example of a medium-weight sparkling wine boasting good complexity gained through extended contact with lees is the prestige cuvée Kreinbacher Égoïste 2017, made from the grapes of Furmint and Chardonnay sourced from Somló, a wine region in northwestern Hungary. Crafted from base wine of which 11% is fermented and matured in wood, this traditional-method medium-bodied wine has an alcohol content of 12% abv and 6 g/l sugar, boasting a rich palette of flavours combined with moderately firm structure, developed as a result of resting in contact with lees for five years.

The category of full-bodied sparkling wines tends to be viewed as composed of sparkling wines with solid and creamy texture, relatively high alcohol content (usually ranging from around 12.5% abv to around 14.5% abv), great complexity, and high flavour intensity, sometimes accompanied by moderate to very high concentrations of sugar. An illustrative example is the vintage-dated wine Schramsberg Crémant Demi-Sec 2019, made from 55% Flora grapes, 26% Chardonnay and 19% Pinot Noir from Napa Valley, with 21% of the base wine intended for the final blend fermented in wood. This sparkling wine, which has rested on lees for two-and-a-half years, has an alcohol content of 13.1% abv, and boasts a firm structure supporting a rich array of fruity and spicy flavours, with a tangy finish. Levels of dry extract, which may also add to the way a wine feels in the mouth, are usually highest in sparkling red wines, owing to their relatively high phenolic contents, which — when combined with wood contact, late disgorgement and high alcohol concentrations — results in very rich mouthfeels. Examples include traditional-method sparkling red wines from Victoria's two winemaking regions: Grampians and Rutherglen. Made by Best Wine's senior winemaker Justin Purser, Best's Great Western Sparkling Shiraz 2019, which has developed its complexity of flavours — dominated by spicy and plummy characters — while resting on lees in bottles for over two years, boasts an alcohol content of 14.5% abv. Made from 99% Shiraz and 1% Viognier, the base blend for Campbells of Rutherglen 'Sparkling Shiraz' NV matured in relatively new French oak for two years and rested in contact with lees in bottles for nine months before disgorgement, resulting in a highly complex wine. With an alcohol content of 14.5%, this wine combines lifted notes of primary fruits and spicy characters with a savoury finish, intensified by dosage liqueur containing a small amount of port-style fortified wine made from the Shiraz grapes, offering hints of mocha and liquorice.

Bell 2015b; Gawel *et al.* 2000: 203–7; Izquierdo-Llopart and Saurina 2019; Paissoni *et al.* 2023;

Schuster 2016; Speller [1994] 2015c: 647–8; Vidal *et al.* 2004: 519–24, Webb [1994] 2015b: 864; WSET 2020: 108–13

bestswines.com
www.campbellswines.com.au
www.kreinbacher.hu
www.marchioriwines.com/en
www.prosecco.it/en
www.schramsberg.com

Sugar

Sparkling wines are made in a number of styles determined by degrees of sweetness. The presence of sugar in sparkling wines can either result from arresting sparkling fermentations before dryness or from addition of dosage liqueur (as is the case with the production of double-fermentation wines, notably traditional-method sparkling wines, including champagnes). One can expect to find the lowest concentrations of sugar in traditional-method wines with no sweetening agents added with dosage liqueur following disgorgement. These sparkling wines, with their sugar contents no higher than 3 g/l, are standardly labelled *brut nature*, but they are also occasionally referred to by other designations, such as *brut zéro*, *dosage zéro*, *non-dosé*, *pas dosé*, *ultra brut* or *zero dosage*. Noteworthy among the most recently released traditional-method sparkling wines with no added sugar are two champagnes: Laurent-Perrier Blanc de Blancs–Brut Nature, a non-vintage wine launched in 2019, and vintage-dated champagne Louis Roederer et Philippe Starck Rosé–Brut Nature 2015, disgorged in spring 2022.

Sparkling wines labelled *extra brut* can be expected to have sugar content equal to or only slightly higher than *non-dosé* sparkling wines, with the amount of sugar allowed in the final wine of this category being in the 0–6 g/l range. An example is the traditional-method sparkling wine House of Arras E.J. Carr Late Disgorged 2007, made from 78% Chardonnay and 22% Pinot Noir grown in cold-climate vineyards in Tasmania and released — following 14 years on lees — with a light dosage of around 2.6 g/l. Another wine with this range of sweetness is Parigot Crémant de Bourgogne 'Le Lumes' 2015, made from 100% Pinot Noir grapes, and released — after resting on lees for three years — with a sugar content of 4 g/l.

Laurent-Perrier Ultra Brut, a wine with no added sugar, was released by the house of Laurent-Perrier for the first time in 1981, and the company's owners continue to use the *ultra brut* name for their zero-dosage wines. The concept of *ultra brut* champagnes at Laurent-Perrier can be traced back to the firm's once popular range known as Grand Vins Sans Sucre. The first wines of this range — intended for the UK market — were launched in London in 1889, at a time when the company was run by Mathilde Emilie-Perrier.

Wines with sugar concentrations of up to 12 g/l are referred to as *brut*, this category allowing a tolerance of up to plus 3 grams of sugar per litre. Sparkling wines from this category remain (by far) the most common among all high-quality sparkling wines, which finds confirmation in annual champagne sales figures. In the period 2020–2023, annual sales of non-vintage *brut* champagnes alone to markets outside France consistently accounted for over 75% of total champagne export sales. Among the most globally recognized sparkling wines classified as *brut* are: Champagne Fleury Blanc de Noir Brut, released with a sugar content of 4.9 g/l, Champagne Bollinger PN Brut AYC 18, containing around 6 g/l of sugar, and Roederer Estate Brut Rosé, with a sugar content of around 8 g/l. Notable quality sparkling wines at the upper end of the *brut* scale include Jansz Tasmania Premium Cuvée, a non-vintage wine with a sugar content of 8.6 g/l, and Pongrácz Méthode Cap Classique Brut, which has 9.06 g/l of sugar. Ridgeview 'Fitzrovia' Rosé is an English sparkling wine with a sugar concentration of 11.2 g/l.

Sparkling wines classified as 'extra dry' (or *extra sec*) have sugar concentrations ranging from 12 g/l to 17 g/l, with a tolerance of up to plus 3 grams of sugar per litre. An extra-dry wine at the lower end of this scale is Alma 4 Bonarda–2017. Made from dark-skinned Bonarda grapes, grown in Valle de Uco (Uco Valley), in the western part of Argentina's wine-producing province of Mendoza, this traditional-method sparkling red wine has a sugar concentration of 14.54 g/l. Among the most acclaimed wines from the *extra sec* category are many Asolo Prosecco wines from the province of Treviso (northeastern Italy). Prominent examples include Giusti Asolo Prosecco Superiore DOCG Extra Dry, with a sugar content of around 15 g/l, and Montelliana '57' Asolo Prosecco Superiore DOCG Extra Dry, which contains around 16 g/l of sugar.

Photo 63. Graham Beck Estate's vineyard in the Robertson area, in the Breede River Valley (South Africa's Western Cape), is the main source of the Pinot Noir and Chardonnay grapes used for Graham Beck 'Bliss Nectar' Rosé. Based on fruit picked in 2022, this non-vintage *demi-sec* cap classique boasts a high sugar content of around 39.5 g/l and a low pH of around 2.95.
Photo: courtesy of Graham Beck Estate, February 2017

It is common for sparkling red wines, especially those produced in Australia, to have over 15 g/l of residual sugar. A number of makers of bottle-fermented and tank-fermented red wines terminate secondary fermentations early to leave appropriate amounts of sugar to balance the effects of tannins and high alcohol content sensed on the palate. A notable example is Taltarni Sparkling Shiraz 2019, with an alcohol concentration of 14.5% abv. Made from base wine kept on lees for five days before pressing, the resultant sparkling wine has been released with a sugar content of around 20 g/l.

In the category 'dry' (or *sec*) are sparkling wines that contain between 17 g/l and 32 g/l of sugar, with a tolerance of up to plus 3 grams of sugar per litre. A notable example at the lower end of this category of sweetness is the vintage-dated wine Leitz 'Dragonstone' Riesling Sekt 2017, made in the winemaking region of Rheingau, in the southwest of Germany. By contrast, Vieni Rosso VQA Ontario, a non-vintage Charmat-method sparkling red wine made from the grapes of Cabernet Franc and Chambourcin grown in Ontario (the southernmost province of

Canada), has been released with around 30 g/l of residual sugar.

Sparkling wines in the '*demi-sec*' category, with sugar content in the range of 32–50 g/l, include the multi-varietal composition Schramsberg Crémant Demi-Sec 2011, made from 85% Flora, 9% Pinot Noir and 6% Chardonnay. Poured on 25 September 2015 during a state dinner hosted by President Barack Obama and held in honour of President Xi Jinping of China, this vintage-dated wine from California contains around 35 g/l of sugar. Sweeter examples are Llopart 'Nèctar Terrenal' 2019, a vintage-dated corpinnat made from 50% Xarello and 50% Parellada, and released with 39 g/l of sugar, and Billecart-Salmon Demi-Sec, a Meunier-dominated champagne (40% Meunier, 30% Chardonnay and 30% Pinot Noir) with a sugar content of 40 g/l. Among wines at the upper end of the range is champagne Veuve Clicquot Demi-Sec, containing 45 g/l of sugar.

The sweetest sparkling wines are classified as *doux* or 'sweet'. These designations are used to refer to wines with sugar concentrations of over 50 g/l. An example is Champagne Doyard Cuvée 'La Libertine'. Released with around 65 g/l of sugar, this wine has been crafted with the intention of replicating the style of champagnes with elevated sweetness, common in the 18th century. Higher sugar contents can be found in Brachetto d'Acqui wines from Piedmont (northwestern Italy), produced from 100% Brachetto grapes. Examples include Tosti 1820 Brachetto d'Acqui DOCG–Dolce and Banfi 'Rosa Regale' Brachetto d'Acqui 2023, with sugar contents of 90 g/l and around 110 g/l respectively. At the far end of the sweetness scale are sparkling ice wines made in Canada, notably those from the grapes of Vidal Blanc. Inniskillin Sparkling Vidal Icewine VQA Niagara Peninsula–2022 has a sugar content of 214 g/l. The sugar concentration in Vieni Sparkling Vidal Icewine VQA Vinemount Ridge–2014 is even higher at 299 g/l.

Earl 2021; Harding 2015; Hemingway 2022; Karlsson and Karlsson [2015] 2016; Kemp *et al.* 2017; Liem 2017: 230; Louch 2023; MacQuitty 2022; Mowery 2023c; OIV 2022: I.1.4–4; Ryan 2020: 437; Schmidt 2024; Stefanowicz 2023; Stelzer 2018: 28–30, 66; Stevenson 2008: 44; WSET 2020: 13, 64–8, 111–12

alma4.com
www.asoloprosecco.com
www.banfi.it/eng
www.brachettodacqui.com/en
www.champagne.fr/en
champagnedoyard.fr
www.champagne-billecart.fr/en
www.champagne-bollinger.com
www.champagne-fleury.fr/en
www.cremant-parigot.com/en
www.giustiwine.com
www.grahambeck.com
houseofarras.com.au
www.inniskillin.com
www.jansz.com.au
www.laurent-perrier.com/en
www.leitz-wein.de
www.llopart.com/en
www.louis-roederer.com
maisons-champagne.com/en
www.montelliana.com
www.pongracz.co.za
www.ridgeview.co.uk
taltarni.com.au
www.tosti.it/en
vieni.ca
wilsondaniels.com

Table 4. Annual exports of non-vintage *brut* champagnes against all champagne shipments outside France by volume and by value for the period 2021–2023, expressed as percentages

Sources: www.champagne.fr/en; maisons-champagne.com/en

Export years	Share of NV *bruts* in champagne exports by volume	Share of NV *bruts* in champagne exports by value
2020	78.5%	64.9%
2021	76.5%	63.8%
2022	75.4%	61.0%
2023	76.0%	60.8%

Acidity

Acidity is regarded as a core quality of sparkling wines, with typical levels ranging from medium to high. The final acid concentration in a sparkling wine depends on a number of factors, the most important being the choice of grape varieties and the climate in which the grapes are grown. Grapes grown in cooler climates tend to yield wines with naturally higher acidity. Given the marked differences in dynamic climatic conditions in the world's areas of sparkling wine production, acid concentrations vary. The typical range is between around 5 g/l and around 11 g/l of total acidity (expressed as tartaric acid), with a pH scale of 2.9–3.4. Crucial for balancing sweetness, acid concentrations tend to be the highest in wines from grapes known for their high acidity, including (among others) Chenin Blanc, Sauvignon Blanc, Riesling and Chambourcin. At the far end of the spectrum is McGregor Vineyard Sparkling Riesling 2019, made using Riesling grapes grown in the region of Finger Lakes, located in the west-central part of the state of New York. Crisp on the palate, this vintage-dated traditional-method sparkling wine boasts an acidity of 10.4 g/l, with a pH of 2.77. Another example is Binah Rosé Sparkling Wine 'Aura' Pennsylvania 2021, a kosher traditional-method wine produced at the boutique urban wine-producing house Binah Winery, founded by the winemaker Kevin Danna in 2019 and based in Allentown (eastern Pennsylvania). Produced in the *extra sec* style from Chambourcin grapes, this vintage-dated wine has a vibrant acidity of 11.2 g/l, with a pH of 3.11.

Production-related factors conducive to preserving sparkling wines' acidity and maintaining their freshness are regarded as crucial in warm-climate areas and desirable during particularly warm growing seasons in cooler wine regions, such as Germany's Mosel Valley and France's Champagne and Loire Valley. These factors include decisions to harvest early, vinifying grapes without malolactic conversion, and blending base wines from grapes from the most recent harvest with higher-acid reserve wines, thus allowing the winemaker to compensate for natural acid deficiencies. Methods applied in the production of double-fermentation sparkling wines include dosage reductions and the storage of reserve wines in magnums. As a result of such practices, wines from Chenin Blanc grapes sourced from the sunny Swan District in Western Australia can show acid concentrations on a par with, almost as high as, or even slightly higher than acidity levels in wines from grapes grown in areas with cooler climate influences. Crafted from grapes grown in the heart of Western Australia's Swan Valley, Sittella Chenin Blanc NV Brut — based on the 2022 vintage and blended with reserve wines from the previous two vintages — has an acidity of 7.15 g/l, with a pH of 3.35. Similarly, the vintage-dated wine Nikola Estate Regional Sparkling Chenin Blanc 2022, made in the same region, using early-harvested grapes, has an acidity of 7.3, with a pH of 3.2. By comparison, the grapes of Chenin Blanc grown in Golden Mile Bench, the subregion of the Okanagan Valley in British Columbia (southeastern Canada), where the sun hides below Mount Kobau from the late afternoon, retain their natural acidity, giving the final wines a bright and fresh taste. Made from the fruit sourced from the Golden Mile Bench area, the traditional-method sparkling wine Road 13 Sparkling Chenin Blanc–2018 has an acidity of 7.7, with a pH of 2.98. The acidity range for the Loire Valley wines is only slightly different. The vintage-dated wine François

Photo 64. Grapes sourced from Ridgeview's estate vines in East Sussex, England, classified as a cool-climate region, produce high-acid wines. Acid concentrations in traditional-method sparkling wines available in the Ridgeview portfolio range from around 7 g/l to slightly below 8 g/l, with typical pH values between 2.9 and 3.05.
Photo: author, March 2024

Pinon Vouvray Brut 2012, crafted using the high-acid grapes of Chenin Blanc, has an acidity of 6.9 g/l (expressed as sulphuric acid), with a pH of 3.27.

Examples of wines from regions classified as relatively cool — increasingly affected by high temperatures in the growing season, diminishing acid concentrations in grapes — in which freshness has been preserved by skilful practices in the winery include champagnes of AR Lenoble. Antoine Malassagne, the house's co-owner and general manager, is among the proponents of making champagnes using partial or no malolactic conversion to preserve high acidity. The non-vintage wine A.R. Lenoble 'Intense' mag 19 contains base wine from grapes collected in 2019 and a proportion of reserve wines matured in magnum bottles under natural cork, which further helps preserve freshness. Made by the Moorilla Estate's winemaker Conor van der Reest from 83% Chardonnay and 17% Pinot Noir sourced from St Matthias Vineyard, located in the relatively cool area of Tamar Valley (in northern Tasmania), the vintage-dated wine Moorilla 'Muse' Extra Brut 2017 has a bright natural acidity. Retained by blocking malolactic conversion in the base wine and reductive handling, the wine's high acidity has been balanced by rich autolytic notes resulting from four years of lees contact. Similarly, the limited-release sparkling wine Clover Hill 'Cuvée Prestige' Late Disgorged Blanc de Blancs–Brut 2008, made from Chardonnay grapes — grown in cool frost-free maritime climate in the Pipers River area (in northeastern Tasmania) — has been marketed with a high acidity of 9 g/l, and a low pH of 3.13. The great natural acidity, largely preserved by the use of partial malolactic conversion, offers a desirable counterweight for the sugar content of 6 g/l and autolytic flavours resulting from the over-10-year period in which the wine rested on lees in bottles prior to disgorgement and release.

Ahmed 2024; Caputo 2019; Hodson 2020a; Karlsson and Karlsson [2015] 2016; LeBeau 2019; Puga *et al.* 2022; Robinson 2014; Schmitt 2024; Shaw 2018; Stevenson, 2012: 149–53; Taylor [1994] 2015: 811; Zoecklein 2002: 5

binahwinery.com
cloverhillwines.com.au
francois-pinon-vouvray.com
lenoble-champagne.com/en
mcgregorwinery.com
moorilla.com.au
nikolaestate.com.au
www.ridgeview.co.uk
www.road13vineyards.com
www.sittella.com.au

Alcohol

Sparkling wines reveal a wide spectrum of alcohol content, from low (below or equal to 11% abv) to moderate (in the range of around 11% to almost 14% abv) to high (at least 14% abv). The lowest alcohol concentrations can be found in single-fermentation sweet wines with special status, such as Moscato d'Asti and Brachetto d'Acqui from Piedmont (north-western Italy). Among illustrious examples are Pelissero Moscato d'Asti DOCG–2023, released with around 5.5% abv, Alice Bel Colle Brachetto d'Acqui DOCG, with an alcohol content of around 5.5% abv, and Tosti 1820's Brachetto d'Acqui DOCG–Dolce, containing 6.5% abv. At the upper end of the range of low-alcohol sparkling wines one can find dry styles of fully sparkling Asti wines (also known as Asti Spumante), many Lambrusco wines, and some proseccos. Duchessa Lia Asti Secco, made in the Piedmont region in Santo Stefano Belbo, has an alcohol content of 10.5% abv.

The majority of sparkling wines are characterized by moderate alcoholic strength. At the lower end of the medium scale (from around 11% to almost 12% abv) are some Lambrusco wines, most proseccos (and other Charmat-method wines) and some crémants. The sparkling wine Charles de Fère's Crémant du Jura 'Grands Crêts'–Brut has an alcohol content of 11.5% abv, balanced by fresh acidity and 9 g/l of sugar. The average alcohol content in champagnes is in the range of 12% to 12.5% abv. A typical example is provided by the vintage-dated champagne Dom Ruinart Blanc de Blancs 2010, which matured on lees for more than nine years. This champagne has been released with an alcohol concentration of 12.5% abv. In California, traditional-method sparkling wines with alcohol content higher than 13% are not uncommon. Examples include: Iron Horse 'Wedding Cuvée' 2019 from Sonoma County's Russian River Valley, containing 13.5% abv, balanced by 6.9 g/l of sugar, an acidity of 7.3 g/l and a pH of 3.04; and Decoy Brut Cuvée, a traditional-method wine of California with an alcohol content of 13.5%, 8.5 g/l of sugar, an acid concentration of 7.3 g/l and a pH of 3.12.

The category of high-alcohol sparkling wines is represented mainly by traditional-method sparkling red wines produced in Australia. A notable example is Teusner Sparkling Shiraz MC 2018 from Barossa Valley of South Australia. Made from 100% Shiraz, matured for two years in seasoned French oak, followed by another 30 months on lees in glass, this wine boasts an alcohol content of 14% abv. Taltarni Sparkling Shiraz 2019, made from 97% Shiraz, 1% Mataro, 1% Cabernet Sauvignon and 1% Viognier sourced from the Pyrenees region (Victoria), has an alcohol content of 14.5% abv, balanced by 20 g/l of sugar. Likewise, made from 100% Shiraz, harvested in the McLaren Vale region (in South Australia), the full-bodied vintage-dated Mollydooker 'Miss Molly' Sparkling Shiraz 2021 has been released with an alcohol content of 15% abv.

Cole 2021: 12; Liem 2017: 9–11, 47, 224–5; Stelzer 2018: 61–2, 66; WA 2022: 14, 18; WSET 2020: 23, 57, 72, 85, 100, 115; Zoecklein 2002: 16, 23

Photo 65. Large, wheeled casks and barrels in the vaults of Schloss Gobelsburg (Kamptal, Lower Austria) are filled with base wines, some of which are destined for traditional-method sparkling wines. The alcohol content in these wines typically ranges from around 11.5% abv and 12% abv.
Photo: courtesy of Schloss Gobelsburg (Michael Moosbrugger), June 2021

www.astidocg.it/en
www.brachettodacqui.com/en
www.cantinaalicebc.it/en
www.charlesdefere.com/en
www.decoywines.com
www.duchessalia.it/en
www.gobelsburg.at/en
www.ironhorsevineyards.com
mollydookerwines.com.au
www.pelissero.com/en
www.ruinart.com/en-int
taltarni.com.au
teusner.com.au
www.tosti.it/en

Tannins

The presence of tannins in sparkling wines ranges from very low to medium, with the smallest amounts found in sparkling white wines and the greatest relative concentrations of these polyphenols (derived, *inter alia*, from wood, stems of grapes and grape skins) detected in sparkling red wines. Given that the extraction of tannins in the production of traditional-method sparkling white wines, such as champagnes and cavas, tends to be low, winemakers seeking to improve the wine's structure (which is believed to also enhance its potential for maturation and ageing) enrich their wine with wood tannins. Notably, the presence of these tannins in some high-quality finished sparkling white wines results from the maturation of base wines in wood. An example is Taittinger 'Comtes de Champagne' Grands Crus Blanc de Blancs 2011, in the production of which 5% of base wines was allowed to mature in fresh oak barrels for four months, thus adding not only to the final wine's flavour richness but also to its structure. At the other end of wood-matured champagnes is Billecart-Salmon 'Sous Bois' Brut: base wines intended for this champagne were vinified entirely in wooden casks. The resultant structure of this wine, further improved as a result of its resting in bottles in contact with lees for almost seven years, has thus gained great ageing potential. Champagne Billecart-Salmon's chief winemaker Florent Nys estimates that this age-worthy wine has a cellaring potential of up to eight years.

Sparkling white wines with copper and orange hues derived from prolonged contact between the juice and the skins of white grapes are more tannic than standard sparkling white wines. One illustrative example is Cantina Orsogna's 'Lunaria Ancestrale' Malvasia Orange Spumante–Brut Nature, made in the Abruzzo region in the central-eastern part of Italy. With skin contact lasting between two and three days, tannins were extracted from the skins of Malvasia grapes, giving the wine a solid structure, similar to that found in some red wines. The tannic content as well as the resultant structure of this ancestral-method wine, further enhanced by lees contact, contrasts with the brightness and brilliance typical of quality sparkling wines, such as corpinnats and franciacortas, made using conventional methods.

Sparkling rosés tend to have limited amounts of tannins as a result of temperature-controlled contact between the must and the skins of red grapes. Depending on the technique used to impart colour, the resultant wines can differ in tannin content, ranging from delicate and refreshing to moderately structured and fruity in the case of blended rosés, and from bright and intense to vinous and complex in the case of macerated rosés. A notable example among delicate and refreshing sparkling rosé wines crafted using the blending technique is Domaine Bousquet Charmat Organic Brut Rosé, made using the tank method from the grapes of Pinot Noir and Chardonnay, grown in high-altitude vineyards in the Tupungato area (western Argentina). Palmer & Co Rosé Solera is an elegant champagne with hints of tannins that are fine-grained in texture. A classic blend (composed of Chardonnay, Pinot Noir and Meunier), this sparkling wine, which has rested on lees for the period of three to five years, contains 8% of over-35-year-old perpetual reserve of red wines from Pinot Noir, stored first in oak casks and then in stainless steel. Notable among macerated sparkling rosés showing good structure with hints of tannin is Champagne Serveaux Fils Rosé de Saignée Brut NV, a wine made exclusively from Meunier grapes, characterized by a long vinous finish.

At the upper end of the range for tannin content are sparkling red wines. Noteworthy examples include two traditional-method wines crafted at Williamson Wines in California. These are Williamson Wines 'Bliss' Sparkling Shiraz NV from Dry Creek Valley in Sonoma Count, and Williamson Wines 'Bubbles' Sparkling Malbec 2022 from Sonoma County's Alexander Valley. The respective grape varieties used for these wines, Shiraz and Malbec, are known to be particularly high in tannins. In addition, base wines intended for these sparkling red wines rested in French oak (for two years in the case of Shiraz and for one year for Malbec) before their conversion into sparkling wines, and their subsequent maturation in bottles in contact with lees.

Hodson 2020b; Köhler 2022; Magyarics 2021; Rose 2021: 28–30; Salpico 2021; Stelzer 2018: 27–8, 60–2; Stevenson 2012: 155–6; Taittinger 2020; White 2009; WSET 2020: 4

www.biocantinaorsogna.it
www.champagne-billecart.fr/en
www.champagne-palmer.fr/en
www.champagne-serveaux.com/en
domainebousquet.com/en
www.taittinger.com/en
williamsonwines.com

TASTING AND PAIRING PRINCIPLES

Tasting principles

Wine experts recommend that sparkling wines be chilled to temperatures required for the type of wine being served. The usual serving temperatures for young sparkling wines (such as most proseccos, standard cavas or young crémants) is 6–9°C. When serving wines qualified as mature and complex (including old cavas and vintage champagnes), the wine's temperature ought to be slightly warmer, typically in the range of 10–12°C. This temperature range is also regarded as suitable for sparkling wines with significant sweetness, most notably those categorized as *demi-sec* and sweet, with sugar contents higher than 32 g/l. Sommeliers tend to recommend a similar temperature scale for serving full-bodied sparkling red wines, including sparkling shiraz produced in Australia, commonly regarded as the benchmark for wines of this style. The vast majority of producers suggest serving these wines at temperatures ranging from slightly below 10°C to a little over 12°C. Notably, temperature of pouring ought to be slightly lower than that prescribed for consumption. Wine scientists have discovered that low temperatures help sparkling wines preserve their effervescence when pouring. Because carbon dioxide is more soluble at low temperatures, cold liquids, including sparkling wine, can better retain their dissolved CO_2.

There are several methods for lowering the temperature of sparkling wine to make it suitable for serving. The one preferred by the majority of sommeliers is to chill it in an ice bucket, with the unopened bottle placed in a bucket and covered with ice cubes. The bucket is then filled halfway with cold water. This chilling technique is reportedly over twice as quick as chilling in the fridge, and much safer for preserving a sparkling wine's qualities than cooling in the freezer.

Bottles of sparkling wine sealed with cork stoppers ought to be opened with care. Since chilled bottles can be damp, making a firm grip difficult, wine waiters use a clean towel or cloth to dry the bottle before opening. On removing the foil from the bottle neck, some sommeliers, such as William Edwards, beverage director at the New York City restaurant Manhatta, drape a serviette over the cork and cage. Then the wire cage is removed by half-turning the circular tab six times with the wrist. With one hand holding the bottle securely at the base, and the thumb of the other hand placed firmly on top of the cork, one should tilt the bottle at an angle, making sure that no one is standing in the direct line of the cork, and then twist the bottle gently

Photo 66. Dom Ruinart Blanc de Blancs 2010 poured from the bottle into a glass during a champagne tasting at Maison Ruinart in Reims.
Photo: author, July 2024

until the cork loosens and comes out of the neck. The ascent of the cork, caused by the pressure inside the bottle, can be controlled by continued slow twisting of the bottle.

Some of the most experienced sommeliers can also open bottles of champagne using the blunt edge of a sabre. Using this method, the sommelier chills the whole bottle, ideally to around 3°C, removes the foil and wire cage, holds the bottle firmly in one hand by placing the thumb inside the punt at the base, and tilts the bottle slightly away from themselves and everyone around them, with the cork still pointing upwards. Then the wine waiter typically runs their sabre up and down the seam of the bottle neck several times — keeping their elbow high — before finally sliding the sabre forcefully along the seam towards the cork in one swift motion, striking the bottom lip of the bottle, and thus removing the cork together with the glass lip. At this stage, some sommeliers will sacrifice a very small amount of wine, allowing it to flow, to ensure that potential glass particles are expelled from the neck. It is also not uncommon for wine waiters to inspect the first glass of champagne poured to ensure that there are no glass shards floating in the wine or on its surface.

While decantation can be beneficial for very mature sparkling wines, known for their complexity, it is usually not recommended for very old quality sparkling wines, which have matured and aged in various conditions for over 15 years. With the bubbles softened as a result of these wines' long rest in bottles, the main objective is to preserve whatever freshness they have retained and protect their delicacy. Hugo Bensimon, sommelier and wine director at Grill 23 & Bar in Boston, advises that the oldest champagnes should be allowed to reveal their features slowly for an optimal tasting experience. Thomas Calder, an export agent and wine connoisseur, observes that decanting mature quality sparkling wine may be the most practical when serving to a large group, as in such a case it will be consumed quickly. However, when a bottle is shared between two or three people, decanting may miss the point, as this wine will have sufficient time to breathe.

Most sparkling wines are poured into glasses directly from the bottle. Decanting is regarded as optional for pétillant naturel wines with sediment. With such wines it is performed mainly for aesthetic reasons. The decantation process is considered beneficial with very mature traditional-method sparkling wines with great complexity, such as 10- to 15-year-old champagnes, including vintage champagnes. Some sommeliers also view the process as desirable with macerated rosés with a significant degree of vinosity. Allowing such wines to breathe can bring forward their taste and perfume while softening their bubbles. Among the most vocal proponents of decanting sparkling wines for the purpose of revealing their complexity of flavours are Florent Nys, *chef de cave* at Champagne Billecart-Salmon, and Cédric Bouchard, owner of small-scale champagne house Roses de Jeanne. Because the flavours of sparkling wines tend to unfold more quickly than those of still wines, decanting — if justified — is usually performed no more than 30 minutes before serving.

The practice of opening bottles of champagne with the blunt edge of a sabre can be traced back to the Napoleonic era, when cavalry officers serving in the French Army would slice the corks off the bottles with their sabres. With time, the practice gained popularity among the French aristocracy. Today, this practice, known by the French word *sabrage*, has many ambassadors. Among the most important are members of Confrérie du Sabre d'Or. This association of wine professionals, founded in France in 1986 by Jean-Claude Jalloux, is dedicated to promoting the sparkling wine of Champagne and the art of *sabrage*.

It is common practice among sommeliers to chill the carafe (or decanter) in the refrigerator prior to decantation so that it is the same temperature as the wine. The practical objective behind this practice is to extend the duration of effervescence. With the same aim in mind, it is often recommended that sparkling wines should be poured gently down the side of the decanter (or carafe). Before pouring sparkling wine into glasses (either from the bottle or from the decanter or carafe), it is recommended to carefully consider their choice.
Flutes — tall, narrow glasses, tapered at the rim, designed to showcase sparkling wine's bubbles and its freshness — are viewed as suitable for toasts at large banquets, at which the host will want the sparkling wines (often young and simple) to remain bubbly for as long as possible. Nowadays, a growing number of wine professionals favour standard (tulip-shaped) wine glasses when the aim is to maximize the appreciation of the wine being served — such as at tastings or gourmet dinners, considering that quality sparkling wines, such as champagne or cava, are considered valuable companions to food.

Wine professionals wishing to pour sparkling wine with minimal loss of effervescence and to prolong the evolution of bubbles from sparkling wines at tastings pay attention to several factors. One of their key concerns is adequate rinsing and drying of glassware to remove soap and detergent residues, as these notorious contaminant surfactants will otherwise coat nucleation sites, thus hindering the formation of bubbles. Another noteworthy factor is the duration of effervescence, which is likely to extend when sparkling wine is poured into glasses slowly in a manner similar to the way that beer is poured into mugs. Oenologists studying champagnes — using infrared imaging — have confirmed that the least CO_2 is lost in the pouring process when the wine is poured against the side of the glass, especially when the glass is held at a 45° angle.

Filling glasses more than half-full ought to be avoided. The aim is for the sparkling wine to have sufficient space for volatile

Photo 67. A sparkling wine stopper for opened bottles, equipped with a built-in activated carbon filter. The filter has been designed to prevent the oxidation process and preserve the wine's organoleptic features.
Photo: author, October 2024

flavour compounds to accumulate above its surface. Moreover, since wine warms up in the glass relatively quickly, it is advisable to pour only 100–150 ml of wine per glass at a time, being ready to refill each glass as often as required, so that the sparkling wines being served can be appreciated at optimum temperatures.

Bottles with leftover sparkling wine can be resealed with especially designed sparkling wine stoppers, which have become increasingly available. They prevent the remaining pressure from escaping the bottle and keep oxygen out, maintaining the wine's effervescence potential and preserving its signature flavours. Airtight seals are also available that prevent spills if the bottle is stored on its side. Storing an opened bottle of champagne or other sparkling wine in the refrigerator keeps the pressure low and allows them to retain their dissolved CO_2 and stay fresh for up to several days. This may be extended to around a week by using stoppers with built-in activated carbon filters.

Cavaleri 2019; Chandra 2022; Cravero 2023; Dara 2022; Denig 2016; Douglas 2016a, 2016b; Dupont 2023b; Gallagher 2023; Gellie 2016; Jackson [2002] 2009: 309–10; Juhlin 2024; Liem 2017: 200–1; Liger-Belair [2004] 2022: 219, 221–2; Liger-Belair *et al.* 2024; McKay 2022; Robinson [1983] 2000: 73; Saladino 2023; Schiessl 2016; Walters 2022: 81–3; Wilson *et al.* [1994] 2015: 788–91; Zielinski 2022

maisons-champagne.com/en
sabra.ge/sabrage
brockenchack.nz
www.champagne.fr/en
www.confreriedusabredorusa.com
eightatthegate.com.au
www.thomascalderselections.com

Pairing principles

Perceived for centuries primarily as celebratory beverages, sparkling wines have for the past several decades been increasingly appreciated for their versatility as accompaniments to a wide variety of dishes. Many producers of high-quality sparkling wines, such as Schramsberg, Veuve Clicquot or Taittinger, offer practical advice on pairing their wines with foods. There are three fundamental principles that underpin most pairing recommendations. These rules, taken together or separately, are: common origins and traditions, similar mouthfeels and textures, and affinity of flavours.

The idea of pairing regional wines with dishes from the same area has a great number of supporters among sommeliers. Matthew Whitaker, sommelier and founder of Woodthorpe Wines, suggests that most English sparkling white wines could pair well with the traditional English dish of (beer-battered) fish and chips. Another illustrative example is the classic combination of champagne and oysters, immortalized in the famous painting *Le Déjeuner d'Huîtres* ('Oyster Lunch'), created in 1735 by Jean-François de Troy, now in the collection of Musée Condé.

The principle of similar mouthfeels relies on finding an optimal match between the weight of food as sensed on the palate, resulting (among other factors) from the fat and protein content of the dish and the methods of heat treatment used for its preparation, and the body of sparkling wine, usually enhanced by the presence of alcohol, tannin and sugar. It is viewed as crucial to match food and wine according to weight in such a way that neither one feels dominant. For instance, a simple prosecco would be too light-bodied to be paired with classic lasagne. Rich in fat and protein, this staple of the Italian cuisine could benefit from the synergistic effect of being served with well-structured, intense and acidic rosé franciacortas. Two prominent examples are Montina Franciacorta Rosé Extra Brut, a macerated wine from 85% Pinot Noir and 15% Chardonnay, matured on lees for almost three years, with an alcohol content of 12% abv, an acidity of 6.3 g/l and a low pH of 3.07; and Lo Sparviere Franciacorta Rosé 'Monique', made from 100% Pinot Noir, characterized by a solid body, matured on lees for more than two years, and with an alcohol content of 12.5% abv, an acidity of 6 g/l and a pH of 3.18.

Publication of technical sheets on wine producers' websites is now an increasingly common practice, an attempt to respond to consumers' quests for ever more precise information. The champagne producer Krug offers post-sales information service linking the ID number provided on the back label of every bottle, which can be keyed in on the producer's website, with detailed information related to the bottle's content and suitable pairings. For instance, entering the six-digit code 111011 provides access to a wealth of information on Krug Grande Cuvée 160-ème Édition–Brut, including the history of the range that the blend represents, the wine's composition, facts about its production and serving suggestions. Interestingly, along with recommendations about pairings, the producer offers suggestions for musical accompaniments to the tasting, an outcome of Krug's cooperation with renowned artists who have created compositions known as 'Krug Echoes'. Among these are violinist Iwao Furusawa (whose piece was inspired by the Krug Grande Cuvée 162ème Édition), the Italian pianist Dario Faini and the Hong Kong-born harmonicist Cy Leo (both of whom have created a different 'Echo' for Krug Grande Cuvée 167ème Édition).

Finally, wine and food professionals distinguish between two main pairing methods based on the so-called affinity of flavours: congruent pairing and contrasting (complementary) pairing. A congruent pairing relies on many flavourants shared between sparkling wine and food, while complementary pairings will rely on few shared flavourants, with flavour harmonies created by contrasting flavours. Versatile (congenial) flavours, such as sweet, fat and salt, can be paired not only with one another but also with less adaptive flavours, such as spicy, bitter and acidic. When some of these maladaptive attributes are found in a sparkling wine or a dish, they tend to be matched with versatile flavours.

Congruent pairings have a loyal following among North Americans and western Europeans, as attested by shared flavour compounds identified in one of the most celebrated food–beverage pairings in the

Photo 68. Some professionals point to the benefits of pairing sparkling wine with radish. This choice is largely due to the crispy and slightly spicy character of both, with the radish usually served with butter and *fleur de sel*. Among the proponents of this combination are Didier Depond, head of champagne houses Delamotte and Salon since 1997, and Christian Holthausen, a strategic consultant and international communications specialist in the wine trade with 25 years of experience in the champagne industry.
Photo: author, August 2023

history of gastronomy: champagne and oysters. Researchers from the University of Copenhagen have recently succeeded in explaining the widespread appreciation of this combination by the synergistic enhancement of the umami taste when partnering European oysters, which contain free glutamate and are particularly rich in nucleotides (inosinate and guanylate), with champagnes with very long lees contact times, containing marked amounts of free glutamate. Notable champagnes recommended for serving with oysters include: Champagne Duval-Leroy 'Femme de Champagne' Brut Nature 2002, which has rested on lees for 20 years, and Champagne Duval-Leroy 'Femme de Champagne' Grand Cru–Brut, NV, released following over 14 years of lees contact. If the level of acidity is considered a key reference point when matching a wine with a dish, experts tend to pair acidic dishes with sparkling wines with acid concentrations equal to or slightly higher than those of the dish. Paired with lower-acidity wines, acidic foods with high intensity, such as salads dressed with a balsamic vinaigrette or a lemon-based sauce, will often make these wines taste flat. Likewise, when creating a congruent pairing between a sweet-tasting dessert and sparkling wines with considerable sweetness, care is usually taken to ensure that the wine is at least as sweet (or intense) as the dessert. For instance, Pavlova — a light sweet dish consisting of a meringue and fresh fruit — can make a congruent pairing with both Brachetto d'Acqui wines and *demi-sec* champagnes.

Contrasting pairings include a number of matches, both tried and tested and creative. For instance, the crisp acidity associated with *crémants de Loire* provides a highly desirable contrast to the fatty and umami flavours detectable in an oily fish, fried onion rings or calamari. Another example is provided by *demi-sec* sparkling wines with subtle sweetness, such as Schramsberg Crémant Demi-Sec 2020, containing 38 g/l of sugar, served with an array of spicy Asian foods. This wine also features good acidity of 7.5 g/l (with a low pH of 3.13), which manifests itself in flavours of peach, lychee and orange zest, acting as a foil for the fatty and salty notes of *foie gras*, served warm with black pepper and *fleur de sel*.

Beckett 2024; Clarke [2000] 2003: 60–3; Dublin 2023; Hallinan 2023; McNeill 2024; Mercer 2020a, 2020c; Schmidt *et al.* 2020; Sohm and Muhlke 2019: 248–51; Whitaker 2017

www.duval-leroy.store/en
franciacorta.wine/en
www.krug.com
losparviere.it/en
montinafranciacorta.it/en
www.schramsberg.com
www.veuveclicquot.com/en-int

WINE AND FOOD PAIRINGS

Sparkling white wines

Light-bodied

Light-bodied sparkling white wines, containing from around 5.5% abv to around 11.5% abv, and rarely more than 12% abv, are usually partnered with delicate dishes, with similar weight and intensity. Asti wines from the region of Piedmont are light-bodied despite their sweet flavours. Produced from the aromatic grapes of

Moscato Bianco, these wines tend to reveal notes of acacia, wisteria and fresh grapes, with hints of sage and lemon. Apart from flavours found in the majority of Asti wines, the low-alcohol wines Pelissero Moscato d'Asti, with 5.5% abv and over 100 g/l of unfermented sugar, balanced with good acidity, and Michele Chiarlo Moscato d'Asti 'Nivole', with an alcohol content of 5% abv, demonstrate their own distinct features. The former shows notes of thyme and mint, while the latter is characterized by flavours of peach and apricot. Owing to their fruity attributes, Moscato d'Asti wines are usually considered suitable accompaniments to the Italian dessert semifreddo, often made with fresh fruits, such as raspberries or peaches. The tank-method fully sparkling wine Acquesi Asti DOCG, with an alcohol content of 7% abv, displays aromas evoking honeycomb and peaches. This sparkling wine is recommended — also by its makers — as an accompaniment to either light, slightly spicy Asian foods, or dry pastries, fruit tarts and nougats.

Photo 69. A wire cage removed from one of the bottles of Marchiori 'Integrale' Sui Lieviti Valdobbiadene Prosecco Superiore–Brut Nature 2020. Characterized by moderate structure with herbal and apple freshness, this light-bodied wine is sometimes served as an accompaniment to risottos, notably those flavoured with herbs and grated Parmesan.
Photo: author, October 2024

Typical examples of light-bodied sparkling wines are provided by most proseccos and Italy's many (other) Martinotti-method wines. Bortolomiol 'Miòl' Prosecco DOC Treviso Brut–2021, with 11% abv and 8 g/l of sugar, shows flavour notes of apple and tropical fruits, with hints of thyme and oregano. If not served on its own as an aperitif, this sparkling wine can be enjoyed with light fish starters, simple risottos sprinkled with seasonal herbs, or fresh and medium-aged cheeses made in the Treviso province. The wine Marchiori 'Integrale' Sui Lieviti Valdobbiadene Prosecco Superiore–Brut Nature 2020, which owes its structure to cold pre-fermentative maceration, fermentation on skins and refermentation in the bottle, has been released with a natural sediment and an alcohol content of 11% abv. The structure of this wine, together with its decent fixed acidity of 5.5 g/l and fresh fragrances of camomile, dandelion, pear and Reinette du Canada apples, enables wine and food professionals to pair it with heavier and richer risottos.

Slightly higher in alcohol (11.5% abv), and hence also a little weightier, is the traditional-method sparkling wine Casa Valduga 'Arte Tradicional' Brut 2019. Made from 60% Chardonnay and 40% Pinot Noir, sourced from the area of Vale dos Vinhedos (located in the wine-producing region of Serra Gaúcha in southern Brazil), this wine reveals aromas of tropical fruits accompanied by notes of toasted bread resulting from having rested in bottles on lees for 12 months. The moderate structure of this wine, its flavour profile and its lively acidity, with a pH of 3.26, warrant its pairing with the casseroled chicken pie known as *empadão de frango*, filled with shredded chicken and vegetables and topped with a buttery crust.

An interesting example of a light-bodied wine with 11.5% abv and a great pairing potential is Karanika Cuvée Spéciale Extra Brut 2020 made by Laurens Hartman and Annette van Kampen in the area of Amyntaio, in the region of Macedonia (northern Greece). Crafted from 90% Xinomavro and 10% Assyrtiko, and released after 12 months on lees, followed by two months of post-disgorgement bottle ageing, this traditional-method sparkling wine is characterized by discreet saltiness accompanying flavours of acidic fruits and roses. With a refreshing acidity of 7.8 g/l and a very low pH of 2.83, this wine could create a contrasting pairing with many dishes with a strong umami flavour, such as salty mussels with pasta.

Another noteworthy example of a relatively light-bodied wine with good pairing potential is Chateau Chizay Carpathian Sekt Cherzegi Extra Dry. Made from the Cserszegi Fűszeres grapes grown in the region of Zakarpattia in western Ukraine, this Charmat-method sparkling wine, with an alcohol content of 11.5% abv and 18 g/l of sugar, can be described as fruit-forward with refreshing acidity, which justifies pairing it with fresh fruits. This wine is also reported as providing a pleasant contrast with salty and umami flavours associated with seafood, and also as pairing well with *banosh*, a cooked dish made from cream and cornmeal, seasoned with herbs and sprinkled with pieces of fried bacon (or fatback) and crumbs of *brynza*, a salty crumbly cheese made from sheep's milk.

The aromatic grapes of Viognier grown at high altitudes in Valle de Uco (Uco Valley), located in Argentina's winemaking province of Mendoza in close proximity to the Andes, produce wines with high acidity and concentrated flavours. The vintage-dated sparkling wine Alma 4 Valle de Uco Viognier 2018, made using the traditional method without malolactic conversion, has been released after 10 months of maturation in contact with lees with 12% abv and 5 g/l of sugar. With an acidity of 7.5 g/l, expressed by fresh fruit flavours, such as those of pineapple, peach and apricot, and found in floral aromas of honeysuckle and jasmine, this wine also reveals notes of dried apricot and toasted bread. This rich palette of flavours enables wine and food professionals to pair it with some intensely flavoured dishes such as chicken salads served with peach or mango sauce, or dishes dominated by flavours of ginger, saffron or coconut. Given its flavour profile, this sparkling wine can also be expected to pair well with sweet root vegetables, notably cooked carrots, parsnips, sweet potatoes and butternut squash.

An interesting example of light-weight sparkling wine comes from the region of Lower Silesia (southwestern Poland), which has given birth to ancestral-method monovarietal vintage-dated wine Winnica Jadwiga Johanniter Orange Pét-Nat 2019. Crafted from Johanniter grapes, this amber-hued pétillant naturel wine has been obtained by macerating fully ripened fruit for around 10 days, and fermenting the grape must using wild yeasts. With an alcohol level of 11.5% abv, an acidity of 6.2 g/l, and with noticeable tannicity and aromas evoking bread crust, apple pie, apricots and tea leaves, the wine is appreciated for creating optimal pairings with local pork sausages seasoned with hand-cut garlic and herbs.

The vintage-dated cap classique Le Lude Vintage Cuvée 2013 deserves to be named among the most complex sparkling wines in its category. Although this multi-regional blend, produced using 57% Chardonnay, 41% Pinot Noir and 2% Meunier, is relatively light in terms of its alcohol concentration (11.56% abv) and sweetness (5 g/l of sugar), it boasts rich flavours resulting from the use of reserve wines stored in relatively young French oak barrels and its subsequent maturation on lees in magnums for nine years. The flavour profile of this wine is dominated by rich notes of roasted nuts, honey and brioche, while hints of bruised apple and grapefruit express the wine's high acidity of 8.5 g/l and its low pH of 3.17. Highlighted by creamy effervescence, these flavour characteristics enable the wine to be paired with a variety of foods, ranging from simple glutamate-rich dishes based on seafood — such as fresh oysters, shrimps or fried calamari — to traditional South African dishes characterized by greater complexity and richness, for instance Cape Malay *bobotie*, offering a combination of savoury and sweet flavours.

Medium-bodied

At the lowest end of the category of medium-bodied sparkling white wines are sparkling ice wines, which are lower in alcohol concentration than the rest of the medium-bodied wines, but their sugar contents are significantly higher, averaging above 200 grams of sugar per litre. A notable example is Inniskillin Sparkling Vidal VQA Niagara Peninsula–2021, crafted from the grapes of Vidal Blanc by Nikolas Gizuk. With around 10% abv and with a sugar content of 214 g/l balanced by an acidity of 8.1 g/l, this wine reveals flavours of fresh peach, orange and mango, with notes of lychee and pineapple. Wine tasters at Inniskillin recommend teaming their wine with spicy dishes, such as tuna tartare with fresh chilli, while also suggesting that its high acid concentration can help balance the creaminess of blue-veined cheeses as well as the fatty and salty character of *foie gras* (*poutine*). Matt Steeves, an experienced sommelier and respected wine writer, suggests that sweet and acidic flavours of sparkling ice wines made from Vidal grapes at Inniskillin can produce memorable pairings with *crème brûlée*.

An English *brut*-style traditional-method sparkling wine made from Chardonnay grapes, Gusbourne Blanc de Blancs 2019 has an alcohol content of 12% abv, which is relatively low for a sparkling wine with a medium body. Among key factors that have added to this wine's complexity, and hence also its weight, are the fact that a small part of the juice was fermented in old oak barrels, and that it was stored after bottling on lees for over 39 months. Described by its makers as brimming with bright citrus notes of ripe lemon and orange peel, along with hints of green apple, Williams pear and white-fleshed peach, this wine boasts a refreshing acidity of 9 g/l, with a low pH of 2.96. This *blanc de blancs* displays classic autolytic character, manifesting itself on the palate in flavours of toasted nuts, brioche and lemon meringue pie. The wine's zesty characteristics can offer a pleasant contrast with oily fish, such as salmon or trout, as it can with beer-battered cod or haddock served alongside salty chips in classic 'fish and chips'. The wine's toasty and buttery flavours match those of the deep-fried ingredients in many dishes.

Grüner Veltliner, regarded as Austria's greatest asset, and Furmint and Hárslevelű, grape varieties widely grown in Hungary, produce wines that may be similar in weight but are distinct in flavour. The traditional-method sparkling wine Jurtschitsch Grüner Veltliner Große Reserve Brut Nature–2018, matured in Austrian oak for several months and stored in bottles on lees for more than three years, has an alcohol content of 12% abv. It shows mouth-watering acidity (expressed in aromas of lime and apple), with aromas of baked bread, and elegant fresh flavours of black pepper and herbs. These flavours are given a lift when pairing it with herbal dishes, such as those containing tarragon, mint or parsley leaves. Its flavours also allow complementary pairings with the umami flavours found in dishes based on fish or veal. Made from Furmint and Hárslevelű, Sauska Tokaj Extra Brut has an alcohol concentration of 12% abv and an acidity of 7.1 g/l. Stored in bottles in contact with lees for over three years, this traditional-method wine demonstrates intense flavours of peach, dried apricots and linden honey, with hints of tobacco. The flavour profile of this wine is reported as pairing well with spicy Hungarian dishes, a notable of which example is *rakott krumpli*, a layered-potato casserole with slices of hard-boiled eggs smothered in sour cream and chunks of smoked *Gyulai kolbász* sausage, made from pork and lard and seasoned with ground paprika, garlic, pepper and ground caraway seeds.

Made in the *extra brut* style, Roger Goulart Gran Reserva Josep Valls 2018 is a vintage-dated cava with an alcohol content of 12% abv and an acidity of 6 g/l. Made from 35% Chardonnay, 35% Xarel-lo, 15% Macabeo and 15% Parellada, and released following its maturation in bottles on lees for more than 36 months, this wine exhibits the fragrance of fleshy tropical fruits with delicate notes of toasts, cream and pastry. The wine is regarded as an apt companion to spicy tapas, such as *patatas bravas* or *bacallà gratinat*, a dish of baked cod served with *allioli*, a thick Catalan sauce made from garlic, salt and olive oil. Toasty notes found in this wine will amplify the baked flavour notes of the dish, while the wine's juicy and creamy acidity can be expected to balance the rich, strongly flavoured sauce.

Photo 70. Medium-bodied champagnes can make multi-level pairings with various types of cheese. Wine and food experts at the champagne house G.H. Mumm offer multiple pairing suggestions, recommending fresh, salty goat's or sheep's cheeses with their G.H. Mumm RSRV Blanc de Blancs Grand Cru Brut 2013, which displays a creamy texture, with notes of fresh flowers and lemon. When the cheese is served with black cherry jam, G.H. Mumm RSRV Rosé Foujita is recommended, with its vinous character, flavours of brioche, pastry and vanilla, and hints of strawberry coulis.
Photo: Aldona Szafrańska, June 2013

Kreinbacher Égoïste 2017 from Somló (northwestern Hungary) is a medium-bodied effervescent wine with increased complexity and a slightly fuller mouthfeel as a result of 11% of the juice intended for the base wine having been fermented in wood and its period of maturation in bottles on lees having been extended to five years. It exhibits pronounced autolytic flavours with intense notes of citrus fruits. Made from 50% Chardonnay and 50% Furmint, this wine — with 12% abv and 6 g/l of sugar — is often recommended as an accompaniment to various starters and main dishes, including the Hungarian pasta dish *túrós csusza*, which contains curds of cottage cheese and bits of crispy bacon, and *paprikás csirke*, a chicken stew flavoured with onions, sweet peppers, tomatoes, paprika, sour cream and seasonings, and served with pasta.

Noteworthy among medium-bodied sparkling wines are those subjected to prolonged maturation both before and after bottling. One example is the English wine Wiston Estate South Downs Blanc de Noirs 2014. Crafted from 65% Pinot Noir and 35% Meunier, this wine has an alcohol content of 12% abv and 6 grams of sugar per litre. Before being made into sparkling wine by the traditional method, the base wine was allowed to rest in used oak on lees for nine months. Stored in bottles in contact with lees after the second fermentation for more than six years, the wine shows flavours of baked and caramelized apples, toast and Seville oranges, with its weighty texture also exhibiting notes of rhubarb, greengage, dried orange and marmalade, with delicate hints of spice, honey and Braeburn apples on the finish. Great intensity of flavours combined with its high acidity has enabled the producer to recommend this wine as a suitable accompaniment to wild mushroom risotto sprinkled with Parmesan. The wine writer Tamlyn Currin considers it an optimal match for Comté, a nutty-tasting and slightly tangy cheese from the Jura region (eastern France).

Two other medium-bodied traditional-method wines that owe their solid structure and good complexity to resting on lees for extended periods following primary fermentation and also after bottling are exemplified by quality sparkling wines crafted by Austrian producers: (Familienweingut) Tement from Styria (southeastern part of Austria) and Schloss Gobelsburg, located in the wine-growing region of Kamptal in Lower Austria (northeastern part of the country). The non-vintage sparkling wine Tement Große Reserve Ehrenhausen Blanc de Blancs Brut Nature, made from 85% Chardonnay (locally known as Morillon) and 15% Pinot Blanc (also known by the name Weissburgunder), rested in old oak barrels on lees for around 50 months followed by another 30 months of lees contact in bottles. Released with 12% abv, and aromas of lime zest, Granny Smith apple peel and lemon balm, the wine exhibits notes of salted butter and toast. Pairing recommendations offered by the sommelier and wine writer Sebastian Bordthäuser include *scaloppine al limone* (veal fillets with lemon) and swordfish with orange and sage vinaigrette. Notably, each of these dishes contains elements that match the wine's refreshing acidity. Similarly, Schloss Gobelsburg Langenloiser Sekt Extra Brut 2012, made from Grüner Veltliner, Riesling and Pinot Noir, owes much of its rich bouquet to the maturation of base wines in Austrian oak for one year, and the storage of the sparkling wine in bottles on lees for more than 10 years. Vinous and smoky notes found in this medium-weight sparkling wine with 12% abv, combined with its juicy acidity, manifesting itself in flavours of lemon and lime, with noticeable hints of brioche, mineral saltiness, papaya and honey blossom, justify its pairing with white asparagus served warm with hollandaise sauce and freshly ground black pepper.

A traditional-method sparkling wine from Riesling grapes produced by Reichsrat von Buhl in the region of Pfalz (southwestern Germany) and released with an alcohol content of 12.5% abv following around 32 months of maturation in bottles on lees, Reichsrat von Buhl Riesling Brut–2019 has an acidic profile dominated by ripe apples, juicy pears and subtle notes of brioche, with a creamy and slightly piquant finish. This wine is regarded as an exceptional companion to salmon with horseradish or steak tartare with capers. A high-quality sparkling wine with a similar weight and only slightly greater complexity, Valentin Zusslin Crémant d'Alsace Clos Liebenberg 2012 has been made from 90% Auxerrois and 10% Riesling. This wine boasts a rich flavour profile resulting — among other factors — from an extended maturation in bottles on lees for around 36 months. Released with 12% abv, an extremely low sugar content of 0.3 g/l and a pH of 3.01, the wine is described as vibrant and full of mineral freshness, which enables its pairing with an onion tart filled with a savoury custard of eggs, bacon and onions, often topped with grated nutty-tasting Gruyère cheese.

Photo 71. Clay pots in the cellars of the cava producer Vilarnau (located in Valls d'Anoia-Foix, part of Comtats de Barcelona) contain base wines undergoing maturation under the eye of winemaker Eva Plazas Torné. The aim is for the wine to gain richness and texture without additional flavours.
Photo: Marta Ręgowska, April 2023

Weightier still and slightly more complex is Domaine Slapšak Žametna Črnina 'Penina' Brut Nature. Crafted in the Slovenian region Dolenjska, also called Lower Carniola, using Žametna Črnina grapes, this wine has 12.5% abv. It displays complex flavours resulting in large measure from partial malolactic conversion in the base wine and the maturation of the sparkling wine in bottles on lees for 72 months before its disgorgement. The wine reveals aromas of bitter orange, dried apricot and tangerine. It also displays peppery and smoky flavours on the palate, with gentle notes of hazelnuts on the finish. Given the intensity and depth of this wine, it can create congruent pairings with umami flavours detectable in cured meats and iodine-rich sea foods such as cod or oysters.

At the upper end of the range of medium-bodied sparkling wines are those that stand out for their alcohol concentrations as well as the maturity and complexity of flavours. One example is the most recent editions of Viña Edén Méthode Champenoise Brut Nature 3D Sur Lies, crafted by winemaker Marcelo Breganti in Uruguay's coastal wine region Maldonado, and released with an alcohol content ranging between 12.6% abv and 13% abv. Dominated by Chardonnay grapes, with a small but significant proportion of Pinot Noir (usually 15%), these traditional-method wines mature in bottles on lees for around two years. Undisgorged, these sparkling wines are described as fresh with intense apple and pear aromas, and with toasty and yeasty notes and hints of almonds and walnuts on the palate. Wines of this type tend to be viewed as good companions to a broad spectrum of savoury dishes with meat (such as beef or venison) or white fish (such as cod or pollock) as the main ingredient. They are also regarded as suitable for *chipirones*, grilled small-sized squids stuffed with tomatoes and onions, drizzled with olive oil.

Crafted using Chardonnay grapes at the wine-producing company Shabo, located near the city of Odessa (southwestern Ukraine), the sparkling wine Shabo Grand Reserve Brut, with an alcohol content of 13% abv and an acidity of almost 6 g/l, displays mineral freshness combined with aromas of flowers. Matured in French oak barrels for several months, and then in bottles on lees for three years, this wine exhibits flavours evoking creamy muffins, bread crust and vanilla. Wine and food tasters at Shabo recommend pairing this wine's elegant freshness with the umami notes found in oysters, sashimi or classic carpaccio. Given this wine's considerable weight, it can also be considered a suitable match for various greasy, savoury meat jellies, also called aspics.

Full-bodied

The category of full-bodied sparkling white wines is dominated by vintage-dated wines produced by the traditional method using mainly the classic grape varieties of Chardonnay and Pinot Noir. Sparkling wines in this group can be expected to have alcohol concentrations higher than 13% abv, but it also includes wines with alcohol content slightly lower than 13% abv but which are characterized by great complexity and concentration of flavours, predominantly an effect of storage of base wine in wood and particularly long periods of lees contact. Among noteworthy examples of such wines are two vintage champagnes made from Chardonnay grapes: Charles Heidsieck Blanc des Millénaires 1995, released with around 12% abv, 12.2 g/l of sugar and an acidity of 6.88 g/l after more than 15 years of maturation in bottles on lees at a stable temperature of 10°C; and Taittinger 'Comtes de Champagne' Grand Crus Blanc de Blancs 2011, launched with 12.5% abv, 9.8 g/l of sugar and a low pH of 3.12 following around 10 years of maturation in bottles on lees.

Crafted by winemaker Thierry Roset, the Charles Heidsieck champagne boasts a multi-layered structure and a creamy texture, the latter characteristic resulting partly from the malolactic conversion process to which this wine was subjected at an early stage of production. Given its opulent

Photo 72. A glass of Dom Ruinart Blanc de Blancs 2010 at a private tasting at Maison Ruinart in Reims. Made from Chardonnay, vinified with malolactic conversion and matured for more than nine years on lees under cork, this wine has developed a solid texture and a rich palette of flavours, from notes of sandalwood and nutmeg to roasted hazelnuts and coffee, with hints of black tea leaves and a lingering finish marked by freshness. Maison Ruinart's resident chef Valérie Radou and cellar master Frédéric Panaïotis recommend this cuvée as an accompaniment to roast black pork tenderloin or smoked blue lobster.

Photo: author, July 2024

bouquet, composed of buttery aromas found in pastries and intense spicy notes of cinnamon and vanilla, combined with rich flavours of candied apricots, quince, hazelnuts and toasted almonds, this wine pairs well with many dishes, including seafood, poultry and creamy cheeses. Owing to its well-preserved freshness, it is also considered a suitable companion to dishes in which juicy meats feature as the main component, such as steak tartare, rare and medium-rare steaks, or venison roasts cooked using the *sous vide* method.

The makers of Taittinger 'Comtes de Champagne' Grand Crus Blanc de Blancs 2011 allowed 5% of base wine intended for this champagne to mature for several months in relatively fresh oak barrels, with the effect of imparting a slightly toasty flavour. This vintage-dated sparkling wine is usually described as intensely spicy with flavours evoking gingerbread, liquorice, fruit meringues and mandarins, with hints of cherimoya blended with subtle notes of almond pastry. Fine, creamy bubbles of this wine combined with its optimal expression of freshness can be viewed as a foil for the crunchy textures and salty flavours of deep-fried seafoods served with lemon wedges, as in the popular Italian dish *fritto misto di mare*.

Ca' del Bosco Franciacorta 'Annamaria Clementi' 2014 is an Italian multi-vineyard traditional-method sparkling wine made from 76% Chardonnay, 9% Pinot Blanc and 15% Pinot Noir. Fermented in oak casks, base wines remained on their lees for six months, undergoing malolactic conversion. The cuvée was then allowed to rest in bottles in contact with lees for eight years. Made in the *brut nature* style, this wine has been released with an alcohol content of 13% abv, a good acidity of over 6 g/l and a pH of 3.06. The long production process has resulted in warm flavours reminiscent of freshly baked baguette, honey, citrus fruits and green apple, with hints of vanilla. Recommended pairings that could match this wine's intensity while also creating a pleasant contrast with its acidity include rich and fatty foods, such as stewed veal served with sweet caramelized onions or goose served with gravy. This wine should also create a multi-level congruent pairing with the classic dish *canard à l'orange*, or with one of its numerous variations, notably crispy honey-glazed roast duck with orange sauce.

Many of the most intense and full-bodied sparkling white wines are produced in California. The multi-regional wine Schramsberg Reserve Late Disgorged 2004 has been made from 84% Pinot Noir and 16% Chardonnay, with 37% of the must intended for this wine fermented in wood. This wine owes much of its richness and depth to malolactic conversion and 17 years of maturation in bottles on lees. With an alcohol content of 13.1% abv, 6 g/l of residual sugar, an acidity of 7.6 g/l and a low pH of 3.14, the wine presents a rich bouquet, dominated by notes of Bosc pear, hazelnuts, honey, candied walnuts, brown sugar and vanilla. Schramsberg's winemakers Sean Thompson, Jessica Koga and Hugh Davies describe their wine as showing dense and juicy flavours, including those of lemon peel, baked oranges, dried apricots and apple crumble, blended with notes of ginger biscuits and almond paste. Only dishes with good texture and rich and intense flavours will be able to match the mouth-weight profile of this wine. Among dishes recommended by the producer is fresh pasta with truffles, a dish based on the recipe shared by the late Schramsberg Vineyards owner Jamie Davies, containing angel hair pasta, diced shallots sautéed in butter and cream, fresh truffles with a dose of brandy, egg yolks, *foie gras*, salt and pepper, and grated Parmesan cheese. The salty and umami elements of the dish can be expected to act as a pleasant foil for the wine's vibrantly fresh and fruity characteristics.

The champagne producer Veuve Clicquot marked its 250th anniversary by joining efforts with the luxury hotel and train services operator Belmond to offer its customers a unique train journey on board the art déco train Venice Simplon-Orient-Express, running from Reims to Venice, with the event featuring haute cuisine combined with Veuve Clicquot champagnes. Towards the end of the journey, which took place from 5–7 June 2022, the executive chef Jean Imbert served for lunch Vitello 'Venice' Burger, a variation of a veal-based street food dish. It was paired with a glass of multi-vintage blend Veuve Clicquot 'Extra Brut Extra Old' 2 (Magnum). The wine tends to be described as showing a range of flavours reminiscent of nutmeg, milk roll and brioche, combined with roasted notes, which are also typical of the dish.

An illustrative example of a sparkling white wine with a solid structure and mouth weight is provided by Iron Horse Vineyards Russian (River) Cuvée 2018 (Magnum). Made from 72% Pinot Noir and 28% Chardonnay sourced from Green Valley (in Sonoma County's Russian River Valley), this wine has an alcohol concentration of 13.5% abv, a sugar content of 13.4 g/l, an acidity of 7.7 g/l and a pH of 3.1. Slow primary fermentation at a temperature of around 17°C and blocked malolactic conversion enabled the preservation of vibrant fruit esters. The complexity and creaminess of this wine come from its storage in bottles on lees for around three years. The wine is described as exhibiting rich notes of lemon meringue with hints of brioche intertwined with flavours of blood orange zest and spiced poached pear, with a long, persistent finish. Tasters at Iron Horse Vineyards suggest that their sparkling wine will pair very well with Meyer lemon curd with goat cheese pastry.

📖 Currin 2021a, 2021b; Demmond 2023; Denig 2023; Farrow 2021; Goode 2020a; Holthausen 2020; Krebiehl 2024b; Mason 2004: 77; Mason 2024; Moser 2024; Sealy 2024; Sinclair [1998] 2008: 15, 105, 232, 263, 278, 422; Steeves 2020; Todorovic 2020; Wang 2020

💻 www.acquesi.it/en
alma4.com
www.astidocg.it/en
www.belmond.com
www.cadelbosco.com
www.casavalduga.com.br/en
charlesheidsieck.com/en
www.domaine-slapsak.com/en
www.gobelsburg.at/en
www.gusbourne.com
www.inniskillin.com
www.ironhorsevineyards.com
www.jurtschitsch.com
karanika.com
www.kreinbacher.hu
www.lelude.co.za
www.marchioriwines.com/en
marynissen.com
www.michelechiarlo.it/en
www.mumm.com/en-ww
www.pelissero.com/en
www.rogergoulart.com/en
www.ruinart.com/en-int
sauska.hu/en
www.schramsberg.com
shabo.ua/en
www.taittinger.com/en
www.tement.at/en
www.vinaeden.com/en
von-buhl.de/en
www.wistonestate.com
www.zusslin.com/en

Sparkling Rosé Wines

Light-bodied

At the lowest end of the category of light-bodied sparkling rosés are lightly coloured effervescent wines characterized by simplicity, their alcohol contents no higher than around 11% abv. A typical example is Paltrinieri Lambrusco di Sorbara DOC 'Radice', produced in the province of Modena in the Emilia-Romagna region (northern Italy). Produced from the grapes of Lambrusco Sorbara, this sparkling wine demonstrates refreshing aromas of grapefruit and wild strawberries, with hints of wild roses, green apple peel and pomegranate detectable on the palate. This flavour profile has earned this wine the reputation of an excellent companion to *antipasti*. The wine's fruity freshness can create pleasant contrasting pairings with the fatty and umami flavours of *prosciutto di Parma*, raw tuna or thin pieces of smoked salmon.

Another relatively light-bodied sparkling rosé wine is Chateau Chizay Carpathian Sekt Blaufränkisch Rosé Brut from the region of Zakarpattia (western Ukraine). With an alcohol concentration of 11.5% abv, a sugar content of 5.2 g/l and subtle aromas of peonies and roses, with delicate flavours of rose petals and sour cherries, this Charmat-method wine has been reported as creating an excellent congruent pairing with strawberries. Professional tasters at Chateau Chizay suggest that their wine could also pair well with tuna, halibut or grilled shrimps.

Made from the grapes of Xinomavro (98%) and Limniona (2%) in the region of Macedonia (northern Greece) using the traditional method, Karanika Cuvée Rosé Brut 2021 is a moderately structured light-bodied wine with an alcohol content of 11.5%. With a sugar content of 5 g/l, an acidity of 7.46 g/l and a low pH of 2.92, this wine demonstrates good food-pairing potential. Delicate aromas of wild roses and raspberries, gentle yeasty characteristics gained from its storage in bottles on its lees for almost one-and-a-half years, along with its lively, mineral freshness with notes of citrus fruits on the finish, can make this wine a delectable accompaniment to grilled halloumi, a semi-hard cheese of Cypriot origin with a salty, lactic (and slightly minty) flavour, made from cow's, goat's and sheep's milk.

A light-bodied fully sparkling rosé with a little richer structure, Augustí Torelló Mata 'Celler Kripta' Trepat Cava Reserva Brut 2021 has been produced from the grapes of Trepat, macerated for eight hours. Matured in bottles on lees for upwards of 18 months and released with an alcohol concentration of 11.5% abv and 8 grams of sugar per litre, this cava reserva has developed aromas of strawberries and hay, with hints of toasted bread, as well as delicate and fresh flavour notes of citrus fruits, perceptible on the palate. According to both the producer and expert tasters, this wine combines well with foods exhibiting similar flavours, notably those with strawberries as the main ingredient. Fruit-driven, this wine is sometimes also considered an agreeable accompaniment to Iberian ham or the savoury, umami notes found in a number of Japanese dishes, including sushi and sashimi.

The group of light-weight sparkling rosés includes those with alcohol concentrations close to or even a little higher than 12% abv. One such wine is Gouguenheim Rosé de Malbec Extra Brut. Produced from Malbec grapes sourced from high-altitude vineyards of the Tupungato area, the northernmost subregion of Uco Valley (the western part of Argentina's wine-growing province of Mendoza), this wine has a flavour profile dominated by floral and fruity notes, a result of using the standard tank method, which usually entails short contact with lees following the second fermentation process. This Charmat-method wine presents fresh aromas of wild cherries and strawberries, with its sugar content of 9.11 g/l, a pH of 3.3 and an alcohol concentration of 12.1% abv. The wine is typically served either on its own as a refreshing drink on a warm day or with fresh fruit salads and light finger foods, such as savoury canapés or spring rolls.

Another notable example among light-bodied wines with around 12% abv is provided by the French wine Calmel & Joseph Crémant de Limoux–Rosé Brut, made from 60% Chardonnay, 30% Chenin Blanc and 10% Pinot Noir, and released following 15 months of maturation in bottles in contact with lees. Dominated by flavour notes of ripe strawberries and raspberries, and tropical fruits, its profile appears to justify the suggestion offered by the producer's tasters that this fruit-driven wine would not only make a suitable companion to fresh Bavarian cream with strawberries but also make for complementary pairings with salty flavours found in bacon or pretzels.

Photo 73. A light and fruity sparkling wine crafted in Sussex from the grapes of Pinot Noir and Meunier, The Bolney Estate Cuvée Rosé Brut 2018 has been matured in bottles on lees for 18 months. With an alcohol content of around 12.3% abv, 8.6 g/l of sugar and flavours of fresh flowers, red apples and cranberries, this vintage wine is known to match well with fruit-based desserts. Renowned wine writers Christopher and Oliver Walkey recommend pairing the wine with apple cakes.
Photo: author, October 2023

Trius Brut Rosé VQA Niagara Peninsula is a quality traditional-method sparkling wine from the Canadian province of Ontario, produced from the grapes of Pinot Noir, Chardonnay, Meunier and Gamay. Released with 12% abv and 8 g/l of sugar, following almost two years of maturation in bottles on lees, this wine has a refreshing acidity with notes of yeast, fresh bread, apples and citrus fruits, with a lingering finish cleansing the palate. Tasters at Trius Winery suggest that their sparkling rosé can make an optimal match for a range of foods, including vegetarian spring rolls, pan-fried perch or pickerel, roasted tomatoes served with goat cheese on lightly toasted baguette, and classic fondue (made with Emmental and Gruyère).

Medium-bodied

Medium-bodied sparkling rosés include pink-hued sparkling wines with alcohol content ranging from around 12% abv to almost 13% abv. The low alcohol concentrations in some of these wines are often compensated by their structure and concentration of flavourants, notably as a result of production methods, such as the choice of grapes, contact of base wines with wood, use of reserve wines when preparing cuvées, and extended maturation of sparkling wine on lees. An example is Wiston Estate South Downs Rosé NV, made from 40% Pinot Noir, 30% Chardonnay and 30% Meunier, where reserve wines (dating back to 2008) constitute up to 30% of the final blend. Bottled in 2019, this wine has been kept on its lees for around 42 months to allow it to mature and take on both flavours and structure. Released with an alcohol content of 12% abv, 8 g/l of sugar and a refreshing acidity, it is described as bursting with aromatic notes of rhubarb, sour cherries and redcurrants, with hints of wild roses and lemon zest, and a touch of hibiscus on the finish. Tasters at Wiston Estate recommend partnering it with shellfish, strawberries and cream, and lemon meringue pie.

Another quality traditional-method medium-bodied sparkling wine with an alcohol content no higher than 12% abv comes from Styria (southeastern Austria). Crafted using the grapes of Pinot Noir, Chardonnay and Pinot Blanc, Harkamp 'Sol' Rosé Große Reserve 2016 has rested in bottles in contact with lees for around 55 months, developing a solid structure and a rich palette of flavours. Made in the *brut nature* style, this wine shows mineral acidity with flavour notes of grapefruit, brioche, mint and raspberry, and hints of wild cherries on the finish. The structure of this sparkling wine should make it a suitable companion to *Wiener schnitzel*, as well as some sophisticated dishes, such as roast venison with semolina dumplings stuffed with carrots, yellow beetroot, celeriac and crushed coriander seeds, topped with bits of fried smoked bacon.

Composed of around 70% Cabernet Franc and around 30% Pinot Noir, Langlois Crémant de Loire Brut Rosé NV is a traditional-method sparkling wine from the Loire Valley with an alcohol concentration of 12.5% abv, a sugar content of 14 g/l and a good acidity. The wine, which has matured for 18 months in bottles in contact with lees, exhibits an array of floral and fruit-forward characteristics, dominated by notes of raspberries, strawberries, redcurrants, pomegranate, and also sour cherries, with some perceptible hints of creamy yoghurt, vanilla and white pepper. While this sparkling wine is typically recommended as an aperitif, it has also been reported as pairing well with red fruit desserts or roasted green vegetables (such as broccoli or green asparagus), especially when served with butter.

Lo Sparviere Franciacorta Rosé 'Monique', made exclusively from Pinot Noir — fermented partly in stainless steel and partly in oak — is a medium-bodied sparkling wine, matured in bottles on lees for more than two years and released with an alcohol content of 12.5% abv, 5 g/l of residual sugar, an acidity of 6 g/l and a pH of 3.18. With a fine solid creamy structure and flavour notes of rose water and red berry fruits, this wine is recommended by its producer (Azienda Agricola 'Lo Sparviere') as an apt, delectable companion not only to selected red berry fruits and charcuterie but also to rich, creamy vegetable dishes, notably *parmigiana di melanzane*, or main courses of red meat.

Kreglinger Rosé Brut 2018 is a traditional-method sparkling wine with an alcohol

Photo 74. Wine flavours carried by the Pinot Noir grapes evolve with age. The Pinot Noir component of a cuvée often shows notes of strawberries during the first several years of *sur lie* maturation taking place in the bottle. Matured in bottles on lees for between four and six years, Pinot Noir-based cuvées can exhibit notes of truffles, which tend to evolve into spicy flavours of ground black pepper if the wine is kept on lees for a longer period.
Photo: author, June 2020

content of 12.5% abv and 7.5 grams of sugar per litre. All the must intended for this wine was obtained from Pinot Noir grapes grown in northern Tasmania and fermented in used French oak, in which the new base wine was subsequently also allowed to mature for nine months prior to bottling. After its rest in bottles for four years in contact with lees, this sparkling wine boasts an exceptional creamy texture. Described as a quality wine with elegance, natural freshness and flavours of raspberries, strawberries and cherries, accompanied by notes of sweet spices and marzipan, this wine can create delectable multi-level pairings with both fresh sashimi with pickled ginger and smoked salmon on blinis with sour cream.

Full-bodied

Full-bodied sparkling rosé wines can be expected to have alcohol content higher than 13% abv. However, some sparkling rosés classified as full-bodied may have slightly lower alcohol content if they stand out for their high complexity and concentration of flavours. One example is the traditional-method wine Ca' del Bosco Franciacorta 'Annamaria Clementi' Rosé 2013, made from Pinot Noir grapes. Sourced from 33-year-old vines grown in the province of Brescia (northern Italy), the grapes were macerated briefly at 14°C and vinified in small oak barrels. New base wines intended for this wine were allowed to mature in oak for three years before being bottled. Stored in bottles on lees for eight years, this franciacorta has been released with an alcohol concentration of 12.8% abv, 1 g/l of sugar, an acidity of 5.8 g/l and a low pH of 3.03. This bottling boasts great complexity and depth, with flavours of wild strawberries, raspberries and citrus fruits, and hints of violets and pastry. This wine is viewed as a suitable match for sophisticated dishes, for instance pumpkin risotto topped with Robiola cheese and thin layers of duck speck.

Jean-Michele Novelle 'Réserve Personnelle' Rosé de Noirs 2018 is a full-bodied traditional-method sparkling wine made exclusively from Pinot Noir grapes in the region of Genève (westernmost part of Switzerland, close to the French border). Released with 13.5% abv and 20 g/l of sugar, this wine has rested in bottles in contact with lees for around three years. It reveals a balance between sweet and sour flavours, with clear notes of sour cherries, raspberries and grapefruit. Jean-Michele Novelle's tasters suggest that their wine pairs best with *foie gras*, lamb tagine with prunes, and cherry pie. It is sometimes recommended as an accompaniment to toasted Raclette served alongside boiled potatoes and pickles, and *Älplermagronen*, a traditional dish from the Alpine regions of Switzerland, which requires six basic ingredients: potatoes, macaroni, cheese, onions, bacon and apple sauce. The juicy flavour profile of this wine, enhanced by its crisp acidity, can balance the rich and creamy texture of either of these two dishes.

A sparkling wine's ageing potential relies to a great extent on the format of the bottle in which the wine resides. Champagne Billecart-Salmon's chief winemaker Florent Nys suggests that his Billecart-Salmon Brut Réserve could benefit from ageing for up to two years if the wine is in a half-bottle (37.5 cl) and for over two to four years in a standard bottle (75 cl). He further recommends that connoisseurs that have purchased the wine in magnums (1.5-litre bottles) or jéroboams (3-litre bottles) age it for four to eight years.

Another notable example is Iron Horse Brut Rosé 2019, a traditional-method sparkling wine from Sonoma County's Russian River Valley. Crafted by Iron Horse Vineyards' winemaker David Munksgard using 62% Chardonnay and 38% Pinot Noir, and matured in bottles on lees for three-and-a-half years, this wine has been released with a sugar concentration of 5.4 g/l and an acidity of 7.3 g/l, with a low pH of 3.06. The wine shows aromas of ripe raspberries and dark cherries, and flavours of strawberries and limes. Tasters at Iron Horse Vineyards suggest that, balanced as it is, the wine could be served on its own as an aperitif or in a pairing with steak.

Among the most full-bodied sparkling rosé wines is the traditional-method House of Arras Vintage Rosé 2015. Made by Ed Carr from 61.7% Pinot Noir, 37.7% Chardonnay and 0.6% Meunier sourced from Tasmanian cool-climate vineyards, this wine has an alcohol content of 13.6% abv. Notably, 10% of the juice destined for this wine was fermented in new French oak barriques and the wine rested in bottles on lees for around six years. Released with 2.4 g/l of sugar, an acidity of 6.7 g/l and a pH of 3.17, it exhibits lifted aromas of raspberries, Turkish delight and lychee, and also seamlessly integrated flavours described as evoking maraschino cherries, dried blood orange and red berry yoghurt. Expert tasters from House of Arras propose pairing this complex wine with fresh seafoods, poultry and game meats. Its natural freshness has also been reported as offering a welcome contrast to the umami flavours of Flinders Island saltgrass lamb, braised or roasted, often previously marinated in olive oil, white wine and herbs, typically sage and rosemary. The salty and herbal elements of the dish offer a pleasant foil to the fruity components of the wine's bouquet.

Beckett 2015, 2023; Bredahl 2017; D'Agata 2024; Farrer *et al.* 2023; Hunt 2022; Hyland 2019; Johnson 2019; Lewis 2024; May 2021; McCarthy 2017; Poggio 2021; Sinclair [1998] 2008: 27, 269, 359; Walkey and Walkey 2022; Woodard 2023

bolneywineestate.com
agustitorellomata.com/en
www.cadelbosco.com/en
www.calmel-joseph.com
cantinapaltrinieri.it/en
chizay.com/en
gouguenheim.com.ar
www.harkamp.at/en
houseofarras.com.au
www.ironhorsevineyards.com
karanika.com
www.kreglingerwineestates.com
losparviere.it/en
www.wistonestate.com

Sparkling Red Wines

Light-bodied

The category of light-bodied sparkling red wines is dominated by wines made in northern Italy and Canada. An illustrative example is the low-alcohol wine Giacomo Bologna 'Braida' Brachetto d'Acqui DOCG 2023, made in Piedmont (the northwestern part of Italy). This sweet vintage-dated wine, released with an alcohol content of 5.5% abv, over 120 g/l of sugar and a lively acidity of around 7.3 g/l, shows flavours reminiscent of ripe red fruit, violets and roses, hence pairing well with fresh strawberries and redcurrants, and also with a host of fruit-based desserts. It is also widely recommended as an accompaniment to Christmas panettone, jam tarts, hazelnuts, walnuts, pistachios and dried figs. With its fresh fruitiness, this wine is also sometimes considered as a suitable complement to the creamy taste of chocolate soufflés.

Vieni Rosso VQA Ontario is a non-vintage Charmat-method sparkling red wine made from the grapes of Cabernet

Franc and Chambourcin grown in Ontario (the southernmost province of Canada). The wine has been released with 10% abv and around 30 g/l of residual sugar. Fragrant and fruity, with distinct blackberry and mulberry flavours, hints of rhubarb and a touch of dried herbs, this wine could make an interesting pairing with roasted vegetables served with barbecue sauce. The sweet flavour features of this wine will correspond with the smoky-sweet flavour characteristics of the sauce, while balancing its tangy character. When the wine is partnered with barbecue sauce, its bright acidity also serves to cleanse and refresh the palate.

Another example of a light-bodied sparkling red wine is offered by Venturini Baldini, a wine-producing estate located in the province Reggio Emilia in the region Emilia-Romagna (northern Italy). Made from Lambrusco Montericco, the wine Venturini Baldini T.E.R.S. Ancestrale Lambrusco Emilia IGP has been released with 11% abv. Made in the *brut nature* style and stored in bottles in contact with lees for 18 months, this wine shows aromas of blackcurrants, raspberries and rhubarb, with pleasant freshness and well-integrated tannins. Tasters at Venturini Baldini suggest the wine should taste excellent with fried gnocchi and mortadella or tortellini in well-flavoured broth, or egg pasta with meat sauces, salami and Parmesan.

Medium-bodied

The lightest among the medium-bodied sparkling red wines are sparkling ice wines, the production of which remains overwhelmingly the preserve of Canadian winemakers, mainly those from the wine province of Ontario. Canadian sparkling red wines are usually made from the grapes of Cabernet Franc. Noteworthy examples of these wines can be found in the portfolios of Inniskillin Niagara Estate Wines and Vieni Estates. Inniskillin Sparkling Cabernet Franc Icewine VQA Niagara Peninsula can be expected to have an alcohol concentration in the range of 9–10% abv and a sugar content from around 240–290 g/l. The vintage-dated wine Vieni Sparkling Cabernet Icewine VQA Vinemount Ridge Doux 2013, produced using 64% Cabernet Franc and 36% Cabernet Sauvignon, has been released with an alcohol content of 10% abv and 243.1 g/l of sugar. Each of these wines has a crisp acidity, but they differ slightly in their expression of flavours. The profile of the Inniskillin bottlings is dominated by aromas of cherries and strawberries, with flavours of cranberries, raspberries and rhubarb, which can be perceived as a justification for pairing these wines with flavours of chocolate and fresh red fruits. The Vieni wine shows aromas of blackberry and plums, with hints of *crème brûlée* and flavour notes of candied red apple, tamarind fruit and honey. The sommelier Edward Rasiulis considers it a delectable accompaniment to duck confit with apricot relish.

A non-vintage traditional-method sparkling red wine with a little more pronounced mouthfeel, Ridgeview Sparkling Red Reserve–Limited Edition has been made from the ripest Pinot Noir grapes sourced from vineyards in southern England. Around 15% of the base wine intended for this sparkling wine was allowed to mature in 10-year-old French oak barrels. Bottle-fermented, it has benefited from its 18-month contact with lees. Released with 12% abv, 6.8 g/l of sugar, an acidity of 7.1 g/l, and a pH of 3.19, the wine shows vibrant aromas of sour cherry blended with notes of vanilla. The mouthfeel is dominated by rich flavours of wild blackberry and pomegranate, accompanied by sweet and peppery spice notes. Tasters at the Ridgeview company consider their wine a match for Sussex charcuterie, blue-veined cheeses and Asian flavours.

The Bolney Estate 'Cuvée Noir' NV Brut is an English medium-bodied traditional-method sparkling red wine crafted from Dornfelder grapes. The base wine destined for this wine was put through malolactic conversion and then allowed to mature in wood to achieve greater complexity. With 12% abv, aromas evoking redcurrants, dark cherries and blueberries, a rich creamy flavour resulting from approximately 18 months of maturation in bottles on lees, and with rich notes of summer fruits and hints of sweet spice on the finish, this wine tends to be recommended with barbecue meats. Chris Siegle, sales manager at The Bolney Estate, also proposes pairing this wine with bitter chocolate desserts.

Another notable example of a medium-bodied sparkling red wine is Treveri Sparkling Syrah Brut, a traditional-method wine produced at Treveri Cellars in Columbia Valley (in the American state of Washington). Crafted using Syrah grapes, it was released after around 24 months of rest in contact with lees, with an alcohol

Photo 75. While red champagnes cannot be found among wines offered by the Reims-based champagne producer G.H. Mumm, traditional-method sparkling red wines are in the portfolio of Mumm Napa, G.H. Mumm's United States outpost in the state of California. The wine has rested for over two years in bottles on lees. With its sugar content of 11.4 g/l, 12.5% abv, soft tannins and a bright acidity of 7.1 g/l, with a pH of 3.18, it displays a flavour profile dominated by fruity notes of Bing cherries, blackcurrants and plums. Mumm Napa's employees serve this wine at Thanksgiving dinner as an accompaniment to a dish of turkey, potatoes and cranberry sauce.
Photo: author, October 2024

content of 12% abv, good acidity, and around 12 g/l of sugar. This sparkling Syrah brings together flavours of dark cherries and tobacco, and is regarded as an agreeable companion to a host of light appetizers, salty snacks and chocolate desserts.

Vitteaut-Alberti Bourgogne Mousseux Rouge, a wine from Burgundy made in the *brut* style, is a blend of two grape varieties: Gamay and Pinot Noir. The wine developed its natural flavours during maturation in bottles on lees for 24 months in air-conditioned cellars at a temperature of around 15°C. Released with a little over 12% abv, this traditional-method sparkling wine displays an array of fresh, elegant and intense flavours, dominated by notes of raspberries, strawberries and blackcurrants. It is marketed by its producer as a delectable accompaniment to a wide variety of foods, most notably grilled red meats, red fruit cakes and fruit-flavoured chocolates.

Among the weightiest in the range of medium-bodied sparkling red wines is Chandon Sparkling Cabernet Sauvignon, combining fruitiness with spicy edge and complexity. Made from 100% Cabernet Sauvignon grown in Chandon's two Napa Valley estate vineyards (in Yountville and Mount Veeder), this wine has an alcohol concentration of 13% abv and a sugar content of 20 g/l. The wine was stored in bottles in contact with lees for around 14 months. It exhibits an abundance of dark fruit flavours with layers of spicy complexity and a hint of brioche. Its juiciness is balanced by a combination

Photo 76. A home vineyard of Winnica Miłosz, a small wine producer in the village of Łaz, around eight miles east of the city of Zielona Góra (western Poland). The wine producer has in its portfolio a medium-bodied sparkling red wine: Grempler Sekt–Pinot Noir 2021.

Photo: author, August 2024

of fine tannins, mineral freshness and vivid acidity. The flavour profile of this traditional-method wine is composed of notes of blackberries, cocoa, cola, dark cherries, plums and tobacco. Tasting panels at Chandon California suggest that it could make suitable pairings with intensely flavoured meats, hard cheeses or mushrooms.

Full-bodied

The category of full-bodied sparkling red wines is represented mainly by Australian bottlings. These wines, typically made using the grapes of Shiraz and/or Merlot, tend to have alcohol contents equal to or higher than 14% abv. At the lower end of the alcohol scale are two Charmat-method wines from South Australia: Hollick Estate Sparkling Merlot 2022 (Coonawarra) and the 2023 release of Genista Wines 'Noble' Sparkling Shiraz NV (Barossa Valley). The Hollick Estate wine has been made from Merlot grapes and released with 14% abv and a pH of 3.5. With flavours evoking red cherries, cranberries, raspberries and vanilla cola, this wine is recommended as an aperitif. While it could be enjoyed on its own, it is also sometimes served alongside a charcuterie board. The sparkling Shiraz offered by Genista Wines, with 14% abv, 18.9 g/l of sugar and a pH of 3.38, comprises Shiraz wines from the vintages of 2017, 2022 and 2023, enriched with a 1% addition of 10-year-old Deutschke fortified Shiraz. This wine exhibits sumptuous flavours of plums, chocolate, prunes, dried herbs, tobacco leaves and cloves. It is highly recommended as a delectable companion to barbecue meats, honey-glazed ham or Peking duck.

A unique sparkling red wine can be found in the portfolio of Uruguayan producer Bodega Familia Pisano, located around 20 miles north of Montevideo. Made from the Tannat grapes grown in vineyards that benefit from proximity to the Río de la Plata and the South Atlantic, the traditional-method wine Pisano 'Río de los Pájaros' Reserve Tannat–Brut Nature 2020 has rested in bottles on lees for almost two years. The wine has an alcohol content of 14% abv and shows intense fruity aromas of cherries and blackberries, with hints of mint and black pepper. Rich notes of blackberries, raspberries and sour cherries with a touch of cinnamon detected on the palate, along with its rich texture and well-integrated acidity, enable the wine to be paired with flavours found in grilled sweetbreads. The producer also recommends pairing it with spicy and fatty cheeses, caviar, grilled prawns and shrimps, oysters *au gratin* and grilled salmon.

Majella Sparkling Shiraz Vintage 2022 is a full-bodied sparkling red wine crafted in keeping with Australia's time-honoured winemaking tradition. This wine matured in French oak hogsheads for around five months before being sealed in bottles for the secondary fermentation. It then rested on lees for over a year before being disgorged in September 2023. Replenished with dosage liqueur based on fortified (port-style) wine, this sparkling wine is well-structured with smooth tannins and flavours of ripe fruits, such as redcurrants, blackberries and plums. It is often served alongside game or cured meats. Interestingly, Brian Lynn, a co-founder of Majella Wines, a wine producer from the Coonawarra region (South Australia), recommends it with a hearty breakfast dish of bacon and eggs.

Langmeil Barossa Stelle Nere Sparkling Shiraz NV is a non-vintage sparkling red wine taking its origin from Shiraz grapes collected in Barossa Valley (South Australia) during the 1998 harvest. Stored in bottles on its lees for two years, and liqueured on disgorgement with fortified Shiraz, this wine has evolved to reveal notes of velvety tannin mingling with rich flavours reminiscent of ripe plums and fresh scones with mulberry jam and cream and hints of leather and vanilla. Owing to these flavour characteristics, an alcohol concentration of 14.5%, a sugar content of 13.5 g/l and a pH of 3.5, this wine is often viewed as a suitable companion to meat dishes featuring roast lamb, grilled beef and game meats (especially when served with sweet and savoury sauces) and to richly flavoured desserts, notably Christmas pudding.

Beckett 2015, 2017; Bennett 2005; Biro 2024; Cardillo 2023; D'Agata 2024; MacLean 2017; Rasiulis [2017] 2019; Sinclair [1998] 2008: 586; Spanos 2023

bolneywineestate.com
www.braida.it/en
www.chandon.com
hollick.com
www.inniskillin.com
www.langmeilwinery.com.au
www.majellawines.com.au
www.pisanowines.com
www.ridgeview.co.uk
www.trevericellars.com
venturinibaldini.it/en
vieni.ca
www.vitteaut-alberti.fr/en

CHRONOLOGY OF EVENTS

23 May 1615 King James I bans the use of wood and its products in the manufacture of glass, thereby laying the foundation for the creation of an industrial economy based on coal, the combustion of which enables the production of pressure-resistant bottles.

17 December 1662 Christopher Merret presents to the Royal Society *Some Observations Concerning the Ordering of Wines*, a paper showing evidence that, in the mid-17th century, in still wines imported to England in barrels English coopers were initiating a second fermentation to produce sparkling wines.

1668–1715 In this long period Benedictine monk Dom Pérignon worked at the Hautvillers Abbey (Abbaye Saint-Pierre d'Hautvillers), perfecting the art of blending to obtain wines with great harmony, an aspect of wine production still regarded as crucial in crafting high-quality sparkling wines.

1709 A bottle factory using coal is established by Gaspard Thévenot in the region of Picardy (northern France) with the aim of meeting the growing demands of the French wine-producing market.

1723 A new plant in France's bottle-producing industry appears when Peter Mitchell sets up his bottle-manufacturing house in Bordeaux (western France).

25 May 1728 The French monarch Louis XV issues an edict granting producers and merchants the right to transport effervescent wines in bottles.

1729 The cloth merchant Nicolas Ruinart sets up a winery dedicated to sparkling wines. The wine company is now part of the business group Moët Hennessy–Louis Vuitton (LVMH).

1730 Pierre Chanoine establishes a winemaking firm known by the name Chanoine Frères, which currently constitutes an integral part of the business group Lanson-BCC.

1734 The company Forest-Fourneaux, since 1931 known as Taittinger, is founded on the initiative of Jacques Fourneaux.

1735 The unveiling of Jean-François de Troy's painting *Le Déjeuner d'Huîtres* ('Oyster Lunch'), commissioned by the King of France the previous year. The painting presents the opening of a bottle of champagne to be served with oysters.

1743 Wine merchant Claude Moët establishes his own business dedicated to producing and selling sparkling wine.

1757 Foundations are laid for the champagne-producing firm now known by the name Abelé.

1760 The year of birth of the currently world-famous champagne-making company known as Lanson, now part of the business group Lanson-BCC.

Photo 77. A spool of string on display in Musée du vin de Champagne et d'Archéologie régionale in Chateau Perrier in Épernay. The string was for securing corks placed in the necks of bottles of effervescent wine. Pieces of string were used widely in Champagne in the 18th century, but now by very few producers, and mainly only as a reference to their heritage.
Photo: author, July 2024

1764 As reported by the historian Isabelle Pierre, archival documents of the house of Ruinart indicate that in March this year the house shipped to the Roman German Empire 60 bottles of very pale rosé champagne (the *oeil de perdrix* style), which can be viewed as a forerunner of today's macerated rosé champagnes.

1765 Wine merchant Pierre-Joseph Dubois begins cooperation with Jean-Baptiste Dussaulois, putting into effect a concept of a wine-trading company based in Reims.

1772 Banker and vineyard owner Philippe Clicquot begins to commercialize his wines, initially selling champagne mainly to business acquaintances in France.

1776 Pierre-Joseph Dubois splits with his business partner Jean-Baptiste Dussaulois to set up his own firm, which was later named Louis Roederer.

1785 Florens-Louis Heidsieck marries Agathe Perthois and establishes a wine company called Heidsieck & Cie.

1789 Antoine Laurent Lavoisier explains the nature of the carbonic gas that forms during fermentations and causes effervescence in wines.

1799 Napoleon Bonaparte's rise to power marks the beginning of an auspicious era for champagne producers, a time characterized by technical advances in the industry.

1805 After the death of François Clicquot, Phillipe Clicquot's son, his widowed wife Barbe Nicole Clicquot, née Ponsardin, takes control of the wine-producing business that her father-in-law and her husband have developed.

1807 Publication of the treatise *L'Art de faire le vin* by the French scientist Jean-Antoine Chaptal (Minister of the Interior under Napoleon Bonaparte, 1800–1804) justifying the use of sweetening agents in the production of wine.

1808 Apolline Henriot, the young widow of cloth and wine merchant Nicolas Henriot, founds Maison Champagne Henriot.

1810 Headed by Barbe Nicole Clicquot, a champagne-producing company — currently known by the name Veuve Clicquot — makes what seems to be the first vintage champagne in the history of the trade.

1810 An oak *foudre* from the region of Algarve is given by Napoleon Bonaparte (also known as Napoleon I) to the then head of the firm Jean-Rémy Moët during one of the emperor's visits to Épernay.

1811 The champagne house Charles de Cazanove is founded by Charles Gabriel in Avize.

1811 Pierre-Nicolas Perrier and Rose-Adélaïde Jouët found the champagne house Maison Perrier-Jouët.

March 1814 Jean-Rémy Moët, the then head of his family's wine-producing business currently known as Moët & Chandon, is awarded the title Chevalier de la Légion d'Honneur for strengthening France's worldwide reputation as a producer of quality wines

In few sparkling-wine-producing regions have women played such fundamental roles in the development of the industry as they have in Champagne. Among the most recognized figures in the history of the champagne trade are the heads of the winemaking houses in the region. They include the widows: Barbe Nicole Clicquot (née Ponsardin) (who took the reins of the champagne house currently known as Veuve Clicquot in 1805); Apolline Henriot (establishing the house Champagne Henriot in 1808); Louise Pommery (taking over the champagne house now called Champagne Pommery from her husband following his death in 1858); Elisabeth Bollinger (assuming control of the firm Champagne Bollinger after her husband Jacques Bollinger passed away in 1941); and Carol Duval-Leroy (becoming head of the house Champagne Duval-Leroy in 1991).

1816 In or around this year, Antoine Müller, the cellar master at the firm headed by Barbe Nicole Clicquot, uses riddling racks for the first time in the history of the champagne industry.

1818 Barbe Nicole Clicquot, the head of the house now known as Veuve Clicquot, blends red wine made from grapes from Bouzy with white base wines, leading to the development of a new method for crafting rosé champagnes, known today as blended rosés or *rosés d'assemblage*.

1825 The foundations are built for the champagne house Joseph Perrier.

1826 Two producers of traditional-method sparkling wines are formed on German soil: G.C. Kessler & Co., founded by Georg Christian von Kessler in Esslingen am Neckar, and Häusler, Förster & Grempler, set up at the initiative of winemaker Carl Samuel Häusler in Grünberg (the city of Zielona Góra, western Poland).

1827 The champagne house known today by the name G.H. Mumm & Cie is established in Reims by a German winemaking family.

February 1829 At Aÿ, foundations are laid by Athanase-Louis-Emmanuel de Villermont and his partners (Joseph Bollinger and Paul Renaudin) for a new champagne-trading firm, now called Bollinger.

1837 Swiss merchant Henri-Marc de Venoge and his son Joseph set up a wine company in Épernay, named De Venoge & Cie.

28 November 1841 The barque *William Salthouse* sinks at the entrance to Port Phillip Bay, near Melbourne, with a cargo of 20 baskets of champagne, a brand most probably owned by the Gosset family.

1843 The champagne producer Krug, now an integral part of the LVMH group, is established by Joseph Krug in Reims.

1846–1849 The champagne business saw very little economic impact from political upheavals of the period known collectively as the Springtime of Nations, including the February Revolution of 1848, which took place in France.

1849 Pol Roger founds a wine company bearing his name, which moved to Épernay two years later, and prospers to the present day.

1850 Carlo Gancia returns to his native Piedmont, having studied champagne-making in Reims, to set up a company now officially known as Fratelli Gancia & C.

April 1850–March 1851 Sales of sparkling wine produced in the region of Champagne climb significantly, amounting to nearly eight million bottles.

1851 Charles Heidsieck establishes his own wine company; soon afterwards he goes on a series of business trips around the United States in search of reliable markets for his champagnes.

1851 The first shipment of Boizel's champagne to London is made.

1858 Eugène Mercier sets up a wine-producing firm that will soon go down in history for some of the most creative and spectacular promotional campaigns in the champagne industry.

1858 Louise Pommery takes over the champagne house, now called Champagne Pommery, following her husband's death.

late 1850s Cooperation between Edme-Jules Maumené, one of the pioneers of sparkling wine science, and Louis Brunet Jaunay, an inventor and wine merchant operating at Reims, results in scientific and technical foundations for making of sparkling wines using pressure-resistant tanks.

21 January 1860 In a letter to the editor of the *New York Dispatch*, Charles Heidsieck demonstrates that the total amount of counterfeit champagne sold to the US market in a year can be several times higher than the quantity of genuine champagne imported to that country in that period.

23 January 1860 The Cobden-Chevalier Treaty is signed between France and the United Kingdom, one of the champagne industry's top export markets, the result being significant reductions in English tariffs on French wines.

early 1860s The German immigrant Jacob Schram comes to Napa Valley to found a wine-producing estate, known to this day as Schramsberg Vineyards, run by him until his death in 1905.

1862 Formation of the champagne house Vincent Testulat, with the house's first vines purchased during the reign of Napoleon III (1852–1870).

1862 Foundations are laid for Schramsberg, a wine producer from Napa Valley's Diamond Mountain District.

1865 Joseph Best founds Great Western Vineyard, currently owned by Seppelt, one of Australia's most renowned sparkling wine producers.

mid-1860s Carlo Gancia's numerous experiments with locally grown grapes, notably Moscato, lead him to create what appears to be the first quality bottle-fermented sparkling wines in Piedmont, which can be viewed as the forerunners of Asti Spumante wines.

Photo 78. Barrels of the champagne producer Vincent Testulat, established in 1862.
Photo: author, July 2024

mid-1860s By this time, Louis Pasteur has proved that yeast cells are responsible for converting sugar in grape juice (and wine) into alcohol, carbon dioxide and secondary products, such as glycerol.

1869 Gustav Heinrich Martel founds the champagne-producing house known to this day as G.H. Martel.

1869–1870 The number of bottles of champagne sold abroad in that period exceeds 13.8 million, a huge increase on champagne sales in the 1863–1864 period, when exports of this wine were around 9.8 million bottles.

19 July 1870 The French emperor Napoleon III declares war against Prussia, often referred to as the Franco-Prussian War, which results in military occupation of the Champagne region by the Prussian army from early September 1870.

1870–1871 Total champagne sales amount to less than 9.2 million bottles, a sharp decline on the previous year, when over 17.4 million bottles were sold, the decline mainly due to the Franco-Prussian War, which ended with the Treaty of Frankfurt, signed on 10 May 1871.

1871–1874 These post-war years are considered the heyday of the champagne trade, with annual sales exceeding 20 million bottles.

1872 Josep Raventós i Fatjó, a leading winegrower in Spain and head of the family business Codorníu, located in the region of Penedès in Catalonia, makes his first bottles of sparkling wine in the likeness of wines produced in Champagne using grapes native to Catalonia.

1874 Grapes from this year's outstanding harvest in the region of Champagne are used at several champagne houses to make sparkling wines with varying degrees of dryness, one example being Pommery & Greno Brut Nature Champagne 1874, a wine introduced to the British market in the latter half of the 1870s which has gone down in history as one of the first fully dry vintage champagnes.

mid-1870s The history of G.H. Mumm's brand Mumm 'Cordon Rouge' reaches back to this period.

1877 Champagne house Pol Roger & Co. receives its first Royal Warrant from Queen Victoria, thus becoming the official supplier of sparkling wines to the British (royal) court.

4 November 1882 A joint initiative of some champagne houses, primarily Heidsieck & Cie, Giesler & Cie and G.H. Mumm & Cie, leads to the creation of the Syndicat du Commerce des Vins de Champagne, an organization established to support champagne manufacturers in promoting their wines, defend them against unfair competition, and exert pressure on legislators to create or modify laws and regulations in their favour.

1882 The winemaking business Korbel Champagne Cellars (F. Korbel & Bros.) is founded at Guerneville, in Sonoma County, by the Czech brothers Francis, Antone and Josef Korbel.

1884 Champagne Bollinger becomes an official supplier of champagne to the British court after a Royal Warrant is bestowed on this producer by Queen Victoria.

1887 The judgement of the Angers Court of Appeal states that the name 'champagne' must not be used as one belonging to the public domain because it refers to wines from the Champagne region, thus indicating where they were produced. This rewards champagne producers' lobbying efforts to protect the name 'champagne', and leads to the creation of one of the world's first widely recognized territorial brands.

1888 Great Western Vineyard, an Australian wine company based in the state of Victoria, is purchased by the passionate entrepreneur and politician Hans William Henry Irvine.

1889 The first champagnes of Laurent-Perrier's once popular range known by the name Grand Vins Sans Sucre — intended for the UK market — are launched in London.

1890s The plague of phylloxera begins to ravage the region of Champagne, leaving some grape-growers with little to depend on for their livelihood.

1890s In the early years of this period, during one of his business trips around France, Great Western Vineyard owner Hans Irvine hires Charles Pierlot, a French winemaker who had previously worked for Pommery & Greno, helping the company become Australia's leading sparkling wine producer within a decade.

1895 A new method of producing effervescence in wine is invented by Italian oenologist Federico Martinotti, involving fermenting cuvée in tightly sealed pressure-resistant tanks.

1896 Franz Hazek is employed by Francis Korbel at Korbel Champagne Cellars.

4 December 1900 Maison Lanson is awarded a Royal Warrant by Queen Victoria, becoming an official supplier of fine champagne to the Royal Court of England.

1901 The use of the name 'Mercier' on a giant-sized champagne bottle in the film *La Barbe Bleue*, directed by Georges Méliès, is considered both the first portrayal of champagne in a fictional film and one of the first product placements in the history of cinematography.

1904 Pommery's grand *foudre* with decorations made by Emile Gallé is exhibited at the St Louis World's Fair, an event organized to commemorate the United States' purchase of the Louisiana territory from France.

Photo 79. An illuminated staircase at the entrance to the cellars of Veuve Clicquot. Dates of key vintages in the history of the champagne house since 1899 are marked on the riser of each step.
Photo: author, July 2024

late September 1905 Eugène-Aimé Salon pioneers single-vintage single-variety *blanc de blancs* champagnes of a single *cru*, using Chardonnay grapes grown at Le Mesnil-sur-Oger.

1907 The method of making sparkling wine using pressure tanks, an invention credited to the Italian oenologist Federico Martinotti, is perfected by Jean-Eugène Charmat.

1908 The first deliveries of Maison Boizel champagne to New York.

October 1910 Major protests are staged by vignerons in Champagne, affected by a failed grape harvest and divided over the issue of

Champagne's borders, their misery compounded by rumours of fraud by some merchants.

April 1911 Protests in the Champagne region culminate in the Marne riots of April 1911, bringing Champagne to the brink of civil war.

1914–1918 During the Great War, champagne production was maintained at half the pre-war levels, with wine cellars and underground caves in the Champagne region providing shelter from shellfire and air raid, and serving as hospitals, schools and concert venues.

3 November 1916 The Swedish schooner *Jönköping*, carrying 2,400 bottles of Heidsieck & C° Monopole 'Gôut Americain' champagne from the 1907 vintage, is sunk in the Baltic Sea by a German submarine; the wreck remains undiscovered until July 1998.

1918 The wine business Great Western Vineyard is acquired by Seppelt.

16 January 1919 In the United States, ratification of the Eighteenth Amendment leads to the closure of the US market for champagne and other sparkling wines for 14 years.

1919 Wirehooding, a technique developed for securing the cork in the neck of the bottle with sparkling wine, begins to gain general acceptance.

1920 Eugène-Aimé Salon founds the champagne house Champagne Salon, currently an integral part of Laurent-Perrier Group.

early 1920s Champagne consumption plummeted from over 21 million bottles sold in 1920–1921 to fewer than 13 million bottles commercialized in 1921–1922, partly as a result of rising prices in France and trade barriers in France's traditional export markets.

1923 The placement of a bottle of Louis Roederer 'Grand Vin'-Sec in a scene of Charlie Chaplin's film *A Woman of Paris*, in which restaurant guests seated at a table are served truffles cooked in champagne.

1923–1927 Champagne sales improve, owing much to consumption in France, which outstrips demand for the wines of Champagne in foreign markets for four consecutive years.

22 July 1927 A French law ends the long-standing conflict between vignerons from the departments of Marne and Aube, recognizing the Aube as part of the wine-producing area of Champagne.

1928 The family-owned company De Bortoli Wines is founded in Australia by Italian immigrant Vittorio de Bortoli.

October 1929 A long period of economic downturn on a global scale, remembered as the Great Depression, begins in the United States, causing a sharp decline in sales of luxury goods, including champagne.

early 1930s Annual exports of champagne drop to fewer than 4.5 million bottles sold to foreign markets, a serious decline compared with the over 15 million bottles sold abroad during the peak in the mid-1920s.

early 1930s–early 1940s The continuous method of producing effervescence in wine, an invention is attributed to Anton Frolov-Bagreev of the Abrau-Durso winery, is developed; it is widely used by winemakers in the Soviet Union until the mid-1960s.

April 1936–March 1937 Sales figures in the champagne trade for this period suggest that the crisis might be coming to an end, with around 40 million bottles of champagne commercialized, including almost 12 million bottles shipped to external markets.

May 1940 France is invaded by Germany. On 22 June the same year the Armistice Agreement is signed, bringing three-fifths of mainland France, including the region of Champagne, under German occupation. Champagne producers are forced to supply copious amounts of their wine to the German army and German-controlled restaurants in France.

April 1941 The Champenois found Comité Interprofessionnel du Vin de Champagne (CIVC), an institution whose aim is to defend the interests of winegrowers and producers against unrealistic demands from the German authorities of the Champagne region represented by Otto Klaebisch.

1941 Jacques Bollinger passes away, and his wife Elizabeth Bollinger assumes control of the family business Champagne Bollinger.

early 1940s The sparkling wine Korbel Brut is first available on the US market.

1942 Champagne houses become vital centres of the French Resistance,

Photo 80. Long-time employees of the champagne house Veuve Clicquot are honoured with plaques placed on the walls of the producer's underground wine cellars; they include their names and dates of service. This plaque is dedicated to Hélène Gerard, who worked at Veuve Clicquot for 45 years.
Photo: author, July 2024

with producers informing the Allies about the locations to which large consignments of their wine are being sent at the behest of the German authorities, thus indicating the likely direction of German war campaigns. After the Second Battle of El Alamein, in November 1942, thousands of empty champagne bottles are found by the Allies in German camps.

1942 Recognition of the champagne brand Mumm 'Cordon Rouge' is cemented by its appearance in the film *Casablanca*, featuring a flashback scene in which the main characters share G.H. Mumm's 'Cordon Rouge' champagne, as the German army approaches Paris.

November 1943 President of the CIVC, Robert-Jean de Vogüé, is arrested on charges of aiding the Allies, and spends the rest of the war as a prisoner.

late 1943 The Champagne region endures severe persecution, with Claude Fourmon, head of the CIVC, deported to the concentration camp Mittelbau, and Gaston Poittevin, former president of Syndicat Général des Vignerons de la Champagne (SGV), in office from 1938 to 1941, taken to the concentration camp Buchenwald, from which he never returns.

late August 1944 Épernay and Reims are liberated from the German occupation by the United States' Third Army, commanded by Lieutenant General George Smith Patton.

February 1945 The Supreme Headquarters of the Allied Expeditionary Force (SHAEF), headed by General Dwight David Eisenhower, are moved to Reims. On 7 May, the authorized representative of the German High Command, Alfred Jodl, signs the Act of Military Surrender at the commanding general's headquarters, an event ending World War II in Europe.

Photo 81. An old wooden case with bottles of G.H. Mumm Cordon Rouge champagne from 1949 on display in the cellars of Maison G.H. Mumm in Reims.
Photo: author, July 2024

5 May 1945 Part of the Fifth Tactical Group of General Philippe Leclerc's Second Armoured Division climbs the Kehlstein summit (near the Bavarian town of Berchtesgaden) and reaches Eagle's Nest, a building often used by Adolf Hitler and the Nazi Party for official and social gatherings, to discover a secret cellar with around half a million bottles, many of them confiscated from champagne houses.

October 1948 Bernard de Nonancourt assumes control of the family business Maison Laurent-Perrier.

early 1950s The idea of a range of Laurent-Perrier unique (multi-vintage) champagnes, now known by the name of Grand Siècle, is born in the mind of Bernard de Nonancourt, head of Laurent-Perrier.

1953–1961 Champagne is a frequent choice at White House dinners during General Eisenhower's two terms as US president.

1954 The company Korbel (Champagne Cellars) (now known as F. Korbel & Bros.) is purchased by the Heck family.

1959 The birth of Chandon Argentina, Moët & Chandon's first vineyard holding in the New World.

1959 The first prestige multi-vintage cuvée of Maison Laurent-Perrier is launched, initiating Laurent-Perrier's Grand Siècle range.

1961 The birth of the now world-famous franciacortas, traditional-method sparkling wines produced in the Franciacorta region, at the northern tip of Italy.

1963 Maison Ruinart is acquired by Moët & Chandon; it is now an integral part of the luxury goods group LVMH.

1965 Jack and Jamie Davies purchase the long-abandoned wine estate called Schramsberg, located in the Napa Valley's Diamond Mountain District; they replant the estate's historic vineyards with Chardonnay and Pinot Noir with the intention of producing quality sparkling wines.

13 August 1969 Korbel Natural', an F. Korbel & Bros. sparkling wine made using the traditional method from fruit sourced from Sonoma County's Russian River Valley, is served at a dinner in the Century Plaza, Los Angeles, held in honour of the Apollo 11 astronauts by President Richard Nixon.

late 1960s The invention of devices called *gyropalettes* (credited to Claude Cazals and Jacques Ducoin), fully developed in the mid-1970s — and used to this day — for the purpose of accelerating the removal of the yeast deposit from bottles of traditional-method sparkling wine following *sur lie* maturation.

1970s–late 1990s The use of sanitized urban waste, locally known as *boues de ville*, as fertilizer in Champagne's vineyards: a practice that is now banned.

1971 The wine-producing business Moët & Chandon, already part of Moët Hennessy, creates two winemaking outposts in Americas: Chandon Brazil (in the Serra Gaúcha region) and Chandon California (in the Napa Valley, US Western Seaboard).

1971 Frans Malan, the owner of Simonsig Wine Estate, located in the Stellenbosch district of the Western Cape's Coastal Region (South Africa), uses Chenin Blanc grapes to create his first natural bottle-fermented sparkling wine, released two years later under the name Kaaps Vonkel.

25 February 1972 The company Schramsberg gains international renown when its Blanc de Blancs 1969 is served by US president Richard Nixon at the Toast to Peace in Beijing during a reciprocal dinner organized for Chinese Premier Zhou Enlai.

December 1973 Andrew Pirie's search for cool-climate areas similar to those of Champagne results in his purchase of land in northern Tasmania, called Pipers Brook Vineyard.

1973 The formation of two Moët & Chandon's sparkling wine-producing outposts: Chandon California and Chandon Brazil.

8 October 1974 The F. Korbel & Bros. wine Korbel Natural' is served at a state dinner organized by US President Gerald Ford for the First Secretary of the Central Committee of the Polish United Workers' Party, Edward Gierek.

1974 The launch of Bollinger's Vieilles Vignes Françaises 1969, the first cuvée in Bollinger's old-vine champagne range Vieilles Vignes Françaises, also commonly referred to as VVF.

13 July 1977 The F. Korbel & Bros. wine Korbel Natural' is served at a state

Photo 82. Winegrowers combatting a heavy frost using burning hay bales (producing heat and a smoke screen) in the Gostchorze vineyard in western Poland in late May 2024. Similar measures have long been applied in vineyards in France, where sub-zero temperatures can occur in some years towards the end of April and the beginning of May. In the Champagne region, spring frosts were particularly severe in 1953, 1955 and 1979.
Photo: Małgorzata Kucikiewicz, 2024

dinner held by US President Jimmy Carter in honour of German Chancellor Helmut Schmidt.

1979 Ernest Émile Rapeneau becomes the owner of the champagne producer G.H. Martel.

1981 The champagne house Bruno Paillard is founded.

1981 Laurent-Perrier Ultra Brut, a wine with no added sugar, is released by the champagne house Laurent-Perrier for the first time.

19 April 1982 The sparkling wine Schramsberg Crémant Demi-Sec 1979 features on the menu for President Ronald Reagan's dinner honouring Queen Beatrix of the Netherlands.

29 July 1982 The F. Korbel & Bros. wine Korbel Blanc de Noirs is served at a state dinner held by US President Ronald Reagan for India's Prime Minister Indira Gandhi.

1982 The National Federation of Crémant Growers and Producers (Fédération Nationale des Producteurs et Élaborateurs de Crémant) is founded in France, an organization that today brings together eight appellations: Alsace, Bordeaux, Burgundy, Die, Jura, Limoux, Loire and Savoy.

1982 Traditional-method sparkling wine is first produced at Roederer Estate, the champagne house Louis Roederer's outpost located in Anderson Valley, California.

1983 The champagne producer Bruno Paillard begins providing information about the date of disgorgement on the back label of all their bottles; they are one of the first houses in Champagne to do so.

1983–1988 Head of Maison Laurent-Perrier Bernard de Nonancourt creates the Laurent-Perrier Group having acquired shares in Maison Champagne de Castellane in 1983 and integrated the brands of Delamotte and Salon into the group's portfolio in 1988; this strategic move is aimed at covering the entire spectrum of the market.

18 June 1984 Bottles of F. Korbel & Bros. Korbel Blanc de Noirs are served at a state dinner held by US President Ronald Reagan in honour of Sri Lanka's President Junius Richard Jayewardene.

mid-1980s Tony Jordan is tasked with finding a suitable location for a Moët & Chandon outpost in Australia.

1986 A new Moët & Chandon sparkling wine-producing estate is set up in Victoria's Yarra Valley, now known widely as Chandon Australia.

1986 Confrérie du Sabre d'Or, an association of wine professionals, is founded in France by Jean-Claude Jalloux to promote the sparkling wine of Champagne and the art of *sabrage*: opening bottles of champagne with the blunt edge of a sabre.

1987 Domaine Carneros is founded by the Taittinger family in the Californian wine district Los Carneros.

1990 The concept of launching the 'Sous Bois' range is born at Maison Boizel as a result of Evelyne and Christophe Roques-Boizel's experimentation with vinifying a proportion of must in old oak.

1990 Joseph Henriot of Champagne Henriot initiates a perpetual reserve: a concept that has enabled the release of the 6th edition of Champagne Henriot 'Cuve 38'–La Réserve Perpétuelle, Blanc de Blancs, a wine made up of 100% Chardonnay wines from every successive harvest from 1990 to 2012.

1990s Major advancements occur in sparkling winemaking in New World countries, notably South Africa and Australia.

1991 Carol Duval-Leroy becomes head of the champagne house Champagne Duval-Leroy.

1991 Artesa (Estate Vineyard and Winery) is set up by the family business Raventós Codorníu in the Los Carneros district, California.

1992 The Cap Classique Producers Association (CCPA) is formed, an organization of South African wine producers making quality bottle-fermented sparkling wines, henceforth commonly referred to as cap classique wines.

1995 The foundations are laid in Tasmania for the award-winning sparkling wine producer House of Arras, known for capitalizing on the concept of blending wines from many areas.

1995 The story of English sparkling wine producer Ridgeview begins, as Mike and Christine Roberts set up the business by planting their first vines in Ditchling, East Sussex.

1996 Vincent Laval takes over as head of the champagne-making family business Georges Laval, one of Champagne's pioneering organic estates.

1997 The wine house Jansz Tasmania, devoted to making quality sparkling wines, becomes part of Hill-Smith Family Estates (HSFE).

1997 Didier Depond becomes head of champagne houses Delamotte and Salon.

July 1998 The wreck of the Swedish schooner *Jönköping* is found at the eastern end of the Baltic Sea with around 2,400 bottles of well-preserved Heidsieck & C° Monopole 'Gôut Americain' champagne from the 1907 vintage.

1998 The grape root aphid phylloxera attacks the Bouzy vineyard with its old ungrafted vines, a source of grapes for the champagne house Bollinger's prestigious VVF range.

1999 Emmanuel Lassaigne takes control of his family's wine-producing estate Jacques Lassaigne (located in Montgueux, near Troyes) to produce champagne with utmost respect for the terroir, using indigenous yeasts and the manual disgorgement method, which is something of a rarity in Champagne.

early 2000s Comité Champagne (CIVC) initiates regular environmental assessments of the champagne industry, which results in Champagne becoming the world's first winegrowing region to measure its carbon footprint.

early 2000s MontRubí, a sparkling wine producer from the Penedès region in Catalonia, revives the cultivation of Sumoll grapes, used recently for the production of the company's lightly sparkling wine MontRubí 'L'Ancestral Rosé'–2021.

2002 François Chartier begins molecular studies on aromatic affinities between traditional-method sparkling wines and food, the results of which show that flavour differences between cavas depend largely on their lees contact times during their maturation, enabling researchers to create aromatic profiles of cava wine categories, now publicly available on the Cava Regulatory Council's official website.

2002–2018 A significant growth in sparkling wine consumption worldwide can be observed, from slightly

Photo 83. Casks with base wine stored in the cellars of Champagne Charles de Cazanove in Reims. The house has been in the hands of the Rapeneau family since 2003.
Photo: author, June 2023

less than 12 million hectolitres to around 19 million hectolitres.

2003 Purchase of the champagne house Champagne Charles de Cazanove by the Rapeneau family.

2003–2022 A collective effort on the part of Champagne's vine-growers and wine producers leads to the reduction of the champagne industry's carbon footprint by around 20%.

2005 The champagne house Champagne Barons de Rothschild is founded by the Roschschild family.

2005 From this year's harvest onward, Aÿ's walled vineyards become the sole source of grapes from old ungrafted vines intended for Bollinger's prestigious VVF range of champagnes.

2005 Benoît Gouez is promoted to the position of *chef de cave* at champagne house Moët & Chandon.

In Champagne, an increasing proportion of women are being promoted to the position of *chef de cave*, a trend that the wine critic and writer Elin McCoy describes as a 21st-century phenomenon. The growing list of female *chefs de cave* in the region includes: Sandrine Logette-Jardin (*chef de cave* at Champagne Duval-Leroy in 2005); Caroline Latrive (head winemaker at Ayala from 2012–2022 and then at Deutz since August 2022); Julie Cavil (*chef de cave* at Krug since 2020, having shadowed the former cellar master Éric Lebel for 13 years); and Séverine Frerson (named chief winemaker at Perrier-Jouët in October 2020).

2005 Sandrine Logette-Jardin is appointed *chef de cave* at Champagne Duval-Leroy.

2006 The acquisition of the Lanson International firm by the group of champagne-making companies Boizel Chanoine Champagne (BCC) renders this business group one of the largest among champagne producers. Currently, the group, known since 2010 as Lanson-BCC, brings together eight champagne houses: Champagne Lanson, Champagne Philipponnat, Champagne De Venoge, Champagne Besserat de Bellefon, Champagne Boizel, Champagne Chanoine Frères, Maison Burtin and Domaine Alexandre Bonnet.

2009 Family-owned champagne-producing business Champagne Francis Boulard et Fille is founded

July 2010 A shipwreck is discovered on the floor of the Baltic Sea near the Finnish Åland archipelago, with 47 of the total 168 bottles found there containing Veuve Clicquot wines from the first half of the 19th century, sealed after 1841.

2010 Beginning this year, entire batches of Maison Ruinart's vintage-dated *blanc de blancs* champagnes have been matured in bottles under natural cork rather than crown closures, a technique found to produce wines with enhanced flavour complexity and depth.

24 May 2011 The English sparkling wine Ridgeview Fitzrovia Rosé 'Cuvée Merret' 2004 is served at a state banquet held at Buckingham Palace by Queen Elizabeth II to

Photo 84. A sculpture of the Madonna with Child in the cellars of Moët & Chandon in Épernay. Christian themes can be found in many wine cellars in Champagne. One of the chalk caves at the house Champagne Pommery in Reims has a similar statue of the Virgin Mary and Child, who are said to watch over Madame Pommery's cellars.
Photo: author, July 2024

honour President Barack Obama of the United States.

2012 De Bortoli Wines, one of Australia's best known producers, launches screw-capped bottles of tank-fermented sparkling wine under the name De Bortoli Willowglen Brut Cuvée.

2012–2022 Caroline Latrive serves as the head winemaker at the champagne house Ayala.

2013 Moët Hennessy, part of the LVMH group, creates Chandon China, the first winegrowing operation in Ningxia Hui Autonomous Region (northern China) to make premium sparkling wines, using mainly the grapes of Chardonnay, Pinot Noir, Marselan and Syrah.

2013 Hervé Dantan joins the champagne house Lanson as cellar master (*chef de cave*).

2014 Moët Hennessy establishes Chandon India, a winemaking estate located at Nashik (in the Indian state Maharashtra), around 200 km northeast of Mumbai, the grapes used by the estate's chief winemaker Kaushal Khairnar being Chenin Blanc, Chardonnay, Pinot Noir and Shiraz.

2014 The owners of Veuve Clicquot embark on a long-term scientific experiment called Cellar in the Sea, involving storing the company's non-vintage Yellow Label champagnes, Vintage Rosé 2004, and its *demi-sec* champagnes, in an underwater cellar called the Åland Vault, located 40 metres below the surface of the Baltic Sea.

2014 The results of research conducted by wine scientist Gérard Liger-Belair of Université de Reims Champagne-Ardenne are published in the *Journal of Physical Chemistry*; they show that around 1 million bubbles are likely to form in 100 ml of champagne served in a flute under standard tasting conditions.

2014 Maison Ruinart, under its managing director Frédéric Dufour, implements a rational delivery policy, with the result that 85% of Ruinart champagne shipments are now made by sea and 15% by land.

2014 Quality sparkling wines from the Penedès region can now be produced under the brand name Clàssic Penedès.

2015 A champagne-making business dedicated exclusively to the production of rosé champagnes, now known as Fleur de Miraval, is conceived at the joint initiative of Brad Pitt and the Perrin and Péters families.

2015 The release of the multi-vintage champagne Moët & Chandon 'MCIII', a special sparkling wine; its production involves blending reserve wines matured in wood, wines matured in stainless steel, and old vintage wines matured in glass.

25 September 2015 Schramsberg Crémant Demi-Sec 2011 is served on at a state dinner held by President Barack Obama for China's President Xi Jinping.

October 2015 Cap classique producer Spier Wine Farm wins the Getaway Award for Leadership in Water Conservation at the Nedbank Green Wine Awards ceremony.

mid-2010s The champagne industry's concerted efforts towards sustainability focus mainly on conserving Champagne's biodiversity and reducing its carbon and water footprints.

2016 Flint Vineyard is set up by Ben Witchell in the village of Earsham (in Norfolk, the east of England).

2017 Catalonia's DO Penedès becomes the world's first appellation to provide a category of premium sparkling wines (known as Classic Penedès) made using a combination of traditional cultivation methods and exclusively organic techniques.

2017 TheAssociació de Viticultors i Elaboradors Corpinnat (Association of Wine Producers and Growers Corpinnat), also called AVEC, is established with the aim of distinguishing quality sparkling wines crafted in the heart of the Penedès region (northeast Spain).

Photo 85. This newly released bottle of Moët & Chandon Grand Vintage 2016, made in the *extra brut* style, rested in bottles on lees until January 2024. It was launched in June the same year.

Photo: author, July 2024

June 2017 The champagne house Leclerc Briant, using the services of Amphoris, a firm specializing in running underwater cellars, have their disgorged wine Leclerc Briant 'Cuvée Abyss' 2013 stored for about a year in the Atlantic Ocean, near the Breton island of Ouessant.

2018 The wine-producing firm Hammerling Wines is set up in Berkeley (California) by the winemaker Josh Hammerling.

2018 Spier Wine Farm, a cap classique producer, launches a series of their own sustainable initiatives, known collectively as Spier Growing for Good, with the aim of fostering environmental sustainability through community engagement and empowerment.

24 April 2018 The sparkling wine Schramsberg Crémant Demi-Sec 2014 is served at US President Donald Trump's state dinner held in honour of France's President Emmanuel Macron.

2018 The Certified Heritage Vineyards seal is first issued by the Old Vine Project initiative in order to certify the dates of plants in South Africa's vineyards that are at least 35 years old.

19 May 2018 The champagne Pol Roger Brut Réserve NV is served at a luncheon reception hosted by Queen Elizabeth II at St George's Hall in Windsor Castle following the marriage ceremony of HRH Prince Henry of Wales and Ms Meghan Markle.

2019 The boutique urban wine-producing house Binah Winery is founded by the winemaker Kevin Danna in Allentown (eastern Pennsylvania).

2019 Maison Ruinart launches recyclable cases as an economical packaging for its bottles; described as 'second skin', they are composed of 99% natural wood fibres and weigh only 40g each.

November 2020 The winemaker Clare Dry joins Seppelt — one of Australia's most renowned producers — as lead winemaker.

Table 5 Champagne sales dynamics in major export markets, 2020–2022

Champagne export markets by country	Quantity in million bottles 2020	2021	2022
United States	20.8	34.1	33.72
United Kingdom	21.3	29.8	28.06
Japan	10.8	13.8	16.57
Germany	10.1	11.2	12.23
Italy	06.9	9.2	10.61
Australia	08.5	9.9	10.53
Belgium	08.9	10.3	10.27

2020 Maison Ruinart's recyclable 'second skin' cases called, launched in the previous year, replace the much heavier single-bottle (unitary) boxes.

October 2020 The winemaker Séverine Frerson is appointed *chef de cave* at the champagne house Perrier-Jouët.

2020 Julie Cavil becomes *chef de cave* at Krug.

October 2020 Fleur de Miraval 'Exclusivement Rosé' ER1 is released, a special cuvée made using the winemaking technique called *remise en cercles*, which involves opening bottles of seven-year-old *blanc de blancs* champagne and adding it to the final blend before bottling the wine and running the secondary fermentation.

2020 Following three years of ageing on lees, the sparkling wine Bacigalupi Vineyards Brillante Blanc de Noir Sparkling-2017 (made in Sonoma County's Russian River Valley) is disgorged before being topped up with expedition liqueur containing honey sourced from a local beekeeper.

2020–2023 Annual sales of non-vintage *brut* champagnes to markets outside France consistently account for over 75% of total champagne export sales.

Champagne producers of the LVMH group exploit their presence among the world's most luxurious brands in many ways. Crucially, they position their products among other luxury goods, sharing with them not only online platforms but also the space of the LVMH-owned grocery store La Grande Épicerie de Paris. Many of the luxury-foods gift boxes offered in that store include bottles of Ruinart champagne. Marketers representing Veuve Clicquot use a combination of strategies, including event marketing, whereby they offer customers special events linked with services provided by other companies within the LVMH group. Veuve Clicquot's gastronomic excursions are known as Solaire Journeys, and are facilitated by luxury train operator Belmond, itself part of the LVMH group since 2019. A recent experience, which took place on 4–6 July 2024, consisted of a gala dinner in Vienna and an overnight journey from Vienna to Paris via Reims on board the art déco Venice Simplon Orient Express. Veuve Clicquot's champagnes remained in the spotlight throughout the package, which included a tour of the champagne producer's historic vineyards and wine cellars during the train's stopover in Reims.

early 2020s Oenological studies conducted by the research group Equipe Effervescence, Champagne et Applications, led by Gérard Liger-Belair of Université de Reims Champagne-Ardenne, confirm that an increase in the size of the bottle in which sparkling wine matures on lees results in the wine's increased capacity to preserve its dissolved CO_2.

November 2021 The quality sparkling wine Bérêche et Fils Aÿ Grand Cru–2014 is disgorged having resting on its lees under cork for around six-and-a-half years.

December 2021 A number of over-30-year-old vintage-dated Krug champagnes from Marina Olssen's collection are tasted at Vinkällaren Grappe in Stockholm.

2022 The Cava Regulatory Council (Consejo Regulador de la Denominación de Origen Cava) commits to producing cava in the Cava de Guarda Superior segment (including cava reserva, cava gran reserva and cava de paraje calificado) using solely organic methods, with this regulation to take full effect from 2025.

March 2022 Fleur de Miraval champagne is served at the Oscars® events: the Oscars Nominees Luncheon (7 March); the 94th Academy (of Motion Picture Arts and Sciences) Awards ceremony in the Dolby Theatre in Los Angeles (27 March); and the Governors Ball (the Academy's official afterparty).

August 2022 Caroline Latrive begins her employment as *chef de cave* at Deutz.

2022 The total amount of France's crémant commercialized reaches 102.3 million standard-sized bottles, surpassing the 100-million-bottle mark for the first time.

June 2023 Seppelt 'Show Sparkling' Limited Release Shiraz 2012, a wine from the range of wines crafted at the Australian wine producer Seppelt only in exceptional vintages, is released.

2023 Drappier, one of few champagne producers to grow all traditional grapes approved for crafting champagne, is among the first to grow the grapes of Voltis, a fungus-resistant variety authorized in the Champagne region on an experimental basis.

2023 Producer of vintage champagnes Champagne Salon celebrates the release of Champagne Salon 2013, marking the 44th release since the first bottles of single-variety *blanc de blanc* vintage-dated champagne were made by the company's founder Eugène-Aimé Salon in 1905.

2023 South Africa's organization Cap Classique Producers Association (CCPA) requires that as from this year's harvest sparkling wines destined for the special category Cap Classique with Extended Lees Ageing be crafted using solely one or more of the (grape) varieties most commonly used for champagne: Chardonnay, Pinot Noir and/or Meunier.

Photo 86. Daniel Arsham's avant-garde wall relief, unveiled for the first time in Autumn 2023 to celebrate Möet et Chandon's 280th anniversary, is on public display in Galerie Impériale in the company's historic cellars in Épernay. The artwork depicts a compilation of symbols relating to the company's rich heritage.
Photo: author, July 2024

22–25 June 2023 The champagne producer Veuve Clicquot offers wine tourists a unique immersive experience. It combines a visit to the house's cellars in Reims, a gala dinner at Hôtel du Marc, gourmet meals, and a chartered flight to the Åland Islands for a comparative tasting of Veuve Clicquot's Cellar in the Sea wines.

ONLINE REFERENCES

www.abbazia.it/en (Casa Vinicola) Abbazia di San Gaudenzio (wine producer, Italy)
www.acquesi.it/en Acquesi (wine producer, Italy)
www.adamiprosecco.it (Adriano) Adami (wine producer, Italy)
adoriavineyards.com Adoria Vineyards (wine producer, Poland)
agustitorellomata.com/en Agustí Torelló (Mata) (wine producer, Spain)
www.alasiavini.it Alasia > Araldica group (wine producer, Italy)
albetinoya.cat/en Albet i Noya (wine producer, Spain)
alma4.com AlmaCuatro (wine producer, Argentina)
amistavineyards.com Amista Vineyards (wine producer, United States)
www.amphoris.fr/en Amphoris (underwater cellar provider, France)
www.artesawinery.com Artesa (Winery) (wine producer, United States)
www.ashtonhills.com.au Ashton Hills (Vineyard) (wine producer, Australia)
www.astidocg.it/en Consorzio Asti DOCG (consortium promoting Asti wines, Italy)
www.astoria.it/en Astoria Wines (wine producer, Italy)
www.australianwine.com Wine Australia (organization promoting Australian wine, Australia)
www.austrianwine.com Austrian Wine Marketing Board (representative body for the Austrian wine industry, Austria)
www.awri.com.au Australian Wine Research Institute (scientific organization supporting the Australian wine industry, Australia)
bacchus.kz/en Bacchus (wine producer, Kazakhstan)
bacigalupivineyards.com Bacigalupi (wine producer, United States)
www.balgownie.com Balgownie Estate (wine producer, Australia)
www.banfi.it/eng Banfi (Piemonte) (wine producer, Italy)
www.bayofshoalswines.com.au Bay of Shoals (Wines) (wine producer, Australia)
www.belmond.com Belmond (hotel and train services operator, United Kingdom)
benjaminbridge.com Benjamin Bridge (wine producer, Canada)
www.bepindeeto.it Bepin de Eto (wine producer, Italy)
www.bereche.com/en Bérêche et Fils (wine producer, France)
bestswines.com Best's Wines Great Western (wine producer, Australia)
beykush.com/en Beykush (Winery) (wine producer, Ukraine)
www.biocantinaorsogna.it (Bio) Cantina Orsogna (wine producer, Italy)
binahwinery.com Binah (Winery) (wine producer, United States)
bodegacruzat.com/en Cruzat (wine producer, Argentina)
www.boizel.com/en (Maison) Boizel (wine producer, France)
bolneywineestate.com Bolney Estate (wine producer, United Kingdom)
www.bordeaux.com/gb CIVB (Conseil Interprofessionnel du Vin de Bordeaux) (organization supporting the Bordeaux wine industry, France)
borgoluce.it/en Borgoluce (producer of agricultural products and processed agricultural products, including wine, Italy)
boschendal.com Boschendal (producer of fresh produce and processed agricultural products, including wine, South Africa)

www.bourgogne-wines.com BIVB (Bureau Interprofessionnel des Vins de Bourgogne) (organization representing the interests of wine producers in Bourgogne, France)
www.brachettodacqui.com/en Consorzio Tutela Vini d'Acqui (organization representing and defending the interests of producers of Brachetto d'Acqui wines, Italy)
www.braida.it/en Braida (di Bologna Giacomo) (wine producer, Italy)
brockenchack.nz Brokenchack Wines (wine producers, Australia)
www.cadelbosco.com Ca' del Bosco (wine producer, Italy)
www.cadirajo.it/en Ca' di Rajo (wine producer, Italy)
www.calmel-joseph.com Calmel & Joseph (wine producer, France)
www.campbellswines.com.au Campbells of Rutherglen (wine producer, Australia)
www.cantinaalicebc.it/en Alice bel Colle (wine producer, Italy)
cantinapaltrinieri.it/en (Cantina) Paltrineri (wine producer, Italy)
capclassique.co.za CCPA (Cap Classique Producers Association) (organization promoting cap classique wines and supporting their producers, South Africa)
carboniste.com Carboniste (wine producer, United States)
www.casavalduga.com.br/en Casa Valduga (wine producer, Brazil)
cascinabo.it/en Cascina Bo (wine producer, Italy)
www.castellane.com/en (Maison) Champagne de Castellane (wine producer, France)
www.castleglenaustralia.com.au Castle Glen Australia (producer of alcoholic beverages, including wine, Australia)
www.cava.wine/en Cava Regulatory Council (organization supporting the cava trade, Spain)
www.cerdon-wojtkowiak.fr/en (Cave) Wojtkowiak (wine producer, France)
champagne-adcoutelas.com A.D. Coutelas (wine producer, France)
champagne-bdr.com/en Barons de Rothschild Champagne (wine producer, France)
www.champagne-billecart.fr/en (Champagne) Billecart-Salmon (wine producer, France)
www.champagne-blin.com (Champagne) H. Blin (wine producer, France)
www.champagne-bollinger.com/en (Champagne) Bollinger (wine producer, France)
www.champagne-cazals.fr/en (Champagne Claude) Cazals (wine producer, France)
champagne-dehours.fr/en (Champagne) Dehours et Fils (wine producer, France)
https://www.champagne-deutz.com/en/intro/o (Champagne) Deutz (wine producer, France)
www.champagne-drappier.com/en Drappier (wine producer, France)
www.champagne-fleury.fr/en Champagne Fleury (wine producer, France)
www.champagne-gimonnet.com Pierre Gimonnet et Fils (wine producer, France)
www.champagne-gosset.com/en Maison Gosset (wine producer, France)
champagne-heucq.com/en André Heucq (wine producer, France)
www.champagne-henriot.com/en Champagne Henriot (wine producer, France)
www.champagne-hure-freres.com Champagne Huré Frères (wine producer, France)
www.champagne-laherte.com Laherte Frères (wine producer, France)
www.champagne-palmer.fr/en Palmer & Co (wine producer, France)
www.champagne-peters.com/en Pierre Péters Champagne (wine producer, France)
champagne-philippe-martin.com/en Champagne Philippe Martin (wine producer, France)
champagne-pouillon.com/en R. Pouillon et Fils (wine producer, France)
www.champagne-salon.fr (Champagne) Salon (wine producer, France)
www.champagne-serveaux.com/en (Champagne) Serveaux Fils (wine producer, France)
champagne-telmont.com (Champagne) Telmont (wine producer, France)
www.champagne-veuve-fourny.com/en Veuve Fourny et Fils (wine producer, France)
www.champagne.fr/en CIVC (Comité Interprofessionnel du Vin de Champagne, also known as Comité Champagne) (organization supporting the champagne industry)
champagneandrechemin.fr André Chemin (wine producer, France)
champagnebrunopaillard.com/en Bruno Paillard (wine producer, France)
champagnedecazanove.com Charles de Cazanove (wine producer, France)
champagnedoyard.fr (Champagne) Doyard (wine producer, France)
www.champagnegastoncollard.org/en Gaston Collard (wine producer, France)
www.champagnejeanvesselle.com/en (Champagne) Jean Vesselle (wine producer, France)
champagnelelarge-pugeot.com/en Lelarge-Pugeot (wine producer, France)
www.champagneveuvedoussot.com (Champagne) Veuve Doussot (wine producer, France)
www.champagnevilmart.fr/en Vilmart & Cie. (wine producer, France)

www.chandon.com Domaine Chandon (also known as Chandon California) (wine producer, United States)

www.chandon.com.au Chandon Australia (wine producer, Australia)

chapeldown.com Chapel Down (wine producer, United Kingdom)

www.chapoutier.com/en M. Chapoutier (wine producer, France)

www.charlesdefere.com/en Charles de Fère (wine producer, France)

charlesheidsieck.com/en (Champagne) Charles Heidsieck (wine producer, France)

chizay.com/en Chateau Chizay (wine producer, Ukraine)

www.clairette-de-die-lantheaume.fr Patrick Lantheaume (wine producer, France)

clairette-de-die.com/en/ Maison de la Clairette (organization promoting old traditions of crafting wines in the Diois region, France)

www.clairette-mongegranon.com (Cave) Monge Granon (wine producer, France)

cloverhillwines.com.au Clover Hill (wine producer, Australia)

www.coates-wines.com Coates Wines (wine producer, Australia)

collavini.it/en Eugenio Collavini (Viticoltori) (wine producer, Italy)

www.confreriedusabredorusa.com Confrérie du Sabre d'Or (association of wine professionals dedicated to promoting champagne and the art of *sabrage*, United States)

www.coppo.com/en (Cantine) Coppo (wine producer, Italy)

www.corpinnat.com/en Associació de Viticultors i Elaboradors Corpinnat (association of wine producers and growers committed to crafting quality sparkling wines in the historic winegrowing area of Penedès, Spain)

www.costadila.it/en Costadilà (Articoltura) (wine producer, Italy)

www.cremant-parigot.com/en Parigot (wine producer, France)

cremants.com FNPEC (National Federation of Crémant Growers and Producers; Fédération Nationale des Producteurs et Élaborateurs de Crémant) (organization representing and supporting producers of crémant, France)

www.debortoli.com.au De Bortoli (wine producer, Australia)

www.decoywines.com Decoy (wine producer, United States)

www.di-giovanna.com Di Giovanna (wine producer, Italy)

diam-cork.com/en Diam (Bouchage) (producer of cork closures)

www.domaine-long.com/en/ Domaine Long (wine producer, France)

www.domaine-slapsak.com/en Domaine Slapšak (wine producer, Slovenia)

domainebousquet.com/en Domaine Bousquet (wine producer, Argentina)

www.domainecarneros.com Domaine Carneros (wine producer, United States)

www.domainekox.lu/en Domaine Laurent & Rita Kox (wine producer, Luxembourg)

www.domaineladentelle.com Domaine La Dentelle (wine producer, France)

www.domperignon.com Dom Pérignon (flagship brand of Moët & Chandon, France)

dopenedes.cat DO Penedès Regulating Council (supervisory body monitoring the production of Penedès wines, France)

www.duchessalia.it/en Duchessa Lia (wine producer, Italy)

www.duval-leroy.store/en Champagne Duval-Leroy (wine producer, France)

eightatthegate.com.au Eight at the Gate (wine producer, Australia)

en.poderepradarolo.com Podere Pradarolo (wine-producing estate, Italy)

www.enotecaemiliaromagna.it/en Enoteca Regionale Emilia Romagna (organization set up to promote and improve wine production in the Emilia-Romagna region, Italy)

www.famillemoutard.com/en Champagne Moutard Père & Fils (wine producer, France)

fleurdemiraval.com/en Fleur de Miraval (wine producer, France)

flintvineyard.com Flint Vineyard (wine producer, United Kingdom)

franciacorta.wine/en Franciacorta Consortium (organization promoting products made in the Franciacorta region, Italy)

francis-boulard.com Champagne Francis Boulard et Fille (wine producer, France)

francois-pinon-vouvray.com François Pinon (wine producer, France)

francoischartier.ca/en François Chartier (website providing access to information about achievements of François Chartier, conducting research on aromatic compounds and molecular harmonies in his labs, mainly in Barcelona, Spain)

www.gancia.com/en Fratelli Gancia & C. (wine producer, Italy)

giolitalia.com/en (Azienda Agricola) Giol (wine producer, Italy)

www.giustiwine.com (Società Agricola) Giusti (wine producer, Italy)

www.gobelsburg.at/en (Weingut) Schloss Gobelsburg (wine producer, Austria) gouguenheim.com.ar Gouguenheim (wine producer, Argentina)

www.grahambeck.com Graham Beck (Estate) (wine producer, South Africa)

grantburgewines.com.au Grant Burge Wines (wine producer, Australia)

www.greenvalley-russianriver.com Green Valley (website providing information about the Green Valley region, located in Sonoma County's Russian River Valley, United States)

www.gusbourne.com Gusbourne (wine-producing estate, United Kingdom)

www.hammerlingwines.co Hammerling Wines (wine producer, United States)

www.harkamp.at/en Harkamp (wine producer, Austria)

www.heidsieckandco-monopole.com Heidsieck & C° Monopole (wine producer, France)

www.hsfe.com Hill-Smith Family Estates (wine producer with a variety of vineyard estates and a rich portfolio of wine brands, Australia)

hollick.com Hollick (Estates) (wine producer, Australia)

houseofarras.com.au House of Arras (wine producer, Australia)

www.hubertsekt.sk/en Hubert J.E. (wine producer, Slovakia)

www.inniskillin.com Inniskillin (Niagara Estate Wines) (wine producer, Canada)

ioc.eu.com IOC (Institut Œnologique de Champagne) (provider of winemaking services and oenological products, France)

www.ironhorsevineyards.com Iron Horse Vineyards (wine producer, United States)

it.zardettoprosecco.com/en Zardetto (wine producer, Italy)

jaillance.com/en Jaillance (wine producer, France)

www.jansz.com.au Jansz Tasmania (wine producer, Australia)

www.jmf.pt José Maria da Fonseca (wine producer, Portugal)

www.jurtschitsch.com Weingut Jurtschitsch (wine producer, Austria)

karanika.com (Domaine) Karanika (wine producer, Greece)

kenforresterwines.com Ken Forrester Wines (wine producer, South Africa)

www.kleinconstantia.com Klein Constantia (wine-producing estate, South Africa)

www.korbel.com F. Korbel & Bros. (wine producer, United States)

www.kreglingerwineestates.com Kreglinger (Wine Estates) (wine producer, Australia)

www.kreinbacher.hu Kreinbacher (wine producer, Hungary)

lambrusco.net/en Lambrusco Consortium (organization promoting Lambrusco wines and their producers, Italy)

www.langmeilwinery.com.au Langmeil Winery (wine producer, Australia)

lanson.com (Maison) Lanson (wine producer, France)

larmandier.fr/en (Champagne) Larmandier-Bernier (wine producer, France)

www.laurent-perrier.com/en (Maison) Laurent-Perrier (wine producer, France)

www.leclercbriant.fr/en Leclerc Briant (wine producer, France)

leftcoastwine.com Left Coast Estate (wine-producing estate, United States)

www.leitz-wein.de (Weingut) Leitz (wine producer, Germany)

www.lelude.co.za Le Lude (wine producer, South Africa)

www.lemanzane.com/en Le Manzane (wine producer, Italy)

lenoble-champagne.com/en A.R. Lenoble (wine producer, France)

lightningrockwinery.com Lightning Rock Winery (wine producer, Canada)

www.limoux-aoc.com/en Syndicat des Vins AOC de Limoux (organization protecting the wine industry in the region of Limoux and promoting its wines, France)

www.llopart.com/en Llopart (wine producer, Spain)

lormarinswines.com L'Ormarins (wine estate, South Africa)

losparviere.it/en Lo Sparviere, (Azienda Agricola) (wine producer, Italy)

www.louis-roederer.com/en Louis Roederer (wine producer, France)

www.lvmh.com LVMH (Moët Hennessy–Louis Vuitton) (business group holding over 70 Maisons, including producers of quality sparkling wines)

m.plantevigne-dubosquet.com/en Plantevigne Dubosquet (producer of rare wines and spirits, France)

www.magnotta.com Magnotta Winery (wine producer, Canada)

maisons-champagne.com/en UMC (Union des Maisons de Champagne) (professional body defending the interests of its members operating in Champagne, France)

www.majellawines.com.au Majella (Wines) (wine producer, Australia)

www.marchioriwines.com/en (Azienda Agricola) Marchiori (wine producer, Italy)

marynissen.com Marynissen Estates (wine producer, Canada)

www.matteosoria.it Azienda Agricola Soria Matteo (wine producer, Italy) mcgregorwinery.com McGregor Vineyard (wine producer, United States)

www.medici.it/en Medici Ermete & Figli (producer of wine and balsamic vinegar, Italy)

www.merselwine.com Mersel (Wine) (wine producer, Lebanon)

www.michelechiarlo.it/en Michele Chiarlo (wine producer, Italy)

www.moet.com/en-gb Moët & Chandon (wine producer, France)

mollydookerwines.com.au Mollydooker (Wines) (wine producer, Australia)

www.montelliana.com Cantina Montelliana (wine producer, Italy)

www.montgueux.com Jacques Lassaigne (wine producer, France)

montinafranciacorta.it/en La Montina (wine-producing estate, Italy)

montrubi.com/en MontRubí (wine producer, Spain)

moorilla.com.au Moorilla Estate (wine-producing estate, Australia)

mountavoca.com Mount Avoca (wine producer, Australia)

www.mumm.com/en-ww G.H. Mumm (& Cie) (wine producer, France)

www.napavalley.wine Napa Valley (Napa Valley Vintners' guide to Napa Valley American Viticulture Area, United States)

napavintners.com Napa Valley Vintners (non-profit trade association tasked with promoting, protecting and enhancing Napa Valley, United States)

www.newbarnfarm.org Newbarn Farm (agricultural estate, home to cultural and educational events, and to Woodfine Wine Vineyard, United Kingdom)

nikolaestate.com.au Nikola Estate (wine-producing estate, Australia)

www.nzwine.com New Zealand Wine (New Zealand Winegrowers' educational guide to New Zealand's grape and wine sector)

oeneo.com/en OENEO (business group offering to wine producers innovative tools and personalized advice, covering each stage of the winemaking process, France)

oldvineproject.co.za OVP (Old Vine Project) (non-profit organization promoting the preservation of vines that are at least 35 years old, South Africa)

www.oregonwine.org Oregon Wine Board (state agency managing marketing, research and education activities that support and advance the wine and grape industry in the state of Oregon, United States)

www.pelissero.com/en (Azienda Agricola Vitivinicola) Pelissero (wine producer, Italy)

perdeberg.co.za Perdeberg (Group) (wine producer, South Africa)

www.perrier-jouet.com/en-ww (Maison) Perrier-Jouët (wine producer, France)

pfeifferwinesrutherglen.com.au Pfeiffer Wines (wine producer, Australia)

www.philipponnat.com Philipponnat (wine producer, France)

www.pisanowines.com Pisano — Artesanía en Vinos Finos (wine producer, Uruguay)

www.pongracz.co.za (Desiderius) Pongrácz (wine producer, South Africa)

www.prosecco.it/en Prosecco (zone) (authoritative guide to the prosecco wine industry, Italy)

www.prosecco.wine/en Consorzio di Tutela DOC Prosecco (organization safeguarding and promoting Prosecco DOC)

proseccobaldi.it (Azienda Agricola) Baldi (wine producer, Italy)

raventoscodorniu.com/en Raventós Codorníu (wine producer, Spain)

www.recaredo.com/en Recaredo (wine producer, Spain)

www.renardat-fache.com Renardat Fache (wine producer, France)

www.ridgeview.co.uk Ridgeview (wine producer, United Kingdom)

www.road13vineyards.com Road 13 (wine producer, Canada)

roedererestate.com Roederer Estate (wine producer, United States)

www.rogergoulart.com/en Roger Goulart (wine producer, Spain)

www.rothschildarchive.org The Rothschild Archive (source of information about the Rothschild family and ways of accessing online resources and visiting the Reading Room of The Rothschild Archive in London, UK)

www.ruinart.com/en-gb/home (Maison) Ruinart (wine producer, France)

russianrivervalley.org Russian River Valley Winegrowers Association (organization of winemakers and farmers committed to preserving natural resources and the farming legacy of Russian River Valley, United States)

sabra.ge/sabrage Confrérie du Sabre d'Or (association of wine professionals dedicated to promoting champagne and the art of *sabrage*, France)

sauska.hu/en Sauska (wine producer, Hungary)

www.sawis.co.za SAWIS (South Africa Wine Industry Information and Systems) (comprehensive guide to the wine industry in South Africa)

www.schramsberg.com Schramsberg (Vineyards) (wine producer, United States)

seguraviudas.com Segura Viudas (wine producer, Spain)

www.selosse-lesavises.com/en Domain Jacques Selosse & Hôtel Restaurant Les Avisés (wine producer and hotelier, France) www.seppelt.com.au Seppelt (wine producer, Australia)

seppeltgreatwestern.com.au Seppelt Great Western (the historic home of Seppelt wines in the Grampians, Australia)

shabo.ua/en Shabo (wine producer, Ukraine)

www.sittella.com.au Sitella Winery (wine producer, Australia)

www.spier.co.za Spier Wine Farm (wine producer, South Africa)

www.sthuberts.com.au St Huberts (wine producer, Australia)

www.sutherlandestate.com.au Sutherland Estate (wine-producing estate, Australia)

www.taittinger.com/en Taittinger (wine producer, France)

taltarni.com.au Taltarni (wine producer, Australia)

www.tement.at/en (Familienweingut) Tement (wine producer, Austria)

www.tenutafa.com.br Tenuta Foppa & Ambrosi (wine producer, Brazil)

terres-secretes.com/en Terres Secrètes (wine-producing business, France)

teusner.com.au Teusner (wine-producing company, Australia)

www.thomascalderselections.com Thomas Calder (export agent and wine connoisseur, currently based in France)

tojovini.it/en Tojo (Vini) (wine producer, Italy)

torello.com/en Torelló (Viticultors) (wine producer, Spain)

torleypezsgo.hu/en Törley (Pezsgőpincészet) (wine producer, Hungary)

www.tosti.it/en Tosti 1820 (wine producer, Italy)

www.trevericellars.com Treveri (Cellars) (wine producer, United States)

venturinibaldini.it/en Venturini Baldini (wine producer, Italy)

www.veuveclicquot.com/en-int Veuve Clicquot (wine producer, France)

vieni.ca Vieni Estates (wine producer, Canada)

www.vietti.com/en Vietti (wine producer, Italy)

villiera.com Villiera Wines (wine-producing company, South Africa)

www.vinaeden.com/en Viña Edén (wine producer, Uruguay)

www.vinetrail.co.uk Vine Trail (importer, seller and distributor of fine wines from France, Northern Spain and Sicily, based in United Kingdom)

www.vineyardbrands.com Vineyard Brands (wine importer and marketer, United States)

vins-cremants.lu/en Fonds de Solidarité Viticole (Luxembourgian organization founded to support and promote wine growing in the country, Luxembourg)

vinsdubugey.net/en Syndicat des Vins du Bugey (organization established to support the winemaking industry in the area of Bugey and to promote Bugey wines, France)

vinselcep.com Vins El Cep (wine-producing estate, Spain)

www.vinicolagaribaldi.com.br Garibaldi (wine producer, Brazil)

www.vitteaut-alberti.fr/en (Maison) Vitteaut-Alberti (wine producer, France)

von-buhl.de/en Reichsrat von Buhl (wine producer, Germany)

vqaontario.ca Ontario Wine Appellation Authority (organization established to support the Ontario wine industry, Canada)

www.werkstatt.wine Werkstatt (wine producer, Australia)

www.weszeli.at/en Weingut Weszeli (wine producer, Austria)

williamsonwines.com Williamson Wines (wine producer, United States)

wilsondaniels.com Wilson Daniels (luxury wines importer, marketer and brand-builder, United States)

www.wineaustralia.com Wine Australia (organization supporting Australia's wine industry by providing quality information and educational tools, Australia)

www.winegrowerscanada.ca WGC (Wine Growers Canada) (organization founded to support Canada's wine industry)

www.winesofgermany.com DWI (Deutsches Weininstitut) (organization founded to promote the quality of and enhance sales of wine from German winegrowing regions, Germany)

www.winesofsa.co.uk Wines of South Africa (nonprofit organization representing the South African wine industry worldwide, South Africa)

winesofnovascotia.ca WGNS (Wine Growers Nova Scotia) (organization serving the wine community in Nova Scotia by building awareness of Nova Scotia wines and expanding market opportunities for their sale, Canada)

winetasmania.com.au Wine Tasmania (official body representing Tasmania's wine producers, advocating their interests, and helping them market and promote their wines, Australia)

www.wistonestate.com Wiston Estate (wine-producing estate, United Kingdom)

www.woodfinewine.com Woodfine Wine (wine producer, United Kingdom)

www.woodparkwines.com.au Wood Park Wines (wine producer, Australia)

www.wosa.co.za Wines of South Africa (non-profit organization representing and supporting South African producers who export their wine)

www.zusslin.com/en Valentin Zusslin (wine-producing estate, France)

WINE LITERATURE

REFERENCE SOURCES

ATF (Bureau of Alcohol, Tobacco, Firearms and Explosives) (2017). Code of Federal Regulations, Title 27, Chapter I, Subchapter A, Part 24, Subpart G — Production of Effervescent Wine. Washington, DC: Government Printing Office. Available from: www.ecfr.gov/current/title-27/chapter-I/subchapter-A/part-24/subpart-G (accessed 7 November 2023).

Atkins, Susy ([1994] 2015a). Moët & Chandon. In Jancis Robinson (Ed.), *The Oxford Companion to Wine*, 4th edition, 531–2. Oxford: Oxford University Press.

—— ([1994] 2015b). Pérignon, Dom. In Jancis Robinson (Ed.), *The Oxford Companion to Wine*, 4th edition, 605–6. Oxford: Oxford University Press.

Baldy, Marian W. ([1993] 2009). *The University Wine Course. A Wine Appreciation Text and Self Tutorial*, 3rd edition. San Francisco, CA: Wine Appreciation Guild.

Bartowsky, Eveline (2022). Malolactic fermentation in winemaking. In V.K. Joshi (Ed.), *Concise Encyclopedia of Science and Technology of Wine*, 1st edition, 255–68. Boca Raton, FL: CRC Press, Taylor & Francis Group.

Braatz, Dieter, Urlich Sautter, & Ingo Swoboda ([2007] 2014). *Wine Atlas of Germany* (Transl. by Kevin D. Goldberg). Los Angeles, CA: University of California Press.

Campanale, Joe, & Joshua David Stein (2022). *Vino: The Essential Guide to Real Italian Wine*. New York: Clarkson Potter, Penguin Random House LLC.

Campbell, Bob ([1994] 2015). New Zealand. In Jancis Robinson (Ed.), *The Oxford Companion to Wine*, 4th edition, 568–72. Oxford: Oxford University Press.

CIVC (Comité Interprofessionnel du Vin de Champagne) (2013). *Champagne: From Terroir to Wine*. Épernay: Comité Interprofessionnel du Vin de Champagne. Available from: www.champagne.fr/en (accessed 9 November 2023).

—— ([2014] 2022). *Champagne: From Cellar to Table*. Épernay: Comité Interprofessionnel du Vin de Champagne. Available from: www.champagne.fr/en (accessed 9 June 2023).

—— (2019). *Les Expéditions de Vins de Champagne: EN 2018*. Épernay: Comité Interprofessionnel du Vin de Champagne. Available from: www.champagne.fr/en (accessed 27 January 2024).

—— (2020). *Les Expéditions de Vins de Champagne: EN 2019*. Épernay: Comité Interprofessionnel du Vin de Champagne. Available from: www.champagne.fr/en (accessed 27 January 2024).

—— (2021). *Les Expéditions de Vins de Champagne: EN 2020*. Épernay: Comité Interprofessionnel du Vin de Champagne. Available from: www.champagne.fr/en (accessed 27 January 2024).

—— (2022). *Les Expéditions de Vins de Champagne: EN 2021*. Épernay: Comité Interprofessionnel du Vin de Champagne. Available from: www.champagne.fr/en (accessed 27 January 2024).

—— (2023a). *Les Expéditions de Vins de Champagne: EN 2022*. Épernay: Comité Interprofessionnel du Vin de Champagne. Available from: www.champagne.fr/en (accessed 27 January 2024).

—— (2023b). *Glossary*. Épernay: Comité Interprofessionnel du Vin de Champagne. Available

from: www.champagne.fr/en/glossary (accessed 9 March 2023).
Coates, Clive (2000). *An Encyclopedia of the Wines and Domaines of France*. Berkeley & Los Angeles, CA: University of California Press.
Collin, Simon ([2004] 2009). *Oddbins Dictionary of Wine*. London: Bloomsbury Publishing.
Considine, Douglas M., & Glenn D. Considine (1982). *Foods and Food Production Encyclopedia*. New York: Van Nostrand Reinhold Company.
Crawford, James Ludovic Lindsay, Earl of (1910). James I. In *A Bibliography of Royal Proclamations of the Tudor and Stuart Sovereigns and of Others Published under Authority, 1485–1714*, Nos. 934–1401. Oxford: The Clarendon Press.
CRC (Consejo Regulador del Cava) (2022). Explanatory guide on new regulations: Cava de Guarda and Cava de Guarda Superior, Integral Producer's Stamp, Territories of origin, quality labels. Vilafranca del Penedès: Consejo Regulador del Cava. Available from: www.cava.wine/en (accessed 13 April 2023).
De la Serna, Victor ([1994] 2015). Cava. In Jancis Robinson (Ed.), *The Oxford Companion to Wine*, 4th edition, 212–3. Oxford: Oxford University Press.
Dunstan, David ([1983] 2006–2023). Irvine, Hans William. In *Australian Dictionary of Biography*. Canberra: Australian National University. Available from: adb.anu.edu.au (accessed 3 April 2023).
Ellis, George (2012). *The Australian Bar Attendant's Handbook*. Frenchs Forest, NSW: Pearson Australia.
Enartis (2019). *Sparkling Wine Handbook*. Windsor, CA: Enartis. Available from: www.enartis.com (accessed 22 February 2024).
Faith, Nicholas, & Jancis Robinson ([1994] 2015a). Champagne. In Jancis Robinson (Ed.), *The Oxford Companion to Wine*, 4th edition, 241. Oxford: Oxford University Press.
Faith, Nicholas & Jancis Robinson ([1994] 2015b). CIVC. In Jancis Robinson (Ed.), *The Oxford Companion to Wine*, 4th edition, 218–20. Oxford: Oxford University Press.
FNPEC (Fédération Nationale des Producteurs et Élaborateurs de Crémant) (2018/2019). *2018–2019 Press Pack: Dare to Taste the Difference with Crémant*. Colmar, France: Fédération Nationale des Producteurs et Élaborateurs de Crémant.
Gawel, Richard ([1994] 2015). *Mannoproteins*. In Jancis Robinson (Ed.), *The Oxford Companion to Wine*, 4th edition, 509. Oxford: Oxford University Press.
Gawel, Richard, Anita Oberholster, & Ian Leigh Francis (2000). A 'Mouth-feel Wheel': Terminology for communicating the mouth-feel characteristics of red wine. *Australian Journal of Grape and Wine Research* 6(3): 203–7. Available from: https://doi.org/10.1111/j.1755-0238.2000.tb00180.x (accessed 24 July 2024).
GBC (Government of British Columbia) (2018–2022). *Food and Agricultural Products Classification Act: Wines of Marked Quality Regulation*, B.C. Reg. 168/2018–158/2022, Consolidated Regulations of British Columbia. Available from: www.bclaws.gov.bc.ca (accessed 7 March 2023).
Hall, Lisa Shara ([1994] 2015). Oregon. In Jancis Robinson (Ed.), *The Oxford Companion to Wine*, 4th edition, 585–6. Oxford: Oxford University Press.
Halley, Ned (2005). *The Wordsworth Dictionary of Drink. An A–Z of Alcoholic Beverages*. London: Wordsworth Editions Ltd.
Halliday, James, & Hugh Johnson ([1992] 2013). *The Art and Science of Wine*. Buffalo, NY: Firefly Books.
Heidsieck, Charles (1860). Letter from Charles Heidsieck to the Editor of the New York Dispatch. In New York Dispatch, January 21, 1860. Available from: www.loc.gov/collections/chronicling-america (accessed 10 October 2022).
Hubble, Gavin (2023). Wine Terms: 'Grape to Glass'. Available from: https://grape-to-glass.com (accessed 10 January 2024).
Hui, Y.H. ([2004] 2012). Wine production in the United States: Regulatory requirements. In Y.H. Hui (Ed.), *Handbook of Plant-Based Fermented Food and Beverage Technology*, 2nd edition, 339–62. Boca Raton, FL, London/New York: CRC Press, Taylor & Francis Group.
IOC (Institut Œnologique de Champagne) (2018). *Sparkling Wines: Effervescence, a Winning Strategy*. Épernay: IOC.
Jackson, Ronald S. ([2002] 2009). *Wine Tasting: A Professional Handbook*, 2nd edition. San Diego, CA/London: Elsevier/Academic Press.
—— ([2002] 2017). *Wine Tasting: A Professional Handbook*, 3rd edition. San Diego, CA/London: Elsevier/Academic Press.
Jeffery, David W., & Kerry L. Wilkinson (2014). Wine. In Charles W. Bamforth & Robert E. Ward (Eds.), *The Oxford Handbook of Food Fermentations*, 54–147. Oxford/New York: Oxford University Press.
LaVilla, Joseph (2010). *The Wine, Beer, and Spirits Handbook: A Guide to Styles and Service*. Hoboken, NJ: John Wiley & Sons, Inc.
Leader (1901). Viticulture. In *Leader* (Melbourne, Victoria: 1862–1918, 1935), Tuesday, January 1, 1901: 58. Available from: http://nla.gov.au/nla.news-article198078695 (accessed 23 January 2023).
Liem, Peter ([2019] 2022). Glossary of Terms. *ChampagneGuide.net (via Peter Liem)*. Available from www.champagneguide.net/info/glossary (accessed 26 April 2023).

Livingstone-Learmonth, John (2005). *The Wines of the Northern Rhône*. Berkeley & Los Angeles, CA/ London: University of California Press.

Lorch, Wink ([1994] 2015). Bugey, Vins du. In Jancis Robinson (Ed.), *The Oxford Companion to Wine*, 4th edition, 175–6. Oxford: Oxford University Press.

MacCulloch, John ([1816] 1829). *Remarks on the Art of Making Wine (with Suggestions for the Application of its Principles to the Improvement of Domestic Wines)* (4th edition, enlarged and improved). London: Longman, Rees, Orme, Brown & Green.

MacKay, Nicole (2021). The 6 most popular sparkling winemaking methods. *Somm TV Magazine*, 14 May 2021. Available from: https://mag.sommtv.com/2021/05/sparkling-winemaking-methods (accessed 5 January 2024).

Marsden, Reginald Godfrey (1905). English Ships in the Reign of James I. In *Transactions of the Royal Historical Society, New Series*, Vol. 19, 309–42. Cambridge: Cambridge University Press on behalf of the Royal Historical Society. Available from: www.jstor.org/stable/3678236 (accessed 17 September 2022).

Maumené, Edme-Jules, & Louis Brunet Jaunay (1859). A.D. 1858, 14th July. No. 1578. Apparatus for the Production of Sparkling Wines: Provisional Specification. In *English Patents of Inventions, Specifications: 1858, 1563–1656*. London: HM Stationery Office.

Mayson, Richard ([1994] 2015). Lancers. In Jancis Robinson (Ed.), *The Oxford Companion to Wine*, 4th edition, 471. Oxford: Oxford University Press.

McCabe, James Dabney, Jr (1871). *History of the War between Germany and France (with Biographical Sketches of the Principal Personages Engaged in the Contest)*. Philadelphia, PA: The National Publishing Company & Jones, Junkin & Co.

McKay, Marianne, Alan J. Buglass, & Chang Gook Lee (2011). Fermented beverages: beers, ciders, wines and related drinks. In Alan J. Buglass (Ed.), *Handbook of Alcoholic Beverages: Technical, Analytical and Nutritional Aspects*, 63–454. Chichester, UK: John Wiley & Sons.

McMillan, Richard (1943). *Mediterranean Assignment*. New York: Doubleday, Doran & Co., Inc.

Nagórka, Piotr (2013). *Thesaurus of Oenology*. Warsaw: Piotr Nagórka/Linguistic Modelling.

OIV (Organisation Internationale de la Vigne et du Vin) (2020a). *OIV Focus: The Global Sparkling Wine Market (April)*. Paris: OIV.

—— (2020b). *Resolution OIV-OENO 631-2020: Review of Practices for the Reduction of SO_2 Doses used in Winemaking*. Paris: OIV.

—— (2021). *OIV Standard for International Wine and Spirituous Beverages of Vitivinicultural Origin Competitions*. Edition 2021. Paris: OIV.

—— (2022). *International Code of Oenological Practices*. Paris: OIV.

Parker, Robert M., Jr (2008). *Parker's Wine Buyer's Guide*, 7th edition. New York/London/Toronto/ Sydney: Simon & Schuster.

Platter, John, & Michael Fridjhon ([1994] 2015). *South Africa: Wine of Origin and Labelling*. In Jancis Robinson (Ed.), *The Oxford Companion to Wine*, 4th edition, 752–3. Oxford: Oxford University Press.

Pomarici, Eugenio, Luigino Barisan, Vasco Boatto, & Luigi Galletto (2019). The Prosecco Superiore DOCG industry structure: Current status and evolution over time. In Adeline Alonso Ugaglia, Jean-Marie Cardebat, & Alessandro Corsi (Eds.), *The Palgrave Handbook of Wine Industry Economics*, 421–35. Cham: Springer Nature Switzerland.

Pozo-Bayón, María Ángeles, Adolfo Martínez-Rodríguez, Encarnación Pueyo, & María Victoria Moreno-Arribas (2009). Chemical and biochemical features involved in sparkling wine production: From a traditional to an improved winemaking technology. *Trends in Food Science & Technology* 20: 289–99.

PPH (Port Phillip Herald) (1841). Shipping intelligence. In *Port Phillip Herald* (Melbourne, Victoria), Friday, December 10, 1841, No. 207, Vol. II.

Ribéreau-Gayon, Pascal, Denis Dubourdieu, Bernard Donèche, & Aline Lonvaud (2006). *Handbook of Enology. Volume 1: The Microbiology of Wine and Vinifications* (2nd edition) (Transl. Jeffrey M. Branco & Christine Rychlewsky). Chichester, UK: John Wiley & Sons, Ltd.

Robinson, Jancis, & Tony Jordan ([1994] 2015). Fizziness. *The Oxford Companion to Wine*, 4th edition, 345. Oxford: Oxford University Press.

Robinson, Jancis, & Véronique Cheynier ([1994] 2015). Colour of wines. *The Oxford Companion to Wine*, 4th edition, 259–60. Oxford: Oxford University Press.

Rodríguez-Bencomo, Juan Jose, María Ángeles Pozo-Bayón, & María Victoria Moreno-Arribas ([2004] 2012). Wine fermentation and production. In Y.H. Hui (Ed.), *Handbook of Plant-Based Fermented Food and Beverage Technology*, 2nd edition, 179–200. Boca Raton, FL/London/New York: CRC Press, Taylor & Francis Group.

Rosberg, Andrés ([1994] 2015). Argentina. In Jancis Robinson (Ed.), *The Oxford Companion to Wine*, 4th edition, 94–100. Oxford: Oxford University Press.

Ryan, Frederick J., Jr (2020). *Wine and the White House: A History*. Washington, DC: The White House Historical Association.

Shaw, Thomas George (1863). *Wine, the Vine, and the Cellar*. London: Longman, Green, Longman, Roberts, & Green.

Sheridan, Philip Henry (1888). *Personal Memoirs of P.H. Sheridan, Vol. II.* New York: Charles L. Webster & Company.
Sherwood, Greg (2018). Agrafe: Le Lude — the Method Cap Classique producer breaking new quality and style barriers in the South African Sparkling Wine category . . . *A Fine Wine Safari*, 26 May 2018. Available from: https://gregsherwoodmw.com (accessed 18 December 2023).
Skelton, Stephen ([1994] 2015). Modern English wine. In Jancis Robinson (Ed.), *The Oxford Companion to Wine*, 4th edition, 324–5. Oxford: Oxford University Press.
—— (2016). *The UK Vineyards Guide. A Directory of Vineyards in Great Britain, Ireland, and the Channel Islands.* London: S.P. Skelton Ltd.
SL (Scott Laboratories) (2018/2019). *Sparkling Handbook*. Petaluma, CA/Pickering, ON: Scott Laboratories.
Smart, Richard, Bryan Coombe, & Patrick Williams ([1994] 2015). Flavour compounds. Jancis Robinson (Ed.), *The Oxford Companion to Wine*, 4th edition, 346–7. Oxford: Oxford University Press.
Speller, Walter ([1994] 2015a). Franciacorta. In Jancis Robinson (Ed.), *The Oxford Companion to Wine*, 4th edition, 357. Oxford: Oxford University Press.
—— ([1994] 2015b). Moscato d'Asti. In Jancis Robinson (Ed.), *The Oxford Companion to Wine*, 4th edition, 541–2. Oxford: Oxford University Press.
—— ([1994] 2015c). Prosecco. In Jancis Robinson (Ed.), *The Oxford Companion to Wine*, 4th edition, 647–8. Oxford: Oxford University Press.647–8.
Speller, Walter, & Tamlyn Currin ([1994] 2015). Lambrusco. In Jancis Robinson (Ed.), *The Oxford Companion to Wine*, 4th edition, 470–1. Oxford: Oxford University Press.
Sullivan, Charles L. (1998). *A Companion to California Wine: An Encyclopedia of Wine and Winemaking from the Mission Period to the Present*. Berkeley & Los Angeles, CA: University of California Press.
TFB (The Finest Bubble) (2016). *Champagne: Glossary of Terms*. London: *The Finest Bubble Ltd*, 20 August 2016. Available from: https://thefinestbubble.com (accessed 26 April 2023).
The Age (1905). The wine industry: Are our methods wrong? In *The Age* (Melbourne, Victoria: 1854–1954), Tuesday, January 10, 1905, p. 5. Available from: http://nla.gov.au/nla.news-article192238687 (accessed 23 January 2023).
Thomases, Daniel ([1994] 1999). Gancia. In Jancis Robinson (Ed.), *The Oxford Companion to Wine*, 2nd edition, 300. Oxford: Oxford University Press.
Trento DOC (2022). *Glossary*. Trento, IT: Trento DOC Institute. Available from: www.trentodoc.com/en (accessed 24 October 2023).
Unwin, Thomas ([1994] 2015). Chaptal, Jean-Antoine. In Jancis Robinson (Ed.), *The Oxford Companion to Wine*, 4th edition, 221. Oxford: Oxford University Press.
WA (Wine Australia) (2018). Wine Australia Regulations 2018 (F2018L00286). Available from: https://classic.austlii.edu.au/au/legis/cth/num_reg/war2018201800286286 (accessed 15 January 2024).
—— (2022). *The Ultimate Guide to Sparkling Wines*. Adelaide: Wine Australia. Available from www.wineaustralia.com (accessed 21 January 2023).
—— (2023). *Foundations of Australian Wine*. Adelaide: Wine Australia. Available from www.wineaustralia.com (accessed 23 January 2023).
Waldin, Monty ([1994] 2015). Natural wine. In Jancis Robinson (Ed.), *The Oxford Companion to Wine*, 4th edition, 557. Oxford: Oxford University Press.
Webb, Albert Dinsmoor ([1994] 2015a). Protective juice handling. In J. *Robinson (Ed.), The Oxford Companion to Wine*, 4th edition, 648. Oxford: Oxford University Press.
—— ([1994] 2015b). Viscosity. In J. *Robinson (Ed.), The Oxford Companion to Wine*, 4th edition, 864. Oxford: Oxford University Press.
Wheatley, Henry Benjamin (1891). Winchester House, Southwark. In *London Past and Present. Its History, Associations, and Traditions, Vol. III*, 524–5. London: John Murray & New York: Scribner & Welford.
Williams, Patrick ([1994] 2015). Pigments. *The Oxford Companion to Wine*, 4th edition, 622. Oxford: Oxford University Press.
Willson, Nigel, Jancis Robinson, & Hildegarde Heymann ([1994] 2015). Tasting. *The Oxford Companion to Wine*, 4th edition, 788–91. Oxford: Oxford University Press.
WSET (Wine & Spirit Education Trust) (2020). *D4: Sparkling Wines. An Accompaniment to the WSET® Level 4 Diploma in Wines*. London: Wine & Spirit Education Trust.

SCIENTIFIC PUBLICATIONS

Abdallah, Zouleika, Véronique Aguié, Roger Douillard, & Christophe Bliard (2007). Champagne bubbles: Isolation and characterization of amphiphilic macromolecules responsible for the stability of the collar at the champagne/air interface. In Philippe Jeandet, Christophe Clément & Alexandra Conreux (Eds.), *Macromolecules and Secondary Metabolites of Grapevine and Wine*, 377–84. Paris: Lavoisier/Tec & Doc.

Araujo, Marcos Vinícius, M.A.C. da Silva & Kelly Lissandra Bruch (2019). The phenomenon of Moscatel sparkling wine in Brazil. *BIO Web of Conferences* 12(03012) (2019). Available from: https://doi.org/10.1051/bioconf/20191203012 (accessed 3 April 2024).

AWRI (Australian Wine Research Institute) ([2011] 2020). *Fact Sheet — Winemaking: Achieving Successful Malolactic Fermentation*. Adelaide: Australian Wine Research Institute. Available from: www.awri.com.au (accessed 17 August 2023).

——— (2020a). *Fact Sheet — Winemaking: Malolactic Fermentation in White and Sparkling Wines*. Adelaide: Australian Wine Research Institute. Available from: www.awri.com.au (accessed 17 August 2023).

——— (2020b). *Fact Sheet — Winemaking: Malolactic Fermentation in Red Wine*. Adelaide: Australian Wine Research Institute. Available from: www.awri.com.au (accessed 17 August 2023).

Bakker, Jokie, & Ronald J. Clarke ([2004] 2012). *Wine Flavour Chemistry*, 2nd edition. Chichester, UK: John Wiley & Sons, Ltd.

Bargmann, Christopher (2005). Geology and wine in South Africa. *Geoscientist* 15(4) (April): 1–8.

Baldwin, Hanson Weightman (1945). America at war. Victory in Europe. In *Foreign Affairs* 23(4) (July). Available from: www.jstor.org/journal/foreignaffairs (accessed 2 February 2023).

Barros, Ana Paula André, Islaine Santos Silva, Luiz Claudio Correa & Aline Camarão Telles Biasoto (2022). Effect of the cold pre-fermentative maceration and aging on lees times on the phenolic compound profile, antioxidant capacity and color of red sparkling wines. *Journal of Food Science and Technology* 59: 3245–55. Available from: https://doi.org/10.1007/s13197-022-05531-z (accessed 26 June 2023).

Becuwe, Stéphane, Bertrand Blancheton & Christopher M. Meissner ([2018] 2019). The French (Trade) Revolution of 1860: Intra-industry trade and smooth adjustment. *NBER (National Bureau of Economic Research) Working Paper Series*. Available from: www.nber.org/papers/w25173 (accessed 9 October 2023).

Benucci, Ilaria (2020). Impact of post-bottling storage conditions on colour and sensory profile of a rosé sparkling wine. *LWT — Food Science and Technology* 118. Available from: https://doi.org/10.1016/j.lwt.2019.108732 (accessed 4 June 2024).

Bittner, Stephen V. (2021). *Whites and Reds: A History of Wine in the Lands of Tzar and Commissar*. Oxford: Oxford University Press.

Bogdanov, Konstantin (2013). Soviet champagne: A festive history (Transl. Rosie Tweddle). *Forum for Anthropology and Culture* 9: 233–44. Available from: https://anthropologie.kunstkamera.ru/en/09/2013_no_9 (accessed 8 April 2024).

Boulton, Roger B., Vernon L. Singleton, Linda F. Bisson & Ralph E. Kunkee ([1996] 1999). *Principles and Practices of Winemaking*. New York: Springer-Verlag Media.

Buglass, Alan J. (2022). *An Introduction to Viticulture, Winemaking and Wine: From Vineyard to Wine Glass*. Newcastle upon Tyne, UK: Cambridge Scholars Publishing.

Busby, James (1825). *A Treatise on the Culture of the Vine, and the Art of Making Wine; Compiled from the Works of Chaptal, and Other French Writers; and from the Notes of the Compiler, During a Residence in Some of the Wine Provinces of France*. Australia: R. Howe, Government Printer.

Buxaderas, Susana, & Elvira López-Tamames (2012). Sparkling wines: Features and trends from tradition. In Jeyakumar Henry (Ed.), *Advances in Food and Nutrition Research* 66: 1–46. Amsterdam: Academic Press.

Caliari, Vinícius, Carolina Pretto Panceri, Jean Pierre Rosier & Marilde T. Bordignon-Luiz (2015). Effect of the Traditional, Charmat and Asti method production on the volatile composition of Moscato Giallo sparkling wines. *LWT — Food Science and Technology* 61(2). Available from: https://doi.org/10.1016/j.lwt.2014.11.039 (accessed 27 March 2024).

Campo, Eva, Juan Cacho & Vicente Ferreira (2008). The chemical characterization of the aroma of dessert and sparkling white wines (Pedro Ximénez, fino, sauternes, and cava) by gas chromatography-olfactometry and chemical quantitative analysis. *Journal of Agricultural and Food Chemistry* 57(7): 2477–84. Available from: https://doi.org/10.1021/jf0729681 (accessed 18 June 2024).

Carrascosa, Alfonso V., Adolfo Martinez-Rodriguez, Eduardo Cebollero & Ramón González (2011). Saccharomyces yeasts II: Secondary fermentation. In Alfonso V. Carrascosa, Rosario Muñoz & Ramón González (Eds.), *Molecular Wine Microbiology*, 33–50. London: Academic Press.

Charnock, Hannah M., Gary J. Pickering & Belinda S. Kemp (2022). The Maillard reaction in traditional method sparkling wine. *Frontiers in Microbiology* 13. Available from: https://doi.org/10.3389%2Ffmicb.2022.979866 (accessed 7 June 2024).

Charpentier, Claudine (2010). Ageing on lees (*sur lies*) and the use of speciality inactive yeasts during wine fermentation. In Andrew G. Reynolds (Ed.), *Managing Wine Quality. Volume 2: Oenology and Wine Quality*, 164–87. Cambridge, UK: Woodhead Publishing Ltd.

Charters, Steve, & Nathalie Spielmann (2014). Characteristics of strong territorial brands: The case of champagne. *Journal of Business Research* 67(7): 1461–7. Available from https://doi.org/10.1016/j.jbusres.2013.07.020 (accessed 7 November 2022).

Chevet, Jean-Michel, Eva Fernández, Eric Giraud-Héraud & Vicente Pinilla (2018). France. In Kym Anderson & Vicente Pinilla (Eds.), *Wine Globalization: A New Comparative History*, 55–91. Cambridge/New York: Cambridge University Press.

Cilindre, Clara, Céline Henrion, Laure Coquard, Barbara Poty, Jacques-Emmanuel Barbier, Bertrand Robillard & Gérard Liger-Belair (2021). Does the temperature of the *prise de mousse* affect the effervescence and the foam of sparkling wines? *Molecules* 26(15): 4434. Available from: https://doi.org/10.3390/molecules26154434 (accessed 13 June 2024).

Cilindre, Clara, Gérard Liger-Belair, Sandra Villaume, Philippe Jeandet & Richard Marchal (2010). Foaming properties of various Champagne wines depending on several parameters: Grape variety, aging, protein and CO_2 content. *Analytica Chimica Acta* 660(1–2): 164–70. Available from: https://doi.org/10.1016/j.aca.2009.10.021 (accessed 17 June 2024).

Cisilotto, Bruno, Fernando Joel Scariot, Luisa Vivian Schwarz, Ronaldo Kauê Mattos Rocha, Ana Paula Longaray Delamare & Sergio Echeverrigaray (2023). Are the characteristics of sparkling wines obtained by the Traditional or Charmat methods quite different from each other? In *OENO One* 57(1) (2023). Available from: https://doi.org/10.20870/oeno-one.2023.57.1.7313 (accessed 31 October 2023).

Colangelo, Donato, Fabrizio Torchio, Dante Marco de Faveri & Milena Lambri (2018). The use of chitosan as alternative to bentonite for wine fining: Effects on heat-stability, proteins, organic acids, colour, and volatile compounds in an aromatic white wine. *Food Chemistry* 264: 301–9. Available from: https://doi.org/10.1016/j.foodchem.2018.05.005 (accessed 23 February 2024).

Cole, Hugh M. ([1950] 1993). *The Lorraine Campaign: United States Army in World War II. The European Theater of Operations*. Washington, DC: Center of Military History, United States Army.

Cordingley, Ben (2020). Ask the AWRI: Sparkling wine gushing — not a cause for celebration. *Australian & New Zealand Grapegrower & Winemaker* 682: 42–3. Available from: www.awri.com.au/information_services/publications/ask-the-awri (accessed 10 October 2023).

Costantini, Laura, Christian Dominik Kappel, Massimiliano Trenti, Juri Battilana, Francesco Emanuelli, Maddalena Sordo, Marco Moretto, Céline Camps, Roberto Larcher, Serge Delrot & Maria S. Grando (2017). Drawing links from transcriptome to metabolites: The evolution of aroma in the ripening berry of Moscato Bianco (*Vitis vinifera* L.). *Frontiers in Plant Science* 8(780) (May). Available from: www.frontiersin.org/journals/plant-science/articles/10.3389/fpls.2017.00780/full (accessed 18 June 2024).

Cravero, Maria Carla (2023). Innovations in sparkling wine production: A review on the Sensory aspects and the consumer's point of view. *Beverages* 9(3): 80. Available from: https://doi.org/10.3390/beverages9030080 (accessed 27 March 2024).

Croguennec, Thomas (2016). From grape to wine. In Romain Jeantet, Thomas Croguennec, Pierre Schuck & Gérard Brulé (Eds.), *Handbook of Food Science and Technology. 3: Food Biochemistry and Technology* (Transl. Geraldine Brodkorb). London/Hoboken, NJ: ISTE Ltd/John Wiley & Sons.

Crouvisier-Urion, Kevin, Jean-Pierre Bellat, Gérard Liger-Belair, Régis D. Gougeon & Thomas Karbowiak (2021). Unravelling CO_2 transfer through cork stoppers for Champagne and sparkling wines. In *Food Packaging and Shelf Life* 27 (March). Available from: https://doi.org/10.1016/j.fpsl.2020.100618 (accessed 26 July 2023).

Crowden, James (2021). Who invented *méthode champenoise*? *Petits Propos Culinaires* 120: 72–87. Available from: https://doi.org/10.1558/ppc.27829 (accessed 4 May 2024).

Day, Martin Peter, Simon A. Schmidt, Wes Pearson, Radka Kolouchova & Paul A. Smith (2019). Effect of passive oxygen exposure during pressing and handling on the chemical and sensory attributes of Chardonnay wine. *Australian Journal of Grape and Wine Research* 25(2): 185–200. Available from: https://doi.org/10.1111/ajgw.12384 (accessed 6 February 2024).

Di Gianvito, Paola, Giuseppe Arfelli, Giovanna Suzzi & Rosanna Tofalo (2019). New trends in sparkling wine production: Yeast rational selection. In Alexandru Mihai Grumezescu & Alina Maria Holban (Eds.), *Alcoholic Beverages. Vol. 7: The Science of Beverages*, 347–86. Woodhead Publishing/Elsevier.

Divies, Charles ([1993] 2002). Bioreactor technology and wine fermentation. In Graham H. Fleet (Ed.), *Wine Microbiology and Biotechnology*, 449–76. London: Taylor & Francis.

Donzé, Pierre-Yves (2018). The birth of luxury big business: LVMH, Richemont and Kering. In Pierre-Yves Donzé & Rika Fujioka (Eds.), *Global Luxury: Organizational Change and Emerging Markets since the 1970s*, 19–38. Singapore: Palgrave Macmillan.

Du Toit, Maret (2020). What do we know about MLF in 2020? [video]. *InfoWine: The Internet Journal of Viticulture and Enology*, 23 July. Available from: www.infowine.com/en (accessed 23 April 2024).

Dungworth, David (2012). Three and a half centuries of bottle manufacture. In *Industrial Archaeology Review* 34(1) (May): 37–50. Available from https://doi.org/10.1179/0309072812Z.0000000002 (accessed 21 October 2022).

Dwyer, Scott (2018). Comparing three bentonite products for efficacy, packing, and color stability. *Winemakers Research Exchange*. Available from: https://winemakersresearchexchange.com (accessed 15 February 2024).

Edwards, Michael (2009). *The Finest Wines of Champagne: A Guide to the Best Cuvées, Houses, and Growers*. Berkeley & Los Angeles, CA: University of California Press.

Escudero, Ana, Monique Charpentier & Patrick Etievant (2000). Characterization of aged champagne wine aroma by GC–O and descriptive profile analyses. *Sciences des Aliments* 20: 331–46.

Farrer, James, Mônica R. de Carvalho & Chuanfei Wang (2023). Reinventing Japanese fine dining in culinary global cities. In James Farrer & David L. Wank (Eds.), *The Global Japanese Restaurant: Mobilities, Imaginaries, and Politics*, 289–327. Honolulu: University of Hawai'i Press.

Fleet, Graham H. ([1985] 1998). *The microbiology of alcoholic beverages*. In Brian J.B. Wood (Ed.), *Microbiology of Fermented Foods*, Vol. 2 (2nd edition), 217–62. London: Blackie Academic & Professional.

Gabbardo, Marcos, Franco Battistutta, Esther Theisen Gabbardo, Lara Tat & Emilio Celotti (2016). Aromatic characterization of Brazilian sparkling wines using olfactometry and a sensory panel. BIO Web of Conferences, Vol. 7, 39th World Congress of Vine and Wine. Available from: https://doi.org/10.1051/bioconf/20160702005 (accessed 19 June 2024).

Garde-Cerdán, Teresa, Pilar Rubio Bretón, Sandra Marín-San Román, Itziar Sáenz de Urturi & Eva P. Pérez-Álvarez (2021). Pre-fermentative maceration with SO_2 enhanced the must aromatic composition. *Food Chemistry* 345: 128870. Available from: https://doi.org/10.1016/j.foodchem.2020.128870 (accessed 16 February 2024).

Garofalo, Carmela, Mattia Pia Arena, Barbara Laddomada, Maria Stella Capello, Gianluca Bleve, Francesco Grieco, Luciano Beneduce, Carmen Berbegal, Giuseppe Spano & Vittorio Capozzi (2016). Starter cultures for sparkling wine. In Ronnie G. Willaert (Ed.), *Yeast Biotechnology*, 111–26. Basel: MDPI.

Garrido-Bañuelo, Gonzalo, Astrid Buica, Brock Kuhlman, Julia Schückel, Anscha J.J. Zietsman, William G.T. Willats, John P. Moore & Wessel J. du Toit (2021). Untangling the impact of red wine maceration times on wine ageing. A multidisciplinary approach focusing on extended maceration in Shiraz wines. *Food Research International* 150 Part A (December). Available from: https://doi.org/10.1016/j.foodres.2021.110697 (accessed 31 May 2024).

Genisheva, Zlatina, José A., Teixeira & José M. Oliveira (2014). Immobilized cell systems for batch and continuous winemaking. *Trends in Food Science & Technology* 40: 33–47. Available from: https://doi.org/10.1016/j.tifs.2014.07.009 (accessed 12 April 2024).

Gilinsky, Armand Jr, Tom Atkin & Stephanie Bayar (2021). The champagne industry in 2020. *Wine Business Journal* 4(1).

González-Lázaro, Miriam, Leticia Martínez-Lapuente, Antonio Palacios, Zenaida Guadalupe, Belén Ayestarán, Marta Bueno-Herrera, Pedro López de la Cuesta & Silvia Pérez-Magariño (2019). Effects of different oenological techniques on the elaboration of adequate base wines for red sparkling wine production: phenolic composition, sensory properties and foam parameters. *Science of Food and Agriculture* 99(10). Available from: https://doi.org/10.1002/jsfa.9697 (accessed 27 May 2024).

Gough, Jerry B. (1998). Winecraft and chemistry in 18th century France: Chaptal and the invention of chaptalization. *Technology and Culture* 39(1): 74–104. Available from: https://doi.org/10.2307/3107004 (accessed 4 March 2023).

Guy, Kolleen Marie (1999). 'Oiling the wheels of social life': Myths and marketing in Champagne during the Belle Epoque. In *French Historical Studies* 22(2) (Spring): 211–39. Available from: www.jstor.org/stable/286747 (accessed 4 November 2022).

—— (2003). *When Champagne Became French: Wine and the Making of a National Identity*. Baltimore, MD/London: The Johns Hopkins University Press.

Harding, Graham (2015). Laurent-Perrier Ultra Brut — A history of innovation. *Wine as Was*, 3 May 2015. Available from: https://wineaswas.com (accessed 26 July 2024).

—— (2018). *The Establishment of Champagne in Britain, 1860–1914*. PhD thesis, University of Oxford.

Available from: www.ora.ox.ac.uk (accessed 10 October 2022).

Heelan, Avery (2015). Volatile thiols: Sulfur containing compounds have high aroma impact, including key varietal flavors. Waterhouse Lab, 2015. Available from: https://waterhouse.ucdavis.edu (accessed 26 June 2024).

Hogendoorn, Katja, Kay Anantanawat & Cassandra Collins (2015). Cap removal by honey bees leads to higher pollen rewards from grapevine flowers. *Apidologie* 47: 671–8. Available from: https://doi.org/10.1007/s13592-015-0419-1 (accessed 22 June 2023).

Holzapfel, Bruno (2014/2015). Nitrogen nutrition and grape and wine production. In Darren Fahey (Ed.), *Grapevine Management Guide, 2014–2015*. Orange, NSW: Department of Primary Industries.

Hornsey, Ian Spencer (2007). *The Chemistry and Biology of Winemaking*. Cambridge, UK: The Royal Society of Chemistry.

Howe, Patricia (2003). Sparkling wines. In Andrew G.H. Lea & John R. Piggott (Eds.), *Fermented Beverage Production*, 2nd edition, 139–56. New York: Springer Science + Business Media.

Izquierdo-Llopart, Anaïs, & Javier Saurina (2019). Characterization of sparkling wines according to polyphenolic profiles obtained by HPLC-UV/Vis and principal component analysis. *Foods* 8(1): 22. Available from: https://doi.org/10.3390/foods8010022 (accessed 24 July 2024).

Jackson, Ronald S. ([1994] 2008). *Wine Science: Principles and Applications*, 3rd edition. Amsterdam: Academic Press/Elsevier.

—— ([1994] 2014). *Wine Science: Principles and Applications*, 4th edition. London: Academic Press/Elsevier.

Jeandet, Philippe, & Steve Charters (2012). Producing champagne. In Steve Charters (Ed.), *The Business of Champagne: A Delicate Balance*, 15–24. London/New York: Routledge.

Jeandet, Philippe, Silke S. Heinzmann, Chloé Roullier-Gall, Clara Cilindre, Alissa Aron, Marie Alice Deville, Franco Moritz, Thomas Karbowiak, Dominique Demarville, Cyril Brun, Fabienne Moreau, Bernhard Michalke, Gérard Liger-Belair, Michael Witting, Marianna Lucio, Damien Steyer, Régis D. Gougeon & Philippe Schmitt-Kopplin (2015). Chemical messages in 170-year-old champagne bottles from the Baltic Sea: Revealing tastes from the past. *Proceedings of the National Academy of Sciences of the United States of America* 112(19) (20 April): 5893–8. Available from: www.pnas.org/doi/full/10.1073/pnas.1500783112 (accessed 3 July 2023).

Jenster, Per V., David E. Smith, Darryl J. Mitry & Lars V. Jenster (2008). *The Business of Wine: A Global Perspective*. Frederiksberg: Copenhagen Business School Press.

Jolly, Neil, Philip Minnaar, Mardé Booyse & Paul Gerber (2021). Bottle fermented sparkling wine: Cork or crown closures during the second fermentation? *South African Journal of Enology and Viticulture* 42(2). Available from: https://doi.org/10.21548/42-2-4522 (accessed 27 November 2023).

Jones, Joanna E., Fiona L. Kerslake, Dugald C. Close & Robert G. Dambergs (2014). Viticulture for sparkling wine production: A review. In *American Journal of Enology and Viticulture* 65(4): 407–16.

Joshi, Vinod Kumar, S. Sharma & A.D. Thakur (2017). Wines: White, red, sparkling, fortified, and cider. In Ashok Pandey, Maria Ángeles Sanromán, Guocheng Du, Carlos Ricardo Soccol & Claude-Gilles Dussap (Eds.), *Current Developments in Biotechnology and Bioengineering: Food and Beverages Industry*, 353–406. Amsterdam: Elsevier.

Joshi, Vinod Kumar, Vikas Kumar & Jaspreet Kaur (2021). Technology of winemaking. In V.K. Joshi & Ramesh C. Ray (Eds.), *Winemaking: Basics and Applied Aspects*, 547–80. Boca Raton, FL: CRC Press.

Julien, Anne, Didier Theodore, Gordon Specht, Ann Dumont & Cecile Brung (2014). *The State of the Art: Yeast Nutrition and Protection for Good Alcoholic Fermentation*. Petaluma, CA: Lallemand.

Just-Borràs, Arnau, Ekaterina Moroz, Pol Giménez, Pedro Cabanillas, Jordi Gombau, Joan M. Canals & Fernando Zamora (2023). Comparison of ancestral and traditional methods in the elaboration of sparkling wines: Preliminary results. *International Congress on Grapevine and Wine Sciences*, 13 October 2023.

Kemp, Belinda, Hervé Alexandre, Bertrand Robillard & Richard Marchal (2015). Effect of production phase on bottle-fermented sparkling wine quality. *Journal of Agricultural and Food Chemistry* 63(1): 19–38. Available from: https://doi.org/10.1021/jf504268u (accessed 16 September 2023).

Kemp, Belinda, Casey Hogan, Shufen Xu, Lisa Dowling & Debbie Inglis (2017). The impact of wine style and sugar addition in *liqueur d'expédition* (*dosage*) solutions on traditional method sparkling wine composition. *Beverages* 3(1): 7. Available from: https://doi.org/10.3390/beverages3010007 (accessed 31 October 2023).

Kemp, Belinda, Bruna Condé, Sandrine Jégou, Kate Howell, Yann Vasserot & Richard Marshal ([2018] 2019). Chemical compounds and mechanisms involved in the formation and stabilization of foam in sparkling wines. *Critical Reviews in Food Science and Nutrition* 59(13). Available from: https://doi.org/10.1080/10408398.2018.1437535 (accessed 17 June 2024).

Kerslake, Fiona, Rocco Longo, & Robert Dambergs (2018). Discrimination of juice press fractions for sparkling base wines by a UV-Vis spectral phenolic fingerprint and chemometrics. *Beverages* 4(2): 45. Available from https://doi.org/10.3390/beverages4020045 (accessed 6 May 2023).

Kunc, Martin (2012). Creating and sustaining a competitive advantage over time: managing the delicate balance between value creation and value appropriation. *The Business of Champagne: A Delicate Balance*, 137–53. London/New York: Routledge.

Kunkee, Ralph E., & Cornelius S. Ough (1966). Multiplication and fermentation of *Saccharomyces cerevisiae* under carbon dioxide pressure in wine. *Applied Microbiology* 14(4): 643–8.

Lambert-Royo, María Ignacia, Cristina Ubeda, Rubén Del Barrio-Galán, Nathalie Sieczkowski, Joan Miquel Canals, Álvaro Peña-Neira & Mariona Gil i Cortiella (2022). The diversity of effects of yeast derivatives during sparkling wine aging. *Food Chemistry* 390: 133174. Available from: https://doi.org/10.1016/j.foodchem.2022.133174 (accessed 21 November 2023).

Lambri, Milena, Donato Colangelo, Roberta Dordoni, Fabrizio Torchio & Dante Marco De Faveri (2016). Innovations in the use of bentonite in oenology: Interactions with grape and wine proteins, colloids, polyphenols and aroma compounds. In Antonio Morata & Iris Loira (Eds.), *Grape and Wine Biotechnology*, 381–400. London: IntechOpen Ltd.

Lavoisier, Antoine ([1789] 1806). *Elements of Chemistry: In a New Systematic Order, Containing All the Modern Discoveries*, 5th edition (Transl. Robert Kerr). New York: Evert Duyckinck & James and Thomas Ronalds.

Lerm, Elda (2020). Applying our knowledge to determine best MLF practices in 2020 [video]. *InfoWine: The Internet Journal of Viticulture and Enology*, 23 July 2020. Available from: www.infowine.com/en (accessed 23 April 2024).

Liger-Belair, Gérard ([2004] 2022). Blame it on the fizzics. In Susan Keevil (Ed.), *On Champagne*, 216–24. Myddle, UK: Académie du Vin Library Ltd.

—— (2014). How many bubbles in your glass of bubbly? *Journal of Physical Chemistry B* 118: 3156–63. Available from: https://pubs.acs.org/doi/10.1021/jp500295e (accessed 17 June 2024).

—— (2016). Wines: champagne and sparkling wines — production and effervescence. In Benjamin Caballero, Paul M. Finglas & Fidel Toldrá (Eds.), *Encyclopedia of Food and Health*, 526–33. Oxford: Academic Press/Elsevier.

Liger-Belair, Gérard, & Clara Cilindre (2021). Recent progress in the analytical chemistry of champagne and sparkling wines. *Annual Review of Analytical Chemistry* 14: 21–46. Available from: https://doi.org/10.1146/annurev-anchem-061318-115018 (accessed 1 February 2024).

Liger-Belair, Gérard, Guillaume Polidori & Philippe Jeandet (2008). Recent advances in the science of champagne bubbles. *Chemical Society Reviews* 37(11) (November): 2490–2511. Available from: https://doi.org/10.1039/B717798B (accessed 19 March 2024).

Liger-Belair, Gérard, Marielle Bourget, Clara Cilindre, Hervé Pron & Guillaume Polidori (2013). Champagne cork popping revisited through high-speed infrared imaging: The role of temperature. *Journal of Food Engineering* 116(1) (May): 78–85. Available from: https://doi.org/10.1016/j.jfoodeng.2012.11.006 (accessed 20 November 2023).

Liger-Belair, Gérard, Daniel Carvajal-Perez, Clara Cilindre, Jérôme Facque, Marc Brevot, François Litoux-Desrues, Vincent Chaperon & Richard Geoffroy (2018). Evidence for moderate losses of dissolved CO_2 during aging on lees of a champagne prestige cuvee. *Journal of Food Engineering* 233 (September): 40–8. Available from: https://doi.org/10.1016/j.jfoodeng.2018.03.026 (accessed 20 November 2023).

Liger-Belair, Gérard, Chloé Khenniche, Clara Poteau, Carine Bailleul, Virginie Thollin & Clara Cilindre (2023). Losses of yeast-fermented carbon dioxide during prolonged champagne aging: Yes, the bottle size does matter! *ACS Omega* 8(25): 22844–53. Available from: https://doi.org/10.1021/acsomega.3c01812 (accessed 16 November 2023).

Liger-Belair, Gérard, Clara Cilindre, Fabien Beaumont & Guillaume Polidori (2024). Understanding the tasting of champagne and other sparkling wines from a scientific perspective. *Food Research International* 191 (September). Available from: https://doi.org/10.1016/j.foodres.2024.114678 (accessed 29 September 2024).

Lira, Eugenio, Fernando Noé Salazar, Simone Vincenzi, Andrea Curioni & Francisco López (2013). Effect of bentonite fining during fermentation on protein content in macabeu wines: Comparison of pilot- and industrial-scale experiments. *Journal International des Sciences de la Vigne et du Vin* 47(3): 221–6.

Makarov, Alexander S., & Igor P. Lutkov (2021). Yeast race effect on the quality of base and young sparkling wines. *Foods and Raw Materials* 9(2). Available from: https://doi.org/10.21603/2308-4057-2021-2-290-301 (accessed 19 February 2024).

Marín, Antonio Castro, Claudio Riponi & Fabio Chinnici (2020). Chitosan in sparkling wines produced by the traditional method: Influence of its presence during the secondary fermentation. *Foods* 9(9): 1174. Available from: https://doi.org/10.3390/foods9091174 (accessed 23 February 2024).

Martin, Alexander E., R. John Watling & Garry S. Lee (2012). The multi-element determination and regional discrimination of Australian wines. *Food Chemistry* 133: 1081–9. Available from: https://doi.org/10.1016/j.foodchem.2012.02.013 (accessed 13 January 2024).

Martinelli, Luiz A., Marcelo Z. Moreira, Jean P.H.B. Ometto, André R. Alcarde, Luiz A. Rizzon, Erik Stange & James R. Ehleringer (2003). Stable carbon isotopic composition of the wine and CO_2 bubbles of sparkling wines: Detecting C4 sugar additions. *Journal of Agricultural and Food Chemistry* 51: 2625–31.

Medina-Plaza, Cristina, Aubrey DuBois, Elizabeth Tomasino, Anita Oberholster (2024). Effect of storing conditions (lighting, temperature and bottle color) on rosé wine attributes. *Food Chemistry* 439. Available from: https://doi.org/10.1016/j.foodchem.2023.138032 (accessed 4 June 2024).

Merret, Christopher ([1662] 1669). Some observations concerning the ordering of wines. In Walter Charleton, *Two Discourses*, 201–30. London: Printed by R.W. for William Whitwood at the Sign of the Golden-Lion in Duck-Lane, near Smithfield. Available from: http://name.umdl.umich.edu/A32715.0001.001 (accessed 20 August 2022).

Minnaar, Philip, Paul Gerber, Mardé Booyse & Neil Jolly (2021). Research Note: Phenolic compounds in cork-closed bottle-fermented sparkling wines. *South African Journal of Enology and Viticulture* 42(1): 19–24. Available from: https://doi.org/10.21548/42-1-4336 (accessed 19 December 2023).

Moreno, Juan, Rafael Peinado ([2010] 2012). *Enological Chemistry* (Transl. Anne Murray and Iain Patten). London: Academic Press.

Paissoni, Maria Alessandra, Guilia Motta, Simone Giacosa, Luca Rolle, Vincenzo Gerbi & Susana Río Segade (2023). Mouthfeel subqualities in wines: A current insight on sensory descriptors and physical-chemical markers. *Comprehensive Reviews in Food Science and Food Safety* 22: 3328–65. Available from: https://doi.org/10.1111/1541-4337.13184 (accessed 24 July 2024).

Panesar, Parmijt S., Vinod Kumar Joshi, V. Bali & R. Panesar (2017). Technology for production of fortified and sparkling fruit wines. In Maria R. Kosseva, Vinod Kumar Joshi & Parmijt S. Panesar (Eds.), *Science and Technology of Fruit Wine Production*, 487–530. London: Academic Press.

Peters, Sera Jane (1996). Archaeological wines: Analysis and interpretation of a collection of wines recovered from the 'William Salthouse' shipwreck (1841). *Australasian Historical Archaeology* 14: 63–8.

Phillips, Rod (2016). *French Wine: A History*. Oakland, CA: University of California Press.

—— (2017). *The Wines of Canada*. Oxford, UK: Infinite Ideas Ltd.

Polidori, Guillaume, Philippe Jeandet & Gérard Liger-Belair (2009). Bubble and flow patterns in champagne. *American Scientist*, July 2009. Available from: https://www.americanscientist.org (accessed 13 June 2024).

Pons, Alexandre, & Christophe Loisel (2021). Search for an analytical method adapted to OTR measurement of stoppers after long-term bottle ageing. *Revue des Œnologues* 180.

Pons-Mercadé, Pere, Pol Giménez, Jordi Gombau, Glòria Vilomara, Marta Conde, Antoni Cantos, Nicolas Rozès, Joan-Miquel Canals & Fernando Zamora (2021). Oxygen consumption rate of lees during sparkling wine (Cava) aging; influence of the aging time. *Food Chemistry* 342 (16 April 2021): 128238. Available from: https://doi.org/10.1016/j.foodchem.2020.128238 (accessed 26 June 2024).

Puga, German, Kym Anderson, Gregory Jones, Firmin Doko Tchatoka & Wendy Umberger (2022). A climatic classification of the world's wine regions. *OENO One* 56(2). Available from: https://doi.org/10.20870/oeno-one.2022.56.2.4627 (accessed 10 August 2024).

Reddy, Lebaka Veeranjaneya, & Vinod Kumar Joshi (2021). Additives, adjuvants, packages, closures, and labels in enology. In V.K. Joshi & Ramesh C. Ray (Eds.), *Winemaking: Basics and Applied Aspects*, 420–51. Boca Raton, FL: CRC Press.

Riu-Aumatelle, Montserrat, Jordi Torrens, Susana Buxaderas & Elvira López-Tamames (2013). Cava (Spanish sparkling wine) aroma: Composition and determination methods. In Diego Muñoz-Torrero, Amparo Cortés & Eduardo L. Mariño (Eds.), *Recent Advances in Pharmaceutical Sciences III*, 45–60. Trivandrum, India: Transworld Research Network.

Riu-Aumatelle, Montserrat, Elena Bosch-Fusté, Elvira López-Tamames & Susana Buxaderas (2006). Development of volatile compounds of cava (Spanish sparkling wine) during long ageing time in contact with lees. *Food Chemistry* 95: 237–42. Available from: https://doi.org/10.1016/j.foodchem.2005.01.029 (accessed 19 June 2024).

Robillard, Bertrand (2012). The use of jetting technology for a better quality management at bottle (disgorging) stage. In Jamie Goode (Ed.), *Sparkling Wine 2012: Proceedings of the Second International Sparkling Wine Symposium*, 85–90. Tunbridge Wells, UK: Flavour Press.

Robinson, Jancis (2014). Champagne — now older and drier. *Jancis Robinson*, 1 March 2014. Available from: www.jancisrobinson.com (accessed 10 August 2024).

Roth, Klaus (2010). Sparkling wine, champagne and co — Part 3. *Chemistry Views*, 22 December 2010. Available from: www.chemistryviews.org (accessed 14 June 2024).

Rountree, Meredith (2023). Like a fine (sparkling) wine: How to age champagne without losing the bubbles. *ACS Axial*, 20 July 2023. Available from: https://axial.acs.org/agriculture-and-food-chemistry (accessed 16 November 2023).

Sandler, Merton, & Roger Pinder (2003). *Wine: A Scientific Exploration*. London/New York: Taylor & Francis Group.

Schmidt, Charlotte Vinther, Karsten Olsen & Ole G. Mouritsen (2020). Umami synergy as the scientific principle behind taste-pairing champagne and oysters. *Scientific Reports* 10 (November): 20077. Available from: https://doi.org/10.1038/s41598-020-77107-w (accessed 23 August 2024).

Shepro, Richard Warren (2018). 'Le mariage entre mets et vins': On the geographical and historical origins of pairing a food with a particular wine in France. In Mark McWilliams (Ed.), *Food and Landscape. Proceedings of the Oxford Symposium on Food and Cookery 2017*, 346–58. London: Prospect Books.

Skelton, Stephen (2007). *Viticulture: An Introduction to Commercial Grape Growing for Wine Production*. London: Stephen Skelton.

—— (2021). *The Knight Who Invented Champagne*. London: S.P. Skelton Ltd.

Staniforth, Mark (2007). William Salthouse 1841: Barrels, beef and bottles. In Michael Nash (Ed.), *Shipwreck Archaeology in Australia*, 99–110. Nedlands: University of Western Australia Press.

Stávek, Jan, Barbora Papouskova, Josef Balik & Petr Bednar (2012). Effect of storage conditions on various parameters of colour and the anthocyanin profile of rosé wines. *International Journal of Food Properties* 15(5): 1133–47. Available from: https://doi.org/10.1080/10942912.2010.511751 (accessed 3 June 2024).

Tarko, Tomasz, Aleksandra Duda-Chodak, Paweł Sroka & Małgorzata Siuta (2020). The impact of oxygen at various stages of vinification on the chemical composition and the antioxidant and sensory properties of white and red wines. *International Journal of Food Science (IJFS)*. Available from: https://doi.org/10.1155/2020/7902974 (accessed 17 February 2024).

Tomasi, Diego, & Federica Gaiotti (2020). *The Terroirs of Conegliano Valdobbiadene Prosecco Wines. A Study of the Origins of Quality in the Hills of the UNESCO Heritage Site*. Crocetta del Montello, Italy: Antiga Edizioni.

Torresi, Sara, Maria Teresa Frangipane & Gabriele Anelli (2011). Biotechnologies in sparkling wine production. Interesting approaches for quality improvement. A review. *Food Chemistry* 129(3) (1 December 2011): 1232–41. Available from: https://doi.org/10.1016/j.foodchem.2011.05.006 (accessed 17 November 2023).

Ubeda, Cristina, Ingeborg Kania-Zelada, Rubén del Barrio-Galán, Marcela Medel-Marabolí, Mariona Gil & Álvaro Peña-Neira (2019). Study of the changes in volatile compounds, aroma and sensory attributes during the production process of sparkling wine by traditional method. *Food Research International* 119 (May): 554–63. Available from: https://doi.org/10.1016/j.foodres.2018.10.032 (accessed 26 June 2024).

UGA (University of Georgia) (2019). Methods for sparkling wine production. *UGA Extension Viticulture Blog*, May 2019. Available from: https://viticulture.uga.edu/production/enology (accessed 2 November 2023).

Varlamov, Andrey A., & Lev G. Aslamazov ([2001] 2019). *The Wonders of Physics*, 4th edition. London: World Scientific Publishing.

Vernhet, Aude (2022). Settling: Clarification of musts. In Antonio Morata (Ed.), *White Wine Technology*, 87–98. London: Academic Press.

Vidal, Stéphane, Leigh Francis, Pascale Williams, Mariola Kwiatkowski, Richard Gawel, Véronique Cheynier & Elizabeth Waters (2004). The mouthfeel properties of polysaccharides and anthocyanins in a wine like medium. *Food Chemistry* 85(4) (May): 519–25. Available from: https://doi.org/10.1016/S0308-8146(03)00084-0 (accessed 24 July 2024).

Waterhouse, Andrew L., Gavin L. Sacks & David W. Jeffery (2016). *Understanding Wine Chemistry*. Chichester, UK: John Wiley & Sons.

Welke, Juliane Elisa, Bruna Dachery, Lucas Dal Magro, Karolina Cardoso Hernandes & Cláudia Alcaraz Zini (2022). Volatile compounds formation in sparkling wines. In Felipe Richter Reis & Caroline Mongruel Eleutério dos Santos (Eds.), *Volatile Compounds Formation in Specialty Beverages*. Boca Raton, FL: CRC Press.

Wilson, James E. ([1998] 1999). *Terroir: The Role of Geology, Climate, and Culture in the Making of French Wines*. Berkeley & Los Angeles, CA: University of California Press.

Zoecklein, Bruce (2002). *A Review of Méthode Champenoise Production*. Blacksburg, VA: Virginia Cooperative Extension.

PROFESSIONAL DOCUMENTS

Adams, Paul (2018). The science of Pét-Nat wines. *Seven Fifty Daily*, 6 February. Available from: https://daily.sevenfifty.com (accessed 20 March 2024).

Ahmed, Sarah (2024). Tasmania unlocked. *The World of Fine Wine*, 26 April. Available from: https://worldoffinewine.com (accessed 10 August 2024).

Alvarez, Pablo (2018). Champagne: From the vine to the glass. *Geneva Business News*, 10 September. Available from: www.gbnews.ch (accessed 6 February 2024).

Amatuzzi, Dan (2013). *A First Course in Wine: From Grape to Glass*. New York: Quayside Publishing Group.

Angelosante, Jennifer (2020). A guide to wine phenolics. *GuildSomm International*, 27 March. Available from: www.guildsomm.com (accessed 1 May 2024).

Anson, Jane (2015). Rothschild one of two newcomers in the elite world of champagne. *South China Morning Post*, 12 June. Available from: www.scmp.com (accessed 8 August 2023).

Antonsson, Björnstierne (2022). New cellar master @ Champagne Deutz — Caroline Latrive. *The Champagne Club*, 30 October. Available from: www.champagneclub.com (accessed 29 April 2024).

AO (Anchor Oenology) (2022). *Base Wine to Bubbles*, 2022 edition. Cape Town: Anchor Oenology. Available from: https://za.anchoroenology.com/downloads (accessed 31 October 2023).

Atkins, Susy (2021). English blanc de blancs: Expert's choice. *Decanter*, 7 April. Available from: www.decanter.com (accessed 21 September 2023).

Avellan, Essi (2022a). Krug 1976–2008. *Club Oenologique* 11 (Spring/Summer). Available from: https://cluboenologique.com (accessed 8 June 2024).

——— (2022b). Champagne — the significance of disgorgement. *Club Oenologique*, 16 November. Available from: https://cluboenologique.com (accessed 19 October 2023).

——— (2022c). Organic and biodynamic Champagne: Green fingers. *The World of Fine Wine*, 17 November. Available from: https://worldoffinewine.com (accessed 28 April 2023).

——— (2023a). Top rosé champagne in 2023. *Club Oenologique*, 25 October. Available from: https://cluboenologique.com (accessed 27 January 2024).

——— (2023b). An experimental new era for oak in Champagne. *Club Oenologique*, 1 November. Available from: https://cluboenologique.com (accessed 19 June 2024).

Avery, Richard ([2010] 2012). *Quicklook at Wine*. Grittleton, UK: Quicklook Books Ltd.

Bacigalupi, Katey (2023). Sparkling & honey: A magical combination. Newsletter > Blog. Available from: https://bacigalupivineyards.com/blog (accessed 23 June 2023).

Baker, Isaac (2023). Wine Reviews: Argentina's Domaine Bousquet. *Terroirist*, 24 June. Available from: www.terroirist.com (accessed 11 May 2024).

Barr, Neria (2019). New Italian wines and how they're made. *Jerusalem Post*, 19 December. Available from: www.jpost.com (accessed 4 April 2024).

Bastianich, Joseph, & David Lynch ([2002] 2005). *Vino Italiano: The Regional Wines of Italy*. New York: Clarkson Potter/Publishers.

Beckett, Fiona (2015). What type of wine works best with a barbecue? *Matching Food & Wine*, 27 May. Available from: www.matchingfoodandwine.com (accessed 13 September 2024).

——— (2017). The best food pairings for Lambrusco. *Matching Food & Wine*, 21 June. Available from: www.matchingfoodandwine.com (accessed 13 September 2024).

——— (2023). The best food pairings for rosé. *Matching Food & Wine*, 31 May. Available from: www.matchingfoodandwine.com (accessed 12 September 2024).

——— (2024). Wine basics: A beginner's guide to pairing food and wine. *Matching Food & Wine*, 26 June. Available from: www.matchingfoodandwine.com (accessed 21 August 2024).

Bell, Emily (2015a). What is crémant? *VinePair*, 11 October. Available from: https://vinepair.com (accessed 17 May 2024).

——— (2015b). Blancs de noirs vs. blanc de blancs — explaining the difference. *VinePair*, 10 December. Available from: https://vinepair.com (accessed 22 July 2024).

Bell, Katie Kelly (2024). A wine writer's top 10 rosé champagnes and sparkling wines. *Forbes*, 14 February. Available from: www.forbes.com (accessed 14 May 2024).

Bennett, Lynne Char (2005). What to eat with sparkling reds? *SFGate*, 26 May. Available from: www.sfgate.com (accessed 11 September 2024).

Berrill, Anna (2011). Barack Obama drinks English wine on Buckingham Palace visit. *Decanter*, 25 May. Available from: www.decanter.com (accessed 27 September 2024).

Bertrand, Patrice (2022). French sparkling wines to rival Champagne on price and taste. *The Connexion*, 6 December. Available from: www.connexionfrance.com (accessed 18 May 2024).

Bielka-Vescovi, Monika (2019). Poland for wine lovers. *Jancis Robinson*, 22 August. Available from: www.jancisrobinson.com (accessed 16 August 2024).

Biro, Lidija (2024). Vieni Estates — Ontario's house of sparkling wine. *Vine Routes*, 17 July. Available from: https://vineroutes.com (accessed 11 September 2024).

Brazil, Nancy (2023). Limoux sparkling wine: Bubbles and history in your glass. *Pull That Cork*, 20 November. Available from: https://pullthatcork.com (accessed 25 March 2024).

Bredahl, Michael (2017). Wine with Trepat grapes. *World's Best Wines*, 16 June. Available from: https://worldsbestwines.eu (accessed 7 September 2024).

Briscoe, Stacy (2023). What does 'cuvée' mean in wine? *Wine Enthusiast*, 6 June. Available from: www.wineenthusiast.com (accessed 24 September 2024).

Broggi, Maurizio (2023). Alcoholic fermentation: Monitoring. *Westgarth Wines*, 14 December. Available from: https://westgarthwines.com (accessed 2 February 2024).

Bryan, Casey (2019). Taltarni Vineyards 2017 Sparkling Shiraz. *Travelling Corkscrew*, 16 December. Available from: https://travellingcorkscrew.com.au (accessed 27 May 2024).

Burgess, Laura (2016). Why the world will never run out of Moscato d'Asti. *VinePair*, 5 December. Available from: https://vinepair.com (accessed 29 May 2024).

Burgess, Noel (2022). What does 'riddling' mean in wine? *Wine Enthusiast*, 28 September. Available from: www.wineenthusiast.com (accessed 26 April 2024).

Burton, Neel ([2014] 2016). *The Concise Guide to Wine and Blind Tasting*, 2nd edition. Oxford, UK: Acheron Press.

Burton, Neel, & James Flewellen (2013). Champagne: Method of production. *The Oxford Wine Blog*, 22 May. Available from: https://theoxfordwineblog.wordpress.com (accessed 26 April 2023).

Butler, David (2023). Blanc de blanc champagne: Styles, 10 great wines (2023). *Vinovest*. Available from: www.vinovest.co (accessed 20 April 2023).

Campbell, Bob (2021). Fizz facts: Sparkling winemaking methods. *The Real Review*, 2 September. Available from: www.therealreview.com (accessed 4 November 2023).

Capretti, Lucia (2022). 20 different types of sparkling wines explained. *Tasting Table*, 16 June. Available from: www.tastingtable.com (accessed 20 September 2023).

Caputo, Tina (2019). How champagne producers are preserving acidity as the climate changes. *Seven Fifty Daily*, 24 June. Available from: https://daily.sevenfifty.com (accessed 2 August 2024).

Cardillo, J'aime (2023). 24 sparkling wines for 2024. *Halliday Wine Companion*, 10 December. Available from: https://winecompanion.com.au (accessed 13 September 2024).

Carr, Thomas (2021). The art of ageing wine underwater. *Halliday Wine Companion*, 1 October. Available from: https://winecompanion.com.au (accessed 1 July 2023).

Castaing, Yohan (2015). Champagne and Burgundy great Joseph Henriot dies. *Decanter*, 30 April. Available from: www.decanter.com (accessed 20 May 2024).

Cavaleri, Annalisa Leopolda (2019). How to serve champagne like a master: Top 10 tips. *Fine Dining Lovers*, 8 February. Available from: www.finedininglovers.com (accessed 19 August 2024).

Chandra, Gowri (2022). 5 mistakes people make when drinking champagne. *Food & Wine*, 6 July. Available from: www.foodandwine.com (accessed 20 August 2024).

Cheurlin, Philippe (2021). Composition and dosage of champagne. *Delozey Blog*, 20 May. Available from: https://delozey.com/en-int/blogs/advice (accessed 14 June 2024).

Chroman, Nathan (1986). Sparkling wines you can afford: Lancers' two new entries are produced under patented system. *Los Angeles Times*, 6 February. Available from: www.latimes.com (accessed 15 April 2024).

Claffee, Scott (2013). The perpetual blend: Champagne's Solera SYSTEM. *Terroirist*, 8 December. Available from: www.terroirist.com (accessed 21 May 2024).

Clarke, Oz ([2000] 2003). *Introducing Wine*. London: Time Warner Books.

Cloete, Carlin (2017). The quest to become whole again. *WineLand*, 3 October. Available from: www.wineland.co.za (accessed 6 May 2023).

Cole, Katherine (2021). *Sparkling Wine Anytime: The Best Bottles to Pop for Every Occasion*. New York: Abrams.

Coppola, Francis Ford (2020). An interview with Francis Ford Coppola: Master filmmaker and major wine producer. An interview by John Mariani. *Forbes*, 2 June. Available from: www.forbes.com (accessed 22 July 2024).

Cowderoy, David (2021). Interview with Vineyard Magazine ('Charmat sparkles'). *Vineyard Magazine*, August: 36–41. Available from: www.vineyardmagazine.co.uk (accessed 11 April 2024).

Crawford, Peter (2023). What is lees ageing and why is this important? *Sip Champagnes*, 19 July. Available from: https://sipchampagnes.com/blogs/news (accessed 20 November 2023).

Currin, Tamlyn (2021a). Wiston, Blanc de Noir 2014 England. *Jancis Robinson*, 1 September. Available from: www.jancisrobinson.com (accessed 2 September 2024).

—— (2021b). Sparkling Slovenia. *Jancis Robinson*, 4 November. Available from: www.jancisrobinson.com (accessed 2 September 2024).

—— (2022). L & R Kox, Sans Sulfites Ajoutés NV Crémant de Luxembourg. *Jancis Robinson*, 11 November. Available from: www.jancisrobinson.com (accessed 20 April 2024).

Curtis, Charles (2020). *Vintage Champagne: 1899 to 2019*. New York: WineAlpha.

D'Agata, Ian (2024). British sparkling wines: All you really need to know and recent vintages. *Terroir Sense*, 25 April. Available from: https://terroirsense.com/en (accessed 11 September 2024).

Dara, Jillian (2022). Should you decant champagne? It depends. *Wine Enthusiast*, 21 October. Available from: www.wineenthusiast.com (accessed 17 August 2024).

Decuers, Larry (2020). Million dollar hangover: Victory celebration at Berchtesgaden. *Explore the War*, 9 May. New Orleans: The National WW II Museum. Available from: www.nationalww2museum.org (accessed 27 April 2023).

DeJean, Joan (2005). *The Essence of Style: How the French Invented High Fashion, Fine Food, Chic Cafés, Style, Sophistication, and Glamour*. New York: Free Press.

Demmond, Carson (2023). 4 kinds of sparkling red wine to try. *Food & Wine*, 8 August. Available from: www.foodandwine.com (accessed 2 December 2023).

Denig, Vicky (2016). How many twists does it take to open a bottle of champagne? *VinePair*, 22 December. Available from: https://vinepair.com (accessed 20 August 2024).

—— (2020). Alternatives to chemicals in Champenoise viticulture. *Paris Wine Company*, 9 July. Available from: https://pariswinecompany.com (accessed 28 July 2023).

—— (2023). How producers are returning to winemaking's roots. *Wine Enthusiast*, 4 May. Available from: www.wineenthusiast.com (accessed 5 October 2024).

Dessauve, Thierry ([2014] 2022). Bernard de Nonancourt (1920–2010). *The World of Fine Wine*, 16 February 2014 (updated 2 November 2022). Available from https://worldoffinewine.com (accessed 3 May 2023).

Dhillon, Navpreet (2021). Innovative producers are making sparkling red wines for everyday drinking. *Sunset*, 23 September. Available from: www.sunset.com (accessed 3 June 2024).

Douglas, Ellie (2016a). How to sabre Champagne — and become a 'grand commandeur'. *Decanter*, 24 August. Available from: www.decanter.com (accessed 20 August 2024).

—— (2016b). How long to chill champagne for. *Decanter*, 24 December. Available from: www.decanter.com (accessed 24 December 2024).

—— (2020). When should I open my NV champagne? *Decanter*, 17 December. Available from: www.decanter.com (accessed 16 November 2023).

Dublin, Edwin (2016). Champagne — better late-disgorged? *Berry Bros. & Rudd Wine Blog*, 9 February. Available from: www.blog.bbr.com (accessed 5 June 2024).

—— (2023). Matching champagne with food: a feast with fizz. *Berry Bros. & Rudd Wine Blog*, 1 March. Available from: www.blog.bbr.com (accessed 24 August 2024).

Dupont, Estelle (2023a). Types of champagne: The many faces of Champagne. *Millesima Blog*, 20 February. Available from: www.millesima-usa.com/blog (accessed 21 March 2023).

—— (2023b). Everything you need to know about champagne saber. *Millesima Blog*, 8 March. Available from: www.millesima-usa.com/blog (accessed 20 August 2024).

—— (2023c). Champagne bubbles: What are they and where do they come from? *Millesima Blog*, 3 July. Available from: www.millesima-usa.com/blog (accessed 5 January 2024).

Eads, Lauren (2014). Ruinart lays claim to 'world's first' rosé champagne'. *The Drinks Business*, 10 March. Available from: www.thedrinksbusiness.com (accessed 30 April 2024).

Eales, Barnaby (2019). The sugar question divides Europe. *Meininger's Wine Business International* 02–2019 (16 May). Available from: www.wine-business-international.com (accessed 6 March 2023).

Earl, Natalie (2021). The best non-vintage champagnes to buy. *Decanter*, 16 December. Available from: www.decanter.com (accessed 27 July 2024).

Edwards, Lucy (2022). Taking cellaring to new depths — a list of champagnes aged under the sea. *Champagne Every Day*, 26 May. Available from: www.champagneeveryday.com.au (accessed 9 August 2023).

—— (2023). Riddling revolution: The twists and turns of Veuve Clicquot's champagne invention. *Champagne Every Day*, 11 November. Available from: www.champagneeveryday.com.au (accessed 26 April 2024).

Edwards, Michael ([2015] 2021). How to store champagne at home. *Decanter*, 18 February. Available from: www.decanter.com (accessed 20 November 2023).

Edwards, Paul (2023). Wine: Sparkling reds for Christmas. *The West Australian*, 15 December. Available from: https://thewest.com.au (accessed 20 November 2024).

Eedes, Christian (2019). Colmant appoints Paul Gerber as new winemaker. *Winemag*, 30 January. Available from: https://winemag.co.za (accessed 27 November 2023).

Epstein, Becky Sue (2011). *Champagne: A Global History*. London: Reaktion Books Ltd.

Faith, Nicholas ([1989] 2016). *The Story of Champagne*. Oxford, UK: Infinite Ideas Ltd.

Fallowfield, Giles (2022). What is a 'perpetual reserve'? *Decanter*, 11 May. Available from: www.decanter.com (accessed 20 May 2024).

Farrow, Steve (2021). Perfect food matches for English wine. *The Wine Society*, 6 May. Available from: www.thewinesociety.com (accessed 26 August 2024).

Feilden, Eloise (2024). Waitrose: 'We have never seen crémant sales this big'. *The Drinks Business*, 18 March. Available from: www.thedrinksbusiness.com (accessed 18 May 2024).

Ferreira, Pieter (2020). Cap Classique conversation — Crown Cap vs. Agrafe Closures [video]. Graham Beck Bubbly Official (YouTube), 10 June. Available from: www.youtube.com/@grahambeckbubblyofficial (accessed 18 December 2023).

Field, Simon (2021). Reflections from a year with wine: Krug Clos du Mesnil sparkles in Sweden. *The World of Fine Wine*, 30 December. Available from: https://worldoffinewine.com (accessed 8 June 2024).

——— (2022a). What's the deal with cava's new top tier, Guarda Superior? *Club Oenologique*, 28 February. Available from https://cluboenologique.com (accessed 13 April 2023).

——— (2022b). Uncovering the story behind Laurent-Perrier Grand Siècle Champagne. *Club Oenologique*, 28 November. Available from https://cluboenologique.com (accessed 26 April 2023).

Fielden, Christopher ([1994] 1998). *Exploring Wines and Spirits*. London: The Wine & Spirit Education Trust.

——— ([2001] 2010). *The Wines of Argentina, Chile and Latin America*. London: Octopus Publishing Group Ltd.

Forbes, Patrick ([1967] 1989). *Champagne: The Wine, the Land and the People*. London: Victor Gollancz Ltd.

Forshaw, Irenie (2023). Veuve Clicquot unveils exclusive cellar in the sea experience. *Elite Traveler*, 28 March. Available from: https://elitetraveler.com/travel/experiences-travel (accessed 30 June 2023).

Fraley, Dennis (2021). FMN — January 2021 — Wine with stars. *The Bottle Talk*, 1 January. Available from: https://thebottletalk.com (accessed 9 April 2024).

Frerson, Séverine (2018). Interview with Séverine Frerson, a cellar master in Champagne ('Champagne: a wine to keep your smile, or to win it back'). An interview by Kristel Balcaen. *Wine and Words*, 27 July. Available from: www.wineandwords.be (accessed 16 September 2023).

——— (2021). Interview with Séverine Frerson, Perrier-Jouët's chef de cave ('Champagne: the art of blending'). An interview by Amy Wislocki. *Decanter*, 8 March. Available from: www.decanter.com (accessed 9 January 2024).

Funke, Evan, & Katie Parla (2019). *American Sfoglino. A Master Class in Handmade Pasta*. San Francisco: Chronicle Books LLC.

Gabay, Elizabeth (2018). *Rosé: Understanding the Pink Wine Revolution*. Oxford, UK: Infinite Ideas Ltd.

Gallagher, Katherine (2023). The best champagne stoppers for preserving your leftover bubbly. *The Spruce Eats*, 26 April. Available from: www.thespruceeats.com (accessed 21 August 2024).

Gardner, Denise M. (2015a). The bubbles: Basics about sparkling wine production techniques. *Wine & Grapes U.*, 18 September. Available from: https://psuwineandgrapes.wordpress.com (accessed 16 May 2023).

——— (2015b). Technical information about Pét-Nats (Pétillant Naturels, or sparkling wines produced by Méthode Ancestrale). *Wine & Grapes U.*, 2 October. Available from: https://psuwineandgrapes.wordpress.com (accessed 25 March 2024).

Gargett, Ken (2012). First sparkling screwcap released. *Decanter*, 25 May. Available from: www.decanter.com (accessed 26 July 2023).

——— (2021). Salon Le Mesnil Blanc de Blanc: An original 'Unicorn' Champagne. *Quill & Pad*, 5 January. Available from: https://quillandpad.com (accessed 4 May 2021).

Gellie, Tina (2016). Krug plans next move to 'kill the flute'. *Decanter*, 19 May. Available from: www.decanter.com (accessed 20 August 2024).

Gerling, Chris, Molly Kelly & Beth Chang (2023). Pét-Nat production: An industry roundtable: Webinar on 18 April 2023 [video]. *Eastern Viticulture and Enology Forum*, 18 April. University Park, PA: Pennsylvania State University. Available from: https://psu.mediaspace.kaltura.com/media/1_uoowefbr (accessed 1 September 2023).

Gibb, Rebecca (2012). The pros and cons of disgorgement dates on champagne labels. Rebecca Gibb, 29 July. Available from: https://rebeccagibb.com (accessed 19 October 2023).
Gilby, Caroline (2020). Winemaking: A sparkling transformation. *The Wine Society*, 10 June. Available from: www.thewinesociety.com (accessed 28 March 2023).
Goldstein, Evan (2006). *Perfect Pairings: A Master Sommelier's Practical Advice for Partnering Wine with Food*. Berkeley & Los Angeles, CA/London: University of California Press.
Goldstein, Michael (2023). Martinotti Charmat method: Prosecco's best bubbly secrets. *Prosecco Ventures*, 11 May. Available from: https://prosecco.com (accessed 15 April 2024).
Goode, Jamie (2013). Guide to closure type. *The Wine Society*, 16 May. Available from: www.thewinesociety.com (accessed 26 July 2023).
—— (2016). Benoît Gouez introduces Moët et Chandon MCIII, the new prestige cuvée. *Wine Anorak: Global Wine Journal*, 1 April. Available from: https://wineanorak.com (accessed 16 October 2023).
—— (2020a). In Uruguay: Visiting Viña Edén in Maldonado. *Wine Anorak: Global Wine Journal*, 19 March. Available from: https://wineanorak.com (accessed 4 September 2024).
—— (2020b). Juice stabulation, an interesting winemaking technique. *Wine Anorak: Global Wine Journal*, 28 June. Available from: https://wineanorak.com (accessed 7 September 2023).
—— (2023). Clairette de Die and friends: The Diois wines from the Rhône. *Wine Anorak: Global Wine Journal*, 6 April. Available from: https://wineanorak.com (accessed 29 August 2023).
Gordon, Susan H. (2016). 3 unsung champagne alternatives that aren't cava or prosecco. *Eater*, 26 January. Available from: www.eater.com (accessed 12 April 2024).
Gouez, Benoît (2020). Champagne legends: An interview by Björnstierne Antonsson. *The Champagne Club*, 28 November. Available from: www.champagneclub.com (accessed 16 October 2023).
Grainger, Keith (2009). *Wine Quality: Tasting and Selection*. Chichester, UK: John Wiley & Sons.
—— (2021). *Wine Faults and Flaws: A Practical Guide*. Chichester, UK: John Wiley & Sons Ltd.
Grainger, Keith, & Hazel Tattersall (2005). *Wine Production: Vine to Bottle*. Oxford, UK: Blackwell Publishing Ltd.
Grainger, Keith, & Hazel Tattersall ([2005] 2016). *Wine Production and Quality*, 2nd edition. Chichester, UK: John Wiley & Sons, Ltd.
Gramona, Xavier, & Ton Mata (2021). Corpinnat: Economic and Social Profile. Presentation of Economic Data 2020 [video]. Corpinnat (YouTube), 29 April. Available from: www.youtube.com/@corpinnat2009 (accessed 23 November 2023).
Green, Martin (2023a). Champagne sales increase by 19.4% in the US. *Decanter*, 6 April. Available from: www.decanter.com (accessed 21 August 2023).
—— (2023b). Brad Pitt launches a Champagne cuvée named Petite Fleur. Hollywood star Brad Pitt has unveiled a new cuvée from the Fleur de Miraval Champagne house at Mesnil-sur-Oger. *Decanter*, 3 October. Available from: www.decanter.com (accessed 16 October 2023).
Gustafsson, Jenny (2023). New wine in old bottles. Lebanon seeks to get back on the wine map amid economic crises. *The Caravan*, 31 March. Available from: https://caravanmagazine.in (accessed 9 September 2023).
Hale, Noelle (2023). What is acidity in wine? *Wine Enthusiast*, 5 May. Available from: www.wineenthusiast.com (accessed 29 October 2023).
Hall, Tim (2021). Champagne AR Lenoble: In a class of its own. *The World of Fine Wine* 71 (March): 148–53.
Hallinan, Janne (2023). Sparkling wines — From seafood to steak — The perfect food wines. *Glass of Bubbly*, 4 April. Available from: https://glassofbubbly.com (accessed 24 August 2024).
Haraszthy, Agoston (1862). *Grape Culture, Wines and Wine-making, with Notes upon Agriculture and Horticulture*. New York: Harper & Brothers Publishers.
Hartley, Clive ([1999] 2022). *Australian Wine Guide*, 7th edition. Daylesford, VIC: Premier Cru Hospitality Services.
Hay, Celia (2022). Sparkling wine: How to open and serve? *New Zealand Herald*, 2 December. Available from: www.nzherald.co.nz (accessed 7 November 2023).
Heiligenthal, Pamela (2012). Discovering the effects of post-disgorgement ageing. *EnoBytes.com*, 5 November. Available from: www.enobytes.com (accessed 16 November 2023).
Hemingway, Claudine (2022). Episode 95: Widows of Champagne. *Claudine Hemingway*, 14 March. Available from: www.claudinehemingway.com (accessed 26 July 2024).
Hemming, Richard (2010). Oxygen and wine — the latest. *Jancis Robinson*, 15 July. Available from: www.jancisrobinson.com (accessed 18 September 2023).
Hewson, Tom (2021). The mysteries of the mousse. *Tim Atkin*, 14 July. Available from: https://timatkin.com (accessed 13 June 2024).
Hill, Lauren (2021). Red alert! Why you should be drinking sparkling red wine. *Luxury Defined*, 20 July. Available from: www.christiesrealestate.com (accessed 6 May 2024).

Hodson, Andrew (2020a). Sparkling wine: User's guide. The continuing saga of understanding sparkling wine. *Veritas*, 15 September. Available from: https://veritaswines.com (accessed 10 August 2024).

——— (2020b). The sparkling wine saga — tannins. *Veritas*, 2 October. Available from: https://veritaswines.com (accessed 14 August 2024).

Holthausen, Christian (2020). Champagne AR Lenoble: Trying to celebrate 100 years in a pandemic. *The Buyer*, 27 May. Available from: www.the-buyer.net/people/producer/champagne-ar-lenoble (accessed 16 August 2023).

Hooke, Huon (2012). Pop the screwcap. *The Real Review*, 14 June. Available from: www.therealreview.com (accessed 30 November 2023).

——— (2021). Chandon goes ancestrale. *The Real Review*, 2 August. Available from: www.therealreview.com (accessed 30 August 2023).

Howell, Greg (2013). Sparkling wine: A peculiar lab example. *VintEssential*, 16 August. Available from: www.vintessential.com.au (accessed 17 May 2024).

Hryhorieva, Oleksandra (2021). Champagne wine exports in 2020. *Drinks Plus: Gastronomy and Wine Portal*, 26 May. Available from: https://drinks.ua/analytics-en (accessed 21 August 2023).

Hudin, Miquel (2016). An introduction to Clàssic Penedès. *GuildSomm*, 14 March. Available from: www.guildsomm.com (accessed 6 October 2024).

Hunt, Alex ([2015] 2022). Pinot Précoce. *The World of Fine Wine*, 7 May 2015 (updated 2 November 2022). Available from: https://worldoffinewine.com (accessed 2 February 2024).

——— (2016). The (increasing) problem of lightstrike. *Jancis Robinson*, 7 July. Available from: www.jancisrobinson.com (accessed 3 June 2024).

Hunt, Maria C. (2022). Rosé food and wine pairing ideas. *Total Wine*, 15 April. Available from: www.totalwine.com (accessed 11 September 2024).

Hyland, Tom (2019). The allure of rosé champagne. *Forbes*, 11 February. Available from: www.forbes.com (accessed 27 January 2024).

——— (2024). Rosé and oeil de perdrix: Glorious champagnes of a different shade. *Forbes*, 16 January. Available from: www.forbes.com (accessed 30 April 2024).

Intardonato, John (2009). Thinking about making sparkling wine? *Wine Business*, 15 January. Available from: www.winebusiness.com (accessed 9 April 2024).

James, Tim (2013). *Wines of the New South Africa: Tradition and Revolution*. Los Angeles: University of California Press.

Jeffs, Julian (2006). *The Wines of Spain*. London: Mitchell Beazley.

Jestin, Hervé (2023). Personal interview on vitivinicultural practices, 2 June.

Jiménez, Santiago (2014). The cellar in the sea. *Selectus Wines*, 1 October. Available from: www.selectuswines.com/en/the-cellar-in-the-sea/1555 (accessed 30 June 2023).

JMF (José Maria da Fonseca) (2017). *Press Kit*. Azeitão: José Maria da Fonseca, Quinta da Bassaqueira. Available from: www.jmf.pt (accessed 15 April 2024).

Joe, Melinda (2023). Passe or sparkling, Lambrusco can get you through a Tokyo summer. *The Japan Times*, 20 August. Available from: www.japantimes.co.jp (accessed 9 September 2023).

Johnson, Hugh ([1989] 2021). France perfects it . . . In Susan Keevil (Ed.), *On Champagne*, 106–16. Myddle, UK: Académie du Vin Library.

Johnson, Hugh, & Margaret Rand ([1977] 2022). *Hugh Johnson's Pocket Wine Book 2023*. London: Octopus Publishing Group Ltd.

Johnson, Robert (2019). A wine solution for salty food. *Vinesse Wines*, 22 October.

Juhlin, Richard (2022). A century of grand vintages [collaboration by Moët & Chandon]. *The Champagne Club*, 3 February. Available from: www.champagneclub.com/a-century-of-grand-vintages-collaboration-by-moet-chandon (accessed 4 November 2023).

——— (2023). Veuve Clicquot — cellar in the sea 2023. *The Champagne Club*, 30 June. Available from: www.champagneclub.com/veuve-clicquot-cellar-in-the-sea-2023 (accessed 4 July 2023).

——— (2024). How to serve champagne: Expert tips for the perfect toast. *The Champagne Club*, 7 March. Available from: www.champagneclub.com/how-to-serve-champagne (accessed 20 August 2024).

Jukes, Matthew (2018). 2017 Flint Vineyard, Charmat rosé, England. *Matthew Jukes*, 31 August. Available from: www.matthewjukes.com (accessed 13 May 2024).

Karlsson, Per (2016). This is how you disgorge a sparkling wine. *BK Wine Magazine*, 27 March. Available from: www.bkwine.com/news/disgorge-sparkling-wine (accessed 18 December 2023).

Karlsson, Per, & Britt Karlsson ([2015] 2016). The newest trend in Champagne: Drier than brut or even sugar free. *Forbes*, 24 August. Available from: www.forbes.com (accessed 31 July 2024).

Kermode, David (2019). Staring at the Abyss: Leclerc Briant's underwater masterpiece. *The Buyer*, 5 February. Available from: www.the-buyer.net/tasting/wine/staring-at-the-abyss-david-kermode-on-leclerc-briants-underwater-masterpiece (accessed 3 July 2023).

Kermode, David (2023). Dom Ruinart: Time for change. *Club Oenologique*, 16 January. Available

from: https://cluboenologique.com (accessed 3 October 2024).
Kirkland, Matt (2016). Chemistry behind champagne aromas. *Wine Scholar Guild*, 18 August. Available from: www.winescholarguild.org/blog/chemistry-behind-champagne-aromas (accessed 19 October 2023).
Kirton, David (2023). Celebrating English Wine Week 2023 with Harrow & Hope & Flint Vineyards. *Reserve Wines*, 7 June. Available from: www.reservewines.co.uk (accessed 13 May 2024).
Kladstrup, Don, & Petie Kladstrup (2001). *Wine and War: The French, the Nazis, and the Battle for France's Greatest Treasure*. London: Hodder & Stoughton.
Kladstrup, Don, & Petie Kladstrup ([2005] 2006). *Champagne: How the World's Most Glamorous Wine Triumphed over War and Hard Times*. New York: HarperCollins Publishers.
Köhler, Stefanie (2022). How do you like your rosé champagne: Blended or macerated? *Envie de Champagne: Le Magazine du Champagne*, 10 July. Available from: https://magazine.enviedechamp.com/en (accessed 25 September 2023).
Kojder, Artur (2022). Personal interview on vitivinicultural practices, 28 June.
Kramer, Kimberley (2023). The art of dosage: Transforming sparkling wines with a delicate touch. *Kramer Vineyards Blog*, 19 June. Available from: www.kramervineyards.com (accessed 29 April 2024).
Krebiehl, Anne ([2017] 2022). Marriage of age and youth. In Susan Keevil (Ed.), *On Champagne*, 234–43. Myddle, UK: Académie du Vin Library Ltd.
——— (2019). *The Wines of Germany*. Oxford, UK: Infinite Ideas.
——— (2023). What is whole bunch fermentation and why does it matter? *Wine Enthusiast*, 8 May. Available from: www.wineenthusiast.com (accessed 12 June 2024).
——— (2024a). Anne Krebiehl MW: Dom Pérignon 2015, 2006 P2, and the missing 2023. *The Buyer*, 16 July. Available from: www.the-buyer.net (accessed 24 September 2024).
——— (2024b). Moët & Chandon Trilogie: 2016, 2009, 2000 — partners in turbulence. *The World of Fine Wine*, 9 September. Available from: https://worldoffinewine.com (accessed 15 October 2024).
LA (Lamothe-Abiet) (2020). *The Birth of Bubbles: Technical Booklet*. Canéjan/Bordeaux: Lamothe-Abiet.
Lanson-BCC (2022). *Corporate Brochure 2022*. Reims: Lanson-BCC. Available from: www.lanson-bcc.com/en/finance/downloads (accessed 22 July 2023).
Lascelles, Alice (2021). Bubbling over: Champagne's new experimental mood. *Club Oenologique*, 14 June. Available from: https://cluboenologique.com (accessed 5 February 2024).
Lawrence, James (2022). Multi-vintage wines are becoming a thing. *Pix™*, 21 April. Available from: www.pix.wine (accessed 27 April 2023).
Leahy, Kate (2021). *Wine Style: Discover the Wines You Will Love through 50 Simple Recipes*. New York: Ten Speed Press.
LeBeau, Amber (2019). Champagne and climate change — is blocking MLF the answer for fresher bubbles? *Spit Bucket*, 27 December. Available from: https://spitbucket.net (accessed 10 August 2024).
Lewis, Tom (2024). Ukrainian wines at ultimate CEE wine fair. *Medium*, 7 July. Available from: https://medium.com (accessed 6 September 2024).
Lhôpital, Bertrand (2022). Interview with Bertrand Lhôpital — cellar master [Champagne Telmont]. An interview by Björnstierne Antonsson. *The Champagne Club*, 30 March. Available from: www.champagneclub.com (accessed 17 July 2024).
LHV (Limeburn Hill Vineyard) (2016). An organic anti-fungal spray. *LHV Blog*, 13 July. Available from: www.limeburnhillvineyard.co.uk (accessed 28 June 2023).
Liem, Peter (2017). *Champagne: The Essential Guide to the Wines, Producers, and Terroirs of the Iconic Region*. New York: Ten Speed Press.
Lindgren, Shelley, & Kate Leahy (2023). *Italian Wine: The History, Regions, and Grapes of an Iconic Wine Country*. New York: Ten Speed Press.
Lloyd, Angela ([2017] 2024). *Agrafe*: A pioneering method for méthode cap classique. *WoSA News > Cape Chatter*, December. Available from: www.wosa.co.za (accessed 7 March 2025).
Louch, Toby (2023). Laurent-Perrier impresses with Brut Nature. *The World of Fine Wine*, 21 April. Available from: https://worldoffinewine.com (accessed 26 July 2024).
LP (Laurent-Perrier) (2022). *Laurent-Perrier Group Annual Report 2021–2022*. Tours-sur-Marne: Domaine Laurent-Perrier. Available from: www.finance-groupelp.com (accessed 2 May 2023).
LVMH (Moët Hennessy–Louis Vuitton) (2014). Veuve Clicquot bubbles in the Baltic. *News: Wines & Spirits*, 23 July. Available from: www.lvmh.com/en/news-lvmh (accessed 30 June 2023).
Lynn, Brian (2020). Majella talks — Sparkling Shiraz [video]. Majella Talks (YouTube), April 2020. Available from: www.youtube.com/@proflynn (accessed 13 August 2019).
Machard de Gramont, Bernadette (2022). The difference between multi-vintage and nonvintage champagnes. *Wine Enthusiast*, 17 November.

Available from: www.winemag.com (accessed 21 March 2023).
MacLean, Natalie (2024). 10 best Moscato d'Asti wines to buy now + 5 surprising facts about Moscato d'Asti. *Natalie MacLean*, 21 April. Available from: www.nataliemaclean.com/blog (accessed 29 May 2024).
——— (2017). Inniskillin Sparkling Cabernet Franc Icewine 2012 wine review. *Natalie MacLean*, 7 July. Available from: www.nataliemaclean.com/blog (accessed 29 May 2024).
MacNeil, Karen ([2002] 2015). *The Wine Bible* (revised 2nd edition). New York: Workman Publishing.
——— (2019). Muselet. *Wine Words*, 27 December. Available from: https://winespeed.com (accessed 18 August 2023).
MacQuitty, Jane (2022). Sugar rush. In Susan Keevil (Ed.), *On Champagne*, 30–38. Myddle, UK: Académie du Vin Library Ltd.
Magyarics, Kelly (2021). A decades-old champagne style is trending. Here's why. *Wine Enthusiast*, 21 December. Available from: www.wineenthusiast.com (accessed 14 August 2024).
Maker, Meg (2021). Méthode ancestrale. *Terroir Review > Glossary*, 7 November. Available from: www.terroirreview.com (accessed 2 September 2023).
Mariotti, Didier (2023). Veuve Clicquot's 40-year sea-ageing experiment. An interview by Chris Boiling. *Canopy*, 13 July. Available from www.internationalwinechallenge.com (accessed 22 November 2023).
Mason, Laura (2004). *Food Culture in Great Britain*. Westport, CT: Greenwood Press.
Mason, Olivia (2024). Decanter World Wine Awards 2024 results revealed. *Decanter*, 19 June 2024. Available from: www.decanter.com (accessed 29 August 2024).
May, Melanie (2021). Franciacorta wine guide — Italian sparkling wine. *Melanie May — Travel, Eat, Write, Repeat*, 28 November. Available from: www.melaniemay.com (accessed 10 September 2024).
Mazzeo, Jacopo (2022). Leonardo DiCaprio invests in Champagne Telmont. *Decanter*, 23 February. Available from: www.decanter.com (accessed 22 July 2022).
McCann, Marisa (2023). Defining tirage: Putting the bubbles in the bottle. *Frank Family Vineyards Blog*, 9 May. Available from: www.frankfamilyvineyards.com/blog (accessed 21 December 2023).
McCarthy, Ed. (2017). Rosé champagne. *Wine Review Online*, 26 December. Available from: https://winereviewonline.com (accessed 12 September 2024).
McCoy, Elin (2022). Homage to the widow. In Susan Keevil (Ed.), *On Champagne*, 30–38. Myddle, UK: Académie du Vin Library Ltd.
McCubbing, Gus (2024). Yarra Valley winery unveils $7m refurb. *The Australian Financial Review*, 25 March. Available from: www.afr.com/life-and-luxury/food-and-wine (accessed 31 May 2024).
McGregor, Lindsaye (2016). Water-wise farming for sustainable wine production. *Cape Chatter*, March. Available from: www.wosa.co.za/WOSA-News/Blogs/Cape-Chatter/Archive (accessed 7 April 2023).
McKay, Annie (2022). Is sparkling red wine served chilled? *Love to Know*, 29 April. Available from: www.lovetoknow.com (accessed 19 August 2022).
McNeill, Eleanor (2024). The best champagne and food pairings. *The World of Fine Wine*, 28 June. Available from: https://worldoffinewine.com (accessed 21 August 2024).
Melnik, Peg (2023). What is riddling? Schramsberg Vineyards does it by hand. *The Press Democrat*, 27 November. Available from: www.pressdemocrat.com (accessed 26 April 2024).
Mercer, Chris (2020a). Hints of jazz piano? Krug talks Champagne and music pairing. *Decanter*, 1 June. Available from: www.decanter.com (accessed 16 May 2024).
——— (2020b). What does 'saignée' mean in rosé wine? *Decanter*, 12 September. Available from: www.decanter.com (accessed 1 February 2024).
——— (2020c). Scientists unlock secret of Champagne and oysters pairing. *Decanter*, 2 December. Available from: www.decanter.com (accessed 22 August 2024).
——— (2021). What are lees in wine? *Decanter*, 22 July. Available from: www.decanter.com (accessed 23 November 2023).
Mileham, Arabella (2022). Asti la vista, baby: How Italy's sweet sparkling wines are fighting back. *The Drinks Business*, 8 February. Available from: www.thedrinksbusiness.com (accessed 3 April 2024).
Miller, Caitlin A. (2023). The complex role of dosage in sparkling wine. *Seven Fifty Daily: The Business and Culture of Drinks*, 27 November. Available from: https://daily.sevenfifty.com (accessed 31 May 2024).
Mobbs, John (2022). A taste of Flint Vineyard 2021 vintage. *Great British Wine*, 10 May. Available from: www.greatbritishwine.com (accessed 13 May 2024).
Mocke, Bernard (2017). Le Lude leads the charge with Agrafe method. *WineLand*, 13 November. Available from: www.wineland.co.za (accessed 18 December 2023).
Moore, Victoria (2022). The rise and rise of rosé. In Susan Keevil (Ed.), *On Champagne*, 56–9. Myddle, UK: Académie du Vin Library Ltd.
Morris, Michaela (2019). Prosecco: Aiming high. *Decanter*, 15 June. Available from: www.decanter.com (accessed 17 April 2024).

Moser, Peter (2024). Weingut Schloss Gobelsburg. 2012 Sekt Austria Grosse Reserve Extra Brut Niederösterreich G.U. Langenlois Vintage. *Falstaff*, 24 July. Available from: www.falstaff.com (accessed 3 September 2024).

Mowery, Lauren (2023a). Looking for a new take on champagne? Try Perpetual Reserve. *Wine Enthusiast*, 5 May. Available from: www.wineenthusiast.com (accessed 20 May 2024).

—— (2023b). A six-bottle master class to sparkling wine. *Wine Enthusiast*, 8 May. Available from: www.wineenthusiast.com (accessed 2 April 2024).

—— (2023c). Everything you need to know about prosecco. *Wine Enthusiast*, 29 September. Available from: www.wineenthusiast.com (accessed 31 July 2024).

Napjus, Alison (2014). First rosé champagne? Older than you think. *Wine Spectator*, 10 March. Available from: www.winespectator.com (accessed 7 February 2024).

Nichols, Marnie (2018). The oeil de perdrix . . . A special kind of rosé champagne. *Bubble & Flute*, 6 July. Available from: https://bubbleandflute.com (accessed 30 April 2024).

Nordestgaard, Simon (2017). Pre-fermentation heating of red grapes: A useful tool to manage compressed vintages? *Grapegrower & Winemaker* 637 (February): 54–61.

—— ([2019] 2020). Inspirations from the past and opportunities for the future. Part 2: In-tank fermentation monitoring and continuous processes. *Grapegrower & Winemaker* 677 (June): 50–6.

O'Connor, Lilla (2024). Ukraine: A wine odyssey. *Decanter*, 23 February. Available from: www.decanter.com (accessed 23 September 2024).

O'Donnell, Ben (2020). Brad Pitt unveils Fleur de Miraval Rosé Champagne after five years of production. *Wine Spectator*, 31 August. Available from: www.winespectator.com (accessed 16 October 2023).

O'Keefe, Kerin (2018). Meet col fondo, the bottle-fermented prosecco. *Wine Enthusiast*, 11 June. Available from: www.wineenthusiast.com (accessed 26 August 2023).

—— (2019). Col fondo: The murky case of the cloudy wine. *Kerin O'Keefe*, 22 October. Available from: www.kerinokeefe.com (accessed 26 August 2023).

Paillard, Alice (2015). Wine in France: Alice Paillard of Champagne Bruno Paillard on the delicate effervescence, remarkable purity, true freshness, and silky texture of Champagne style. An interview by *A Woman's Paris*™. In *A Woman's Paris*, 2 March. Available from: barbararedmond.com (accessed 21 November 2023).

Palmer, Kaaren (2024). Blending in: Champagne's reserve wines. *Vine and Bubble*, 14 January. Available from: https://vineandbubble.com (accessed 20 May 2024).

Pambianchi, Daniel (2006). Understanding sparkling wines. *Winemaker (Magazine)*, December 2006/January 2007. Available from: https://winemakermag.com (accessed 6 April 2024).

Pàmies, Jaume (2021). Cava ageing — the cava production process. *D.O. Cava Blog*, 31 August. Available from: www.cava.wine/en/news-articles (accessed 23 November 2023).

Parker, Robert M. ([1995] 2008). *Parker's Wine Buyer's Guide*, 7th edition. New York: Simon & Schuster.

Parzen, Jeremy (2014). Absurdist EU wine marketing regulation and why Prosecco Col Fondo matters more than ever. *Do Bianchi*, 16 December. Available from: https://dobianchi.com (accessed 2 September 2023).

—— (2018). Italian wine world mourns the loss of Ernesto Cattel, 54, Prosecco col fondo pioneer and organic advocate. *Do Bianchi*, 9 August. Available from: https://dobianchi.com (accessed 2 September 2023).

Perrin, Marc, & Rodolphe Péters (2022). Fleur de Miraval with Marc Perrin and Rodolphe Péters [video]. Vineyard Brands LLC (YouTube), February 2022. Available from: www.youtube.com/@vineyardbrandsllc (accessed 13 August 2019).

Person, Dan, & Jacqueline Person (2024). Rouge Brute: A sparkling red wine that keeps you sipping. *Carboniste Blog*, 5 March. Available from: https://carboniste.com/blog (accessed 3 June 2024).

Pessey, Christian (2002). *The Little Book of Champagne* (Transl. Fui Lee Luk). Paris: Flammarion.

Phillips, Rod ([2012] 2014). Moët & Chandon, Maison de Champagne. *NUVO*, 30 December. Available from: https://nuvomagazine.com (accessed 21 September 2024).

Phillips, Rod (2019). The case for crown caps: Why some sparkling wine makers are re-thinking dangerous corks. In *NUVO*, 23 September 2019. Available from: https://nuvomagazine.com (accessed 24 July 2023).

Pickett, Natalie (2023). Yarra Valley sparkling cellar doors. *The Bubbles Review*, 9 March. Available from: https://thebubblesreview.com (accessed 31 May 2024).

Pitcairn, Becky (2022). England's most expensive sparkling wines. *The English Wine Diaries*, 25 September. Available from: https://englishwinediaries.com/2022/09/25/englands-most-expensive-sparkling-wines (accessed 27 February 2026).

Poggio, Danilo (2021). Franciacorta food pairings for the holidays. Advice from Michelin-starred chefs. *La Cucina Italiana*, 24 December. Available from: www.lacucinaitaliana.com (accessed 10 September 2024).

Prichard, Alex (2020). The Charmat method. *Wine Academy*, 24 February. Available from: https://wineacademy.com.au (accessed 11 April 2024).
Quaile, James (2023). *Tales from the Wine Floor. 100 Questions Asked of a Sommelier*. Lanham, MD: Lyons Press/The Rowman & Littlefield Publishing Group, Inc.
Radil, Dan ([2016] 2018). Sparkling wine a passion for Bellingham couple. *Wine Press North West*, 1 September 2016/12 August 2018. Retrieved from: www.winepressnw.com (accessed 24 October 2023).
Rand, Margaret ([2018] 2022). The 'O' word. In Susan Keevil (Ed.), *On Champagne*, 226–33. Myddle, UK: Académie du Vin Library.
—— (2022). The méthode champenoise. In Susan Keevil (Ed.), *On Champagne*, 18–20. Myddle, UK: Académie du Vin Library.
Rasiulis, Edward ([2017] 2019). Vieni Sparkling Cabernet Ice Wine Doux 2013, Vinemount Ridge, Niagara, Ontario. *Twilight Wines*, 1 January. Available from: https://twilightwines.com (accessed 12 September 2024).
Riley, Lisa (2018). Piper-Heidsieck promotes Séverine Freson to Chef de Cave. *Harpers Wine and Spirit*, 15 June. Available from: https://harpers.co.uk (accessed 16 September 2023).
Ring, Treve (2023). Cap Classique, an exciting category for South African wine. *Wine Anorak: Global Wine Journal*, 25 January. Available from: www.wineanorak.com (accessed 6 May 2023).
Robinson, Andrea Immer ([2000] 2005). *Great Wine Made Simple* (revised and updated). New York: Broadway Books.
Robinson, Jancis ([1983] 2000). *How to Taste: A Guide to Enjoying Wine*. New York: Simon & Schuster.
—— (2011). Rosé champagne — the missing ingredient. *Jancis Robinson*, 3 September. Available from: www.jancisrobinson.com (accessed 18 January 2024).
—— (2019a). Champagne — Leclerc Briant and Jacquesson. *Jancis Robinson*, 28 August. Available from: www.jancisrobinson.com (accessed 5 February 2024).
—— (2019b). Pink champagne — a serious wine now. *Jancis Robinson*, 28 September. Available from: www.jancisrobinson.com (accessed 27 January 2024).
Rose, Anthony (2021). *Fizz! Champagne and Sparkling Wines of the World*. Oxford, UK: Infinite Ideas.
Ruth, Noémi (2024). Champagne trends in Japan: Luxury and celebration on the rise. *Trajaan*, 1 March. Available from: https://trajaan.io/blog (accessed 20 September 2024).
RV (Ridgeview) (2023). *Ridgeview Impact Report 2023*. Ditchling Common, UK: Ridgeview Estate Winery. Available from: www.ridgeview.co.uk (accessed 19 March 2024).
Sacchetto, Anna (2023). Moscato d'Asti: Production, questions and curiosities. *Love Langhe/Tasty Reads*, 4 July. Available from: https://langhe.net/en/magazine-en/eat-drink (accessed 2 April 2024).
Saladino, Emily (2023). How to open a champagne bottle without spilling a drop. *Wine Enthusiast*, 21 August. Available from: www.wineenthusiast.com (accessed 20 August 2024).
Salpico, Ines (2021). Tannins — what are they and what do they do? *Decanter*, 1 November. Available from: www.decanter.com (accessed 16 August 2024).
Sanderson, Bruce (2008). A singular vision in Champagne. *Wine Spectator*, 31 October. Available from: www.winespectator.com (accessed 16 January 2024).
Schiessl, Courtney (2016). On the floor: Should you ever decant champagne? *VinePair*, 29 September. Available from: https://vinepair.com (accessed 20 August 2024).
—— (2017a). I would always choose cava over prosecco. Here's why. *VinePair*, 9 August. Available from: https://vinepair.com (accessed 22 April 2024).
—— (2017b). All the ways to make champagne and sparkling wine, explained. *VinePair*, 26 November. Available from: https://vinepair.com (accessed 5 January 2023).
—— (2023). How sparkling wine is made. *Wine Enthusiast*, 5 May. Available from: www.wineenthusiast.com (accessed 3 November 2023).
Schmidt, Benjamin (2024). The race to zero: The story of Brut Nature champagne. *The Interior Review*, 15 February 2024.
Schmitt, Patrick (2016). Champagne Lanson now allowing MLF. *The Drinks Business*, 18 January. Available from: www.thedrinksbusiness.com (accessed 23 April 2024)
—— (2020). Sparkling wine has 'more flavours' when tirage performed under cork, says scientist. *The Drinks Business*, 20 March. Available from: www.thedrinksbusiness.com (accessed 27 November 2020).
—— (2024). Ruinart creates a champagne for climate change. *The Drinks Business*, 31 July. Available from: www.thedrinksbusiness.com (accessed 2 August 2024).
Schneider, Elizabeth (2019). *Wine for Normal People*. San Francisco: Chronicle Books LLC.
Schuster, Michael (2016). Determining a wine's weight. *Decanter*, 16 July. Available from: www.decanter.com (accessed 1 May 2024).
Sealy, Christopher (2024). The sommelier suggests . . . Grüner Veltliner by Christopher Sealy. *Decanter*,

29 January. Available from: www.decanter.com (accessed 2 September 2024).
Seidanova, Laura (2022). Complete guide to sparkling wines. *Wine Tourism*, 31 March. Available from: www.winetourism.com (accessed 13 April 2024).
Shaw, Lucy (2018). Preserving acidity is 'the biggest issue in Champagne'. *The Drinks Business*, 11 October. Available from: www.thedrinksbusiness.com (accessed 10 August 2024).
Sheather, Shawn (2012). Australia's first screw-capped sparkling wines. *Australian Good Food Guide*, 14 May. Available from: www.agfg.com.au (accessed 30 November 2023).
Sheppard, Julie (2023). Brazil's Altos de Pinto Bandeira becomes first DO exclusively for sparkling wines in the New World. *Decanter*, February. Available from: www.decanter.com (accessed 3 February 2023).
Shevela, Nika (2020). The traditional method behind Spain's top sparkling wine. *D.O. Cava Blog*, 2 October. Available from: www.cava.wine/en/news-articles (accessed 26 April 2024).
Simon, André-Louis (1905). *History of Champagne Trade in England*. London: Wyman & Sons, Ltd.
Simon, Joanna ([2020] 2022). Moët's Luxury Cuvée MCIII and the enigmatic non-appearance of its successor. In *Joanna Simon*, 11 May. Available from: www.joannasimon.com (accessed 16 October 2023).
Sinclair, Charles ([1998] 2008). *Dictionary of Food: International Food and Cooking Terms from A to Z*. London: A & C Black.
Slinkard, Stacy (2013). *Wine*. New York: Alpha Books/Penguin Group (USA) Inc.
Small, Robert W., & Michelle Couturier (2011). *Beverage Basics: Understanding and Appreciating Wine, Beer, and Spirits*. Hoboken, NJ: John Wiley & Sons.
Smith, Bridgette (2022). Fizz the season. *Williams Gap Vineyard Blog*, 1 January. Available from: www.williamsgapvineyard.com (accessed 10 April 2024).
Smith, Gavin (2020). Terroir immersion therapy with Sébastien Le Golvet. *Fine and Rare*, 9 April. Available from: www.frw.co.uk (accessed 5 February 2024).
Sohm, Aldo, & Christine Muhlke (2019). *Wine Simple: A Totally Approachable Guide by a World-Class Sommelier*. New York: Clarkson Potter.
Soto, Amaia (2020). Cava — the search for the perfect pairing. *D.O. Cava Blog*, 1 October. Available from: www.cava.wine/en/news-articles (accessed 24 June 2024).
Spanos, Georgia (2023). The uniquely Australian sparkling shiraz. *Halliday Wine Companion*, 22 May. Available from: https://winecompanion.com.au (accessed 14 September 2024).
Spaziani, Raymond (2019). Champagnes and other sparkling wines. *Milford-Orange Times*, 9 January. Available from: https://theorangetimes.com (accessed 9 April 2024).
Steeves, Matt (2020). Wine review: 2018 Inniskillin Sparkling Vidal Icewine — 96 points. A decadent treat for ringing in your 2021 NYE. *Quercus Vino*, 29 December. Available from: https://quercusvino.ca (accessed 29 August 2024).
Stefanowicz, Patricia (2023). The medallists from The Prosecco Masters 2023. *The Drinks Business*, 17 April. Available from: www.thedrinksbusiness.com (accessed 31 July 2024).
Stelzer, Tyson (2012). A new twist for sparkling wines. *Wine Spectator*, 7 May. Available from: www.winespectator.com (accessed 26 July 2023).
Stelzer, Tyson (2014). Champagne disgorgement: It's all in the timing. *Decanter*, 1 August. Available from: www.decanter.com (accessed 24 September 2024).
—— (2016). Champagne bubble size: Does it matter? — ask Decanter. *Decanter*, 26 November. Available from: www.decanter.com (accessed 19 March 2024).
—— (2018). Australian Sparkling Report 2018. Available from: www.tysonstelzer.com (accessed 16 January 2024).
Stevenson, Tom (2003). *The Wine Report 2004*. London: Dorling Kindersley Publishing, Inc.
—— (2008). Ageing gracefully. *The World of Fine Wine* 21: 44.
—— (2011). Blanc de noir. All black — the new white. *The World of Fine Wine* 34: 168–75.
—— (2012). Vintage champagne and English sparkling wine: Lemons and limes. *The World of Fine Wine* 38: 148–59.
—— ([2014] 2022). Quality time. In Susan Keevil (Ed.), *On Champagne*, 244–7. Myddle, UK: Académie du Vin Library Ltd.
—— (2015a). Bollinger Vieilles Vignes Françaises 1969–2005. *The World of Fine Wine*, 13 January. Available from: https://worldoffinewine.com (accessed 19 April).
—— (2015b). Jet-age fizz. *The World of Fine Wine* 48: 62.
—— ([2015] 2022). The magnum effect. In Susan Keevil (Ed.), *On Champagne*, 84–7. Myddle, UK: Académie du Vin Library Ltd.
—— (2019). Dr Tony Jordan OAM (1944–2019): Goodbye to a dear friend, colleague, and mentor. *The World of Fine Wine* 66: 44–5.
—— (2022). Le Déjeuner d'Huîtres: The first sparkling Champagne painting. *The World of Fine Wine*, 2 March. Available from: https://worldoffinewine.com (accessed 25 September 2024).

Stimpfig, John (2018). Spot the difference: Tirage and dosage in champagne — ask Decanter. *Decanter*, 11 January. Available from: www.decanter.com (accessed 31 October 2023).

Sumners, Charles Eugene (2002). *Darkness Visible. Memoir of a World War II Combat Photographer* (Ed. Ann Sumners). Jefferson, NC/London: McFarland & Co.

Sussman, Zachary (2015). What is 'pét-nat,' really? *Punch*, 26 May. Available from: www.punchdrink.com (accessed 23 August 2023).

——— (2020a). Natural Lambrusco is just the beginning. Returning to old traditions, the frizzante wines of Emilia-Romagna are ready for their pét-nat moment. *Punch*, 26 May. Available from: www.punchdrink.com (accessed 7 September 2023).

——— (2020b). Don't call them pét-nat. *Punch*, 18 September. Available from: www.punchdrink.com (accessed 29 August 2023).

——— (2021). *Sparkling Wine for Modern Times. A Drinker's Guide to the Freewheeling World of Bubbles*. New York: Ten Speed Press/Penguin Random House LLC.

Szabo, John (2022). South Africa update: Old vines leading the way out of crisis and into the premium markets. *Wine Align*, 3 August. Available from: www.winealign.com (accessed 14 April 2023).

Taber, George M. ([2007] 2009). *To Cork Or Not To Cork*. New York: Scribner, Simon & Schuster.

Tagliabue, John (2010). Buried treasure in Baltic has vintage taste. *New York Times*, 14 December. Available from: www.nytimes.com/2010/12/15/world/europe/15champagne.html (accessed 30 June 2023).

Taittinger, Vitalie (2020). Are there tannins in Champagne? *Decanter*, 16 February. Available from: www.decanter.com (accessed 16 August 2024).

Tardi, Alan (2016). *Champagne, Uncorked: The House of Krug and the Timeless Allure of the World's Most Celebrated Drink*. New York: PublicAffairs.

Taylor, Geoff ([1994] 2015). Total acidity. In Jancis Robinson (Ed.), *The Oxford Companion to Wine*, 4th edition, 811. Oxford: Oxford University Press.

Teclemariam, Tammie (2023). Your pét-nat primer. *Wine Enthusiast*, 4 May. Available from: www.wineenthusiast.com (accessed 22 August 2023).

Teeter, Adam (2023). Can pét-nats age as long as champagne and other sparkling wines? *VinePair*, 8 May. Available from: https://vinepair.com/articles/ask-adam-pet-nat-aging (accessed 16 November 2023).

TFA (Tenuta Foppa & Ambrosi) (2022). *We Love Enology*. Garibaldi, RS: Tenuta Foppa & Ambrosi. Available from: www.tenutafa.com.br (accessed 17 April 2024).

TFB (The Finest Bubble) (2017a). *The Finest Bubble: June 2017 Portfolio*. Available from: https://thefinestbubble.com (accessed 16 October 2023).

——— (2017b). Moët & Chandon MCIII — the world's first 'ultra-prestige cuvée'? *The Finest Bubble*, 19 July. Available from: https://thefinestbubble.com/news-and-reviews/moet-chandon-mciii-worlds-first-ultra-prestige-cuvee (accessed 16 October 2023).

——— (2018). Disgorgement: What is it and why is the date important? *The Finest Bubble*, 7 June. Available from: https://thefinestbubble.com/news-and-reviews/disgorgement-what-is-it-and-why-is-the-date-important (accessed 18 October 2023).

——— (2019a). Bioenergetic Winemaking at Leclerc Briant. *The Finest Bubble*, 26 April. Available from: https://thefinestbubble.com/news-and-reviews/bioenergetic-winemaking-at-leclerc-briant (accessed 1 July 2023).

——— (2019b). Recently Listed: Leclerc Briant Abyss 2013 and Les Basses Prières 2014. *The Finest Bubble*, 7 May. Available from: https://thefinestbubble.com/news-and-reviews/recently-listed-leclerc-briant-abyss-2013-and-les-basses-prieres-2014 (accessed 1 July 2023).

Tirabassi, Eric (2020). *Champagne and the Best Houses*. Paris: Eric Tirabassi.

Todorovic, Sara (2020). Top Slovenian wine: Žametna črnina/Black velvet. *The Slovenia*, 19 May. Available from: https://the-slovenia.com/en (accessed 3 September 2024).

Vierra, Mary Beth (2021). What's old is new again: Bubbles col fondo. *Vero Vino*, 3 December. Available from: https://verovino.com/vero-blog (accessed 24 August 2023).

Vine, Richard P. ([1981] 2012). *Commercial Winemaking: Processing and Controls*. Westport, CT: Avi Publishing Company.

Visconti, Paola (2022). The Gancia Cellars: Birthplace of Italian sparkling wines. An interview by Irene S. Levine. *Forbes*, 21 July. Available from: www.forbes.com (accessed 1 April 2023).

Vittek, Shelby (2023). The growing movement behind 'long Charmat' sparkling wine. *Seven Fifty Daily: The Business and Culture of Drinks*, 22 May. Available from: https://daily.sevenfifty.com (accessed 17 April 2024).

Vizetelly, Henry (1882). *A History of Champagne (with Notes on the Other Sparkling Wines of France)*. New York: Scribner & Welford.

Waldin, Monty (2013). *Monty Waldin's Best Biodynamic Wines*. Edinburgh, UK: Floris Books.

Walkey, Christopher (2019). Charmat doesn't always mean Prosecco. *Glass of Bubbly*, 15 October. Available from: https://glassofbubbly.com (accessed 19 April 2024).

——— (2021). Pétillant naturel (pét-nat). *Glass of Bubbly*, 9 July. Available from: https://glassofbubbly.com (accessed 22 August 2023).
——— (2022). What does rosé de saignée taste like? *Glass of Bubbly*, 28 April. Available from: https://glassofbubbly.com (accessed 25 September 2023).
——— (2023). Sparkling wine production methods. *Glass of Bubbly*, 15 August. Available from: https://glassofbubbly.com (accessed 29 November 2023).
Walkey, Eve (2014). Champagne Jacquesson. *Glass of Bubbly*, 18 March. Available from: https://glassofbubbly.com (accessed 11 August 2023).
——— (2016). Orange sparkling wine — Vej Brut. *Glass of Bubbly*, 20 June. Available from: https://glassofbubbly.com (accessed 16 May 2023).
Walkey, Oliver, & Christopher Walkey (2022). What sparkling wine to pair with apple pie? *Glass of Bubbly*, 1 April. Available from: https://glassofbubbly.com (accessed 7 October 2024).
Wallner, Anna (2018). *Cava: Spain's Premium Sparkling Wine*. London: Eken Press/Grenadine Publishing.
Walters, Robert ([2016] 2022). Get set, go! In Susan Keevil (Ed.), *On Champagne*, 193–8. Myddle, UK: Académie du Vin Library Ltd.
——— (2022). Death to the flute! In Susan Keevil (Ed.), *On Champagne*, 80–3. Myddle, UK: Académie du Vin Library Ltd.
Wang, Natalie (2020). By the glass: Christian Holthausen of AR Lonoble. *Vino Joy News*, 24 August. Available from: https://vino-joy.com (accessed 14 September 2024).
Warren-Adamson, Robin (2020). Cuvée Noir wine tasting [video]. *Bolney Estate (YouTube)*, 14 May. Available from: https://bolneywineestate.com (accessed 6 May 2024).
Wasserman, Sheldon, & Pauline Wasserman (1984). *Sparkling Wine*. Piscataway, NJ: New Century Publishers.
Waterworth, Shay (2023). Crémant ready to pop. *Drinks International*, 13 April. Available from: https://m.drinksint.com (accessed 18 May 2024).
Wheeler, Teri (2019). Tasmanian sparkling wines. *Glass of Bubbly*, 19 October. Available from: https://glassofbubbly.com (accessed 12 January 2024).
——— (2023). Christmas down under with Australian Sparkling Shiraz. *Glass of Bubbly*, 14 December. Available from: https://glassofbubbly.com (accessed 14 May 2024).
Whitaker, Matthew (2017). Focus on flavour: English wine and food pairings. *The Oxford Wine Company Blog*, 31 August. Available from: https://shop.oxfordwine.co.uk/company-blog (accessed 22 August 2024).
White, Carolyn (2020). Our Italians: Denny Bini of Podere Cipolla. *Casa Mia Tours Blog*, 18 February. Available from https://casamiatours.com/blog (accessed 7 September 2023).
White, David (2018). Tracing the origin story of grower champagne. *Seven Fifty Daily: The Business and Culture of Drinks*, 24 September. Available from: https://daily.sevenfifty.com (accessed 21 February 2024).
White, Tim (2009). Bitter truth behind bubbly. *The Australian Financial Review*, 23 October. Available from: www.afr.com (accessed 16 August 2024).
Whitney, Mike (2022). Personal interview on vitivinicultural practices, 12 September.
Willcox, Kathleen (2020). Champagne's rule-breaking winemakers have a sweet secret. *Wine Enthusiast*, October. Available from: www.wineenthusiast.com (accessed 22 June 2023).
——— (2022). How Tasmania's sparkling wine is bubbling up from down under. *Alcohol Professor*, 22 April. Available: www.alcoholprofessor.com (accessed 20 February 2024).
——— (2023). For sparkling wines' secondary fermentation, closure choice can have a huge impact. *Wine Enthusiast*, December. Available from: www.wineenthusiast.com (accessed 19 December 2023).
Williams, Janice (2022). Exploring the world of red sparkling wines. *Pix™*, 13 January. Available from: www.pix.wine (accessed 9 June 2023).
Wilson, Chris (2018). Bruno Paillard and the trauma of disgorgement. *The Buyer*, 22 January. Available from: www.the-buyer.net/tasting/wine/bruno-paillard-trauma-disgorgement (accessed 21 November 2023).
Wislocki, Amy (2022). Champagne Henri Giraud: Serious about oak. *Decanter*, 24 April. Available from: www.decanter.com (accessed 5 February 2024).
Woodard, Richard (2023). Japan: How the Japanese market bounced back. *The Drinks Business*, 9 August. Available from: www.thedrinksbusiness.com (accessed 21 August 2023).
Wu, Yiyao (2015). Prosecco, Asti and others — Methods of production — WSET Level 2. *Decanter*, 23 December. Available from: www.decanter.com (accessed 14 September 2023).
Zamuner, Marco (2021). Vidal Sparkling Icewine — Niagara Peninsula, Limited Edition, VQA [video]. *Magnotta Winery (YouTube)*, 12 March. Available from: www.magnotta.com (accessed 14 September 2023).
Zielinski, Tyler (2022). How to properly saber champagne. *Food & Wine*, 28 June. Available from: www.foodandwine.com (accessed 20 August 2024).
Zohar, Yvonne (2020). A natural sparkling Muscat wine named Clairette. Confused? Don't be. *Winejus*, 7 April. Available from: https://winejus.com (accessed 25 March 2024).

APPENDIX
THE LANGUAGE OF WINE

THIS GUIDE is a quality-focused product of linguistic work, prepared with the aim of facilitating communication among all stakeholders in the vocational field of sparkling wines. Analyses of the publishing market[1] reveal that this reference work is unprecedented in this subject area in terms of its educational merit. This has been corroborated by the author against digital repositories and catalogues of the Web of Science,[2] as well as the world's most renowned and largest libraries, including the UNESCO Library, the Library of Congress, the British Library, the University of California Libraries, the New York Public Library and Library and Archives Canada. This companion presents cutting-edge specialized knowledge processed and structured in terminological (semantic) meta-language. As a product of advanced research into the current concept systems of sparkling wines (expressed in language), it is a multifunctional work, which has the potential to become a central part of vocational education on the subject. The potential beneficiaries include: winemakers; wine scientists and historians; winemaking trainees; students of viticulture, oenology and related fields; wine educators; wine journalists; sommeliers; wine business investors and marketeers; connoisseurs; regular wine customers; restaurant chefs and home cooks; participants of wine courses and wine workshops; and designers of wine libraries and online catalogues.

While the English-language corpus has been the main source for terminological inquiry, background corpora,

1 A full document search, on 25 May 2022, was undertaken both in Google Books and Amazon. While the search returned a large number of reference works on the subject of sparkling wine, including guides, encyclopaedias and handbooks, most of them on champagne, no field thesauri or fully structured encyclopaedias dedicated solely to the subject of sparkling wines were found.

2 The Web of Science was searched twice: on 2 December 2023 and on 29 October 2024. The final search returned 989 records on the subject of sparkling wines, including 183 documents on champagnes, 83 on cavas, 51 on Australian sparkling wines, 30 on proseccos, nine on franciacortas, seven on Asti wines and five on cap classiques, with no terminologies found.

mainly French but also Spanish, Italian and German (languages spoken in the world's largest sparkling-wine-producing countries), served as control instruments throughout the terminology work. Desktop research resulted in the compilation of the English corpus and the extraction of candidate terms, with the completion of the project of creating a map of sparkling wine terms made possible by means of Microgrant UW, received from the University of Warsaw in 2022. Field research, which began in 2015, was conducted during my visits to sparkling-wine-producing estates in the winemaking regions of Lower Austria (2015), western Poland (2023, 2024), Champagne (2023, 2024) and England's East Sussex (2024). The visits allowed me to test my terminological assumptions in consultation with professionals in the field of sparkling wine production while also providing an opportunity to take photographs. A total of 618 sources have been consulted, including 96 reference sources, 139 scientific publications (the majority of which are multi-authored works) and 383 professional (mainly single-authored) works.

The conceptual structure of the sparkling wine field is manifested in five consecutive displays. These are:

- Factual guide, which discusses the history of the sparkling wine industry
- Map of terms, serving as an introduction to the conceptual markers
- A–Z display, providing definitions of all the terms found on the map
- Wine and food pairing, discussing the principles of matching wines with foods, accompanied by explanations and examples
- Order of events, which offers a chronological list of developments shaping the trade from its origins to the present day.

These displays of professional content have been ordered based on the progressive complexity principle in order to serve educational purposes. Besides its pragmatic use, this dedicated compendium has been designed to appeal to readers on an aesthetic level, with its rich descriptions, tables with clear data and high-quality photographs.

The terminology research has revealed the conceptual links of the microfield of sparkling wine with other major concepts in the field of grape wine. A bird's-eye view of the wine field (see Figure A) shows the main notion of sparkling wine under its parent notion of grape wine and alongside its sibling notions of still wine and fortified wine, with the affinity between the concept of sparkling wine and fortified wine being greater as these two concepts taken together form the category of special wines (see OIV 2020a: 4). This ideogram shows the basic structure of the semantic field of wine to illuminate the connections between the different notions in the winemaking field. These basic divisions result from applying the generic link of parent–child, the rule of pattern formation most typical of scientific thought in advanced fields of knowledge. The arrows run from the superordinate concept marker of parent or genus (e.g. 'grape wine') to the subordinate concept of a child or species (e.g. 'sparkling wine'). In order to ensure terminological compatibility among all the microfields subsumed under the top category of 'grape wine', distinctions between concepts have been made. For example, the descriptor 'vinification of grapes',

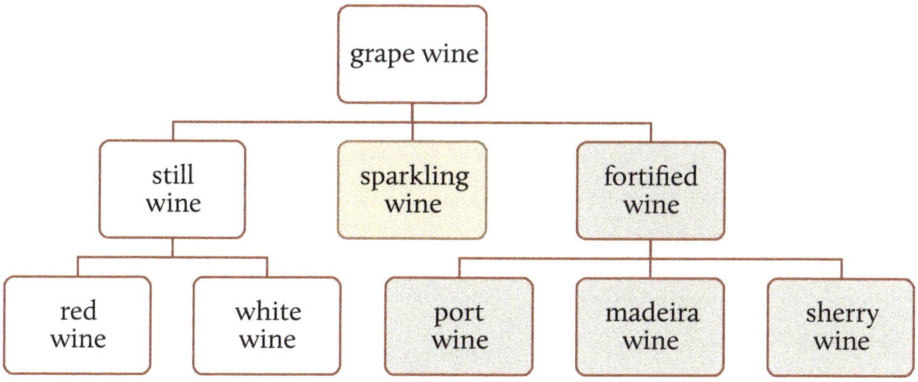

Figure A. The place of *sparkling wines* in the system of wine concepts

which was used in the working version of this companion, needed to be refined to 'vinification of sparkling wine grapes'. In a similar vein, an analogical concept in the area of madeira wine — a subcategory of fortified wine — was marked by the term 'vinification of Madeira grapes'.

The factual guide offers historical background to aid in understanding the sparkling wine trade. No special instructions are needed to appreciate this section. The map of terms captures all the concepts (expressed as concept markers) and all the divisions identified in the field with the exception of the equivalence links. The map uses colour coding to differentiate between types of conceptual link. It is accompanied by a legend to explain its connectors, signifying the relationships between the concepts. The map's connectors mark:

- Performative ('performer–activity') relationship
- Generic ('parent–child') relationship
- Partitive ('whole–part') relationship
- Concomitant ('companion–companion') relationship
- Causative ('cause–effect') relationship
- Sequential ('prequel–sequel') relationship.

The sequential relationship has been marked by ordering two or more concepts of *part* (which form a single *whole*), with each of these concepts (meronyms) given a numerical identifier.

The full semantic depth of the terminological metalanguage finds expression in the A–Z display, which comprises the bulk of this work. Articles in this display are constructed as follows:

1. Title descriptor
2. Definition of term
3. Key literature (and web links)
4. Table with data (optional)
5. Photograph (optional)

A title descriptor opens each of the 170 descriptor articles contained in the A–Z display. Beneath that the reader will find a classical definition with extensive content, with each definition based on a given concept's 'nuclear family', an effect of the natural sets of close relationships at play between relevant conceptual entities. Given the reciprocity of semantic relationships, the whole conceptual system underpinning professional communication in the sparkling wine trade has been found to be reliant on a paradigm of 12 types of conceptual relationship.

The reciprocal linkage between the concepts of *performer* and (the) *activity* (conducted by a specialist) manifests itself in the relationship between the concepts marked by the respective terms 'sparkling wine producers' and 'sparkling winemaking'. This link is existential in character, as the performer referred to by the term is the key factor in the existence of the activity and, thus, also its product. Generic concepts are linked by the relationship of *parent* (genus) to *child* (species), and *child* (species) to *parent* (genus). For example, the term 'ageing of sparkling wine' functions as a parent descriptor for the two terms 'ageing of sparkling wine underwater' and 'ageing of sparkling wine in cellars'. A partitive relationship can be one of *whole* (holonym) to *parts* (meronyms) or one of *part* to *whole*. For instance, the descriptor 'rigorous selection of sparkling wine grapes' functions as the first part of the process known as 'preparation of base wines and partially fermented musts'.

Two types of horizontal relationship have been identified in the sparkling wine conceptual system. One of them is a relationship between two or more concepts of *child* (species) belonging to one class of *parent* (genus). Terms representing the concepts of *child* imply alternative options the winemaker can pursue in selecting a procedure that they consider as the most appropriate for a particular task in the production process. For instance, the *child–child* relationship links the concept identified as 'short ageing of sparkling wine in cellars' and 'extended ageing of sparkling wine in cellars', which become alternatives when the winemaker ponders the broad idea symbolized by the term 'ageing of sparkling wine in cellars'.

The other of the two horizontal relationship types typical of communication in the sparkling wine field is that identified between two or more concepts of *part* (maronym) which constitute a single *whole* (holonym). This type of relationship has been found to assume the character of a *sequence*, which can be subcategorized into the link of *prequel* to *sequel* or *sequel* to *prequel*. For example, the concept marker 'extended maturation on lees' functions as a prequel and the term 'riddling' enacts the defined systemic role of a sequel in the context of the broader term 'traditional method', which marks the comprehensive concept of *whole*.

The (non-hierarchical, reciprocal) causative linkage is divisible into two types of semantic relationship: one that assumes the character of *cause* to *effect*, and one that assumes the character of *effect* to *cause*. The term 'blending of base wines or partially fermented musts' (which functions as *cause*) refers to a winemaking activity resulting in a blend of base wines commonly known as 'cuvée' (which function as *effect*). Similarly, the winemaking activity named 'second pressing of grapes for white must' gives press juice referred to as 'taille'.

The conceptual system underlying communication in the sparkling wine industry is also reliant on concomitance relationship, which can also be referred to as one of partnership or co-occurrence, as it occurs between the concepts of *companions* or *partners*. This relationship is exemplified by the semantic space revolving around the concept expressed as 'sealing of bottles of sparkling wine with corks'. The term's underlying concept creates semantic links with the concepts marked by the descriptors 'securing of final corks' and '*poignettage* for distribution of expedition liqueur'. These descriptors are used to refer to activities which sometimes, often or nearly always co-occur with the activity signalled by the term in question. This link is viewed as the weakest of all the relationships in the system.

A number of diverse conceptual families have been found in the field of sparkling wines. For example, the concept signalled by the term 'extraction of skin components for red must' belongs to a homogeneous family of six concepts, fully taxonomic in character. All the connections between the relevant concepts form either generic links (*parent–child* or *child–parent*) or sibling relationships (*child x–child y*). By contrast, the term 'fermentation of coloured musts' belongs to a heterogeneous family of six, in which concepts are connected not only by taxonomic and partonomic (technical) relationships but also by the partnership linkage (*companion–companion*). In a similar vein, the term 'sparkling winemaking', which forms one of the terminological system's triad of top descriptors, establishes a heterogenous nuclear family of four, with as many as three different types of underlying relationship, namely: a performative link (*activity–performer*), a causative relationship (*cause–effect*), and two partitive relationships (*whole–parts*).

References to key literature are provided at the end of every descriptor article, beneath the definition. The rule governing the acceptance of a concept into the system is that it must have been expressed in at least two independent sources. As in the case of the encyclopaedia *Madeira, Port, Sherry: The Equinox Companion to Fortified Wines*, the choice of sources was concept-oriented, rather than based solely on the formal and statistical aspects of expressions, which is to say that the important thing was that a source was discussing the same thing as the descriptor; it was less important that it was using the same terminology.

Links to websites, placed underneath the standard literature references, offer added value, as do the tables containing numerical and qualitative data, attached to some of the articles. In addition, selected articles are complemented with a photograph. The tables serve to add depth by providing data, whereas the photographs serve a broader purpose. They can provide a rich source of conceptual data, and can be used as substitute objects of observation while providing the reader with a framing of the real-world professional environment. Just as a term specialist employs images to aid understanding of special concepts, images such as these can progress the reader's grasp of the subject matter from (verbal and) systemic to (non-verbal and) concrete.

Equivalence relationships are signalled by the presence of non-descriptor articles (those with no descriptive text to define the term). There are 58 such articles in the A–Z chapter. Each non-descriptor article comprises a non-preferred term followed by the descriptor (or descriptors) proper, indicated by the symbol '≈'. The status of a non-preferred term (non-descriptor) is accorded either to synonymous expressions used less frequently or to quasi-synonyms, the meanings of which have failed to meet the strict requirements of the field's network of concepts. The most important role of non-preferred terms is to function as access points (i.e. lead-ins to the preferred terms), with enhanced access raising the work's educational value.

There can be more than one link from a non-preferred term. A non-preferred term that functions as a quasi-synonym of a descriptor tends to be semantically broader than the latter, in view of the fact that it does not belong to the current system's hierarchy. In such cases, a non-preferred term that marks a parent concept can have more than one reference. One example is the expression '*agrafage*', a non-preferred term with references to 'securing of final corks with *agrafes*' (an activity used when sealing bottles of finished sparkling wine with corks) and 'securing of temporary corks with *agrafes*' (an activity that can accompany sealing bottles of cuvée with corks). Thus, the system brings together concepts which would otherwise remain on distinct branches of a hierarchical tree.

The principle of simplification applies on three basic levels: conceptual (a concept finding its expression in forms with which the terminologist determines its place in the system); semantic (relational grammar reduced to several productive conceptual links); and formal (term specialists limiting the number of forms representing one concept based on their interoperability, while resorting to nominalization to maintain the extent of formality accepted in dictionary-making). The systemic rule has been employed to identify and display the semantic links between the concepts expressed by all interest holders in the winemaking field. The formal aspect is evidenced in the facts that preference is given to those professional language expressions that are statistically most relevant in the subject field.

The entry terms have been checked for correctness and harmonized with French, the subject area's primary language. A number of borrowings from this language (*coeur de cuvée*, *ficelage*, stacking *sur pointes*, *tirage*, etc.) have been observed and documented. French documents were consulted when English-language documents proved to be inconsistent or when target expressions were used in an ambiguous manner. The following instance is described here to serve as a snapshot of the 'backstage' of this terminological work, so that the reader can learn something of the origins of the entry term. In this instance, a candidate term is given and two corresponding quotations from source documents are presented, with key expressions underlined. A 'terminologist's note' is provided which discusses the semantic motivation of the expression's underlying elements and the reason for the candidate term's acceptance.

CHARMAT METHOD

- 'The next cost-effective method is what is known as the Charmat method (or 'tank' or 'bulk' method). This is where the fermentation that produces the bubbles takes place in a closed pressure tank (*cuve close*) and all other downstream operations (clarifying, filtering and bottling) take place under pressure [. . .]. In the case of wines such as Prosecco and Sekt, a still 'base wine' is produced in the normal manner, but it then undergoes a secondary fermentation with additional sugar and yeast added in a tank under pressure' (Skelton 2021: 150–1).
- 'Perhaps most sparkling wine that is drunk in the world is produced by tank fermentation. This process in French is called *cuve close* and is often known as the Charmat method after the Frenchman who developed it at

the beginning of the 20th century. In this method the second fermentation takes place in a sealed tank (*cuve close*) rather than in a bottle. Dry base wine is placed, together with sugar and yeast, in the sealed tank and following the second fermentation the resulting sediment is removed by filtration under pressure before the wine is bottled' (Fielden [1994] 1998: 296).

Terminologist's note: *The expression 'Charmat method' is statistically the most often used for the relevant concept. The expressions 'bulk method', 'tank method' and the French expression 'cuve close' seem too broad, as they can imply both the creation of sparkling wines using a secondary fermentation in a closed tank (typically referred to as Charmat method) and the now rare continuous method, where a second fermentation is conducted using a system of pressure tanks.*

The book contains a chapter on matching sparkling wine with food, composed of three consecutive small parts: 'Sparkling wine profile'; 'Tasting and pairing principles'; and 'Wine and food pairings'. The first of these micro-displays is made up of articles discussing components of evaluation typically used during professional tastings of sparkling wines: colours, mousse, flavours, body, sugar, acidity, alcohol and tannins. The second micro-display covers tasting principles and pairing principles; the third presents cutting-edge knowledge on pairing sparkling wines and foods, with articles organized by wine category.

This guide also has a chapter on chronology of events, which places significant events, most of them mentioned in the previous chapters, into chronological order. This allows the reader to understand the history of sparkling wines against the background of a sequence of developments that have shaped sparkling winemaking businesses around the world. The list of online references and literature on sparkling wines can be found at the very end of this companion.

References to publications are organized into three sections in accordance with the degree of recognized reliability accorded documents of different types; the majority were published in the 21st century. The three sections are: reference sources; scientific publications; and professional documents. The reference section includes encyclopaedias, thesauri, companions, dictionaries, handbooks, history books and glossaries. If articles or chapters in these works are signed with the name of an author, reference has been made to that author and not to the editor. A large proportion of the research publications, primarily scholarly books, scientific reports and scientific journal articles, were accessed via the Web of Science. The section of professional documents comprises books, trade press articles, technical reports, interviews, blog articles, *etc.* The vast majority of documents referred to in this volume of the book are written works. The index at the back of this guide includes company names, geographic place names, names of grape varieties, names of historic events, personal names, and also the names of winemakers' products and chemical compounds.

This guide is the result of extensive qualitative research in the field of sparkling wine terminology as well as consistency (quantity-oriented) checks of the findings. The microfield of sparkling wines is covered by 170 terms, making it the largest subject area in the field of special wines. Its conceptualization is characterized by a greater complexity than that of madeira wines (the largest conceptual area in the microfield of fortified wines). The current state of knowledge on sparkling wines is represented by a system of terms, whose depth reaches the eighth level of subordination in the hierarchy of meanings ('sparkling winemaking' > 'preparation of sparkling wines' > 'sparkling method' > 'secondary fermentation method' > 'bottle method' > 'traditional method' > 'finishing of traditional-method sparkling wines' > '*transversage*' > 'transfer of disgorged bottle-fermented sparkling wine to non-standard-sized bottles'). These results indicate the vast amount of work involved in the sparkling winemaking process.

The reliability of these conceptual schemes is underpinned by the recency of the sources and by the consultations in the field with winemakers during my visits at Adoria Vineyards, Winnice Kojder and Winnica Gostchorze (western parts of Poland); Ridgeview Wine Estate (England's East Sussex); and Leclerc Briant (Champagne). During my field research in Champagne, consultations were also conducted with wine educators from Lanson, Champagne Charles de Cazanove, Maison Ruinart, Veuve Clicquot, Maison Mumm, Maison Boizel and Moët & Chandon. Historical aspects relating to local production of sparkling wines in western Poland have been consulted with Tomasz Kowalski, the curator at Wine Museum in Zielona Góra.

The compilation of terms and the creation of the descriptor system has taken almost five years. The most recent update of this book was made in the last days of September 2024. Preparation for publication ended on 31 October the same year. In order to maintain this guide's relevance, it is planned that this encyclopaedia will be revised every five years. Terminologies develop according to the direction in which the industry is heading. New descriptors for the coverage of the conceptual microfield of sparkling wines could result from the rapid development of its potential for expanding wine classifications and from the ever-growing trend of sustainable practices in the winery. The area of sparkling wine and food pairing will also be closely monitored. This subject area is expected to continue to provide a stimulus to the entire sparkling wine industry, and can therefore be considered as a likely source of new terms.

INDEX

www.ingramcontent.com/pod-product-compliance
Ingram Content Group UK Ltd.
Pitfield, Milton Keynes, MK11 3LW, UK
UKHW060220061026
15625UKWH00007B/419

9 781800 506848